Discrete Geometry and Algebraic Combinatorics

CONTEMPORARY MATHEMATICS

625

Discrete Geometry and Algebraic Combinatorics

AMS Special Session
Discrete Geometry and Algebraic Combinatorics
January 11, 2013
San Diego, CA

Alexander Barg
Oleg R. Musin
Editors

American Mathematical Society
Providence, Rhode Island

2010 *Mathematics Subject Classification.* Primary 52C35, 52C17, 05B40, 52C10, 05C10, 37F20, 94B40, 58E17.

Library of Congress Cataloging-in-Publication Data

Discrete geometry and algebraic combinatorics / Alexander Barg, Oleg R. Musin, editors.
pages cm. – (Contemporary mathematics ; volume 625)
"AMS Special Session on Discrete Geometry and Algebraic Combinatorics, January 11, 2013."
Includes bibliographical references.
ISBN 978-1-4704-0905-0 (alk. paper)
1. Discrete geometry–Congresses. 2. Combinatorial analysis–Congresses. I. Barg, Alexander, 1960- editor of compilation. II. Musin, O. R. (Oleg Rustamovich) editor of compilation.

QA640.7.D575 2014
516′.116–dc23 2014007424

Contemporary Mathematics ISSN: 0271-4132 (print); ISSN: 1098-3627 (online)

DOI: http://dx.doi.org/10.1090/conm/625

∞ The paper used in this book is acid-free and falls within the guidelines
established to ensure permanence and durability.
Visit the AMS home page at http://www.ams.org/

10 9 8 7 6 5 4 3 2 1 19 18 17 16 15 14

Contents

Preface

This volume contains a collection of papers presented at, or closely related to the topics of, the Special Session on "Discrete Geometry and Algebraic Combinatorics" (January 11, 2013) held as a part of 2013 Joint Mathematics Meetings in San Diego, CA. The papers in the volume belong to one of the two related subjects in the session's title, and can be divided into two groups: distance geometry with applications in combinatorial optimization, and algebraic combinatorics, including applications in coding theory.

In the first area, the paper by K. Bezdek discusses the affine plank conjecture of T. Bang. Bezdek gives a short survey on the status of this problem and proves some partial results for the successive inradii of the convex bodies involved. The underlying geometric structures are successive hyperplane cuts introduced several years ago by J. Conway and inductive tilings introduced recently by A. Akopyan and R. Karasev.

Transportation polytopes arise in optimization and statistics, and also are of interest for discrete mathematics because permutation matrices, Latin squares, and magic squares appear naturally as lattice points of these polytopes. The survey by J.A. De Loera and E.D. Kim is devoted to combinatorial and geometric properties of transportation polytopes. This paper also includes some recent unpublished results on the diameter of graphs of these polytopes and discusses the status of several open questions in this field.

The paper by A. Ivanov and A. Tuzhilin presents an overview of a new branch of the one-dimensional geometric optimization problem, the minimal fillings theory. This theory is closely related to the generalized Steiner problem and offers an opportunity to look at many classical questions appearing in optimal connection theory from a new point of view. The paper is essentially a survey, which serves as a useful introduction to a new theory that so far has been scattered in multiple papers mostly appearing in the Russian literature.

A.M. Raigorodskii presents a survey of recent advances in many classical open problems related to the notion of a geometric graph. He discuss some properties ofdistance graphs and graphs of diameters. The study of such graphs is motivated by famous problems of combinatorial geometry going back to Erdős, Hadwiger, Nelson, and Borsuk.

The paper by A. Niedermaier, D. Rizzolo and F.E. Su extends the famous Sperner lemma to finite labellings of trees. In this paper the authors prove 15 theorems around a tree Sperner lemma. In particular they show that any proper labelling of a tree contains a fully-labelled edge and prove that this theorem is equivalent to a theorem for finite covers of metric trees and a fixed point theorem on metric trees. They also exhibit connections to Knaster-Kuratowski-Mazurkiewicz-type theorems and discuss interesting applications to voting theory.

In the second area (algebraic combinatorics), A. Barg and W.-H. Yu use semidefinite programming to obtain new bounds on the maximum cardinality of equiangular line sets in $\mathbb{R}^n$. They obtain some new exact answers, resolving in part a 1972 conjecture made by Lemmens and Seidel.

The Poisson summation formula underlies a number of fundamental results of the theory of codes, lattices, and sphere packings. In their paper, H. Cohn, A. Kumar, C. Reiher, and A. Schürmann address the notion of formal duality introduced earlier in the work on energy-minimizing configurations. Formal duality is well known in coding theory where several classes of nonlinear codes are formal duals of each other. The authors attempt to formalize this notion for the case of packings relying on the Poisson summation formula.

The paper by G. Cohen and S. Mesnager is devoted to the classical problem of constructing bent and semi-bent functions. This problem has been the focus of attention in computer science in particular because of aplications in cryptography including correlation attacks and linear cryptanalysis. The authors construct new families of semi-bent functions and reveal new links between such functions and bent functions.

In his paper, S. Kopparty studies so-called multiplicity codes; i.e., codes obtained by evaluating polynomials at the points of a finite field whereby at each point one computes not just the value of the polynomial but also values of the first few derivatives. Such codes were known for about 15 years in the case of univariate polynomials, while recently these ideas were extended to the multivariate case. It turns out that these constructions are well suited for local decoding including list decoding procedures.

O.R. Musin presents a new approach to the well-known semidefinite programming bounds on spherical codes. Previously these bounds were derived using positive definite matrices, while this paper defines a new class of multivariate orthogonal polynomials that can be used to give a direct proof of the bounds. These polynomials satisfy the addition formula as well as positivity conditions generalizing the conditions given the classical Schoenberg theorem for univariate Gegenbauer polynomials.

A part of the special session was dedicated to the 60th birthday of our friend and colleague Professor Ilya Dumer (UC Riverside). Several authors, including the present editors, also dedicate their papers to Ilya with affection and admiration.

Alexander Barg
University of Maryland

Oleg R. Musin
University of Texas at Brownsville

Contemporary Mathematics
Volume **625**, 2014
http://dx.doi.org/10.1090/conm/625/12489

Plank theorems via successive inradii

Károly Bezdek

ABSTRACT. In the 1930's, Tarski introduced his plank problem at a time when the field discrete geometry was about to born. It is quite remarkable that Tarski's question and its variants continue to generate interest in the geometric as well as analytic aspects of coverings by planks in the present time as well. Besides giving a short survey on the status of the affine plank conjecture of Bang (1950) we prove some new partial results for the successive inradii of the convex bodies involved. The underlying geometric structures are successive hyperplane cuts introduced several years ago by Conway and inductive tilings introduced recently by Akopyan and Karasev.

1. Introduction

As usual, a *convex body* of the Euclidean space $\mathbb{E}^d$ is a compact convex set with non-empty interior. Let $\mathbf{C} \subset \mathbb{E}^d$ be a convex body, and let $H \subset \mathbb{E}^d$ be a hyperplane. Then the distance $w(\mathbf{C}, H)$ between the two supporting hyperplanes of $\mathbf{C}$ parallel to H is called the *width of* $\mathbf{C}$ *parallel to* H. Moreover, the smallest width of $\mathbf{C}$ parallel to hyperplanes of $\mathbb{E}^d$ is called the *minimal width* of $\mathbf{C}$ and is denoted by $w(\mathbf{C})$.

Recall that in the 1930's, Tarski posed what came to be known as the plank problem. A *plank* $\mathbf{P}$ in $\mathbb{E}^d$ is the (closed) set of points between two distinct parallel hyperplanes. The *width* $w(\mathbf{P})$ of $\mathbf{P}$ is simply the distance between the two boundary hyperplanes of $\mathbf{P}$. Tarski conjectured that if a convex body of minimal width w is covered by a collection of planks in $\mathbb{E}^d$, then the sum of the widths of these planks is at least w. This conjecture was proved by Bang in his memorable paper [**5**]. (In fact, the proof presented in that paper is a simplification and generalization of the proof published by Bang somewhat earlier in [**4**].) Thus, we call the following statement Bang's plank theorem.

THEOREM 1.1. *If the convex body* $\mathbf{C}$ *is covered by the planks* $\mathbf{P}_1, \mathbf{P}_2, \ldots, \mathbf{P}_n$ *in* $\mathbb{E}^d, d \geq 2$ *(i.e.,* $\mathbf{C} \subset \mathbf{P}_1 \cup \mathbf{P}_2 \cup \cdots \cup \mathbf{P}_n \subset \mathbb{E}^d$*), then* $\sum_{i=1}^{n} w(\mathbf{P}_i) \geq w(\mathbf{C})$.

In [**5**], Bang raised the following stronger version of Tarski's plank problem called the affine plank problem. We phrase it via the following definition. Let $\mathbf{C}$ be a convex body and let $\mathbf{P}$ be a plank with boundary hyperplanes parallel to the

---

2010 *Mathematics Subject Classification.* Primary 52C17, 05B40, 11H31, and 52C45.

Partially supported by a Natural Sciences and Engineering Research Council of Canada Discovery Grant.

hyperplane H in $\mathbb{E}^d$. We define the **C**-*width* of the plank $\mathbf{P}$ as $\frac{w(\mathbf{P})}{w(\mathbf{C},H)}$ and label it $w_{\mathbf{C}}(\mathbf{P})$. (This notion was introduced by Bang [**5**] under the name "relative width".)

CONJECTURE 1.2. *If the convex body* $\mathbf{C}$ *is covered by the planks* $\mathbf{P}_1, \mathbf{P}_2, \ldots, \mathbf{P}_n$ *in* $\mathbb{E}^d, d \geq 2$, *then* $\sum_{i=1}^{n} w_{\mathbf{C}}(\mathbf{P}_i) \geq 1$.

The special case of Conjecture 1.2, when the convex body to be covered is centrally symmetric, has been proved by Ball in [**3**]. Thus, the following is Ball's plank theorem.

THEOREM 1.3. *If the centrally symmetric convex body* $\mathbf{C}$ *is covered by the planks* $\mathbf{P}_1, \mathbf{P}_2, \ldots, \mathbf{P}_n$ *in* $\mathbb{E}^d, d \geq 2$, *then* $\sum_{i=1}^{n} w_{\mathbf{C}}(\mathbf{P}_i) \geq 1$.

It was Alexander [**2**] who noticed that Conjecture 1.2 is equivalent to the following generalization of a problem of Davenport.

CONJECTURE 1.4. *If a convex body* $\mathbf{C}$ *in* $\mathbb{E}^d, d \geq 2$ *is sliced by* $n-1$ *hyperplane cuts, then there exists a piece that covers a translate of* $\frac{1}{n}\mathbf{C}$.

We note that the paper [**7**] of A. Bezdek and the author proves Conjecture 1.4 for successive hyperplane cuts (i.e., for hyperplane cuts when each cut divides one piece). Also, the same paper ([**7**]) introduced two additional equivalent versions of Conjecture 1.2. As they seem to be of independent interest we recall them following the terminology used in [**7**].

Let $\mathbf{C}$ and $\mathbf{K}$ be convex bodies in $\mathbb{E}^d$ and let H be a hyperplane of $\mathbb{E}^d$. The **C**-*width of* **K** *parallel to* H is denoted by $w_{\mathbf{C}}(\mathbf{K}, H)$ and is defined as $\frac{w(\mathbf{K},H)}{w(\mathbf{C},H)}$. The *minimal* **C**-*width of* **K** is denoted by $w_{\mathbf{C}}(\mathbf{K})$ and is defined as the minimum of $w_{\mathbf{C}}(\mathbf{K}, H)$, where the minimum is taken over all possible hyperplanes H of $\mathbb{E}^d$. Recall that the inradius of $\mathbf{K}$ is the radius of the largest ball contained in $\mathbf{K}$. It is quite natural then to introduce the **C**-*inradius of* **K** as the factor of the largest positive homothetic copy of $\mathbf{C}$, a translate of which is contained in $\mathbf{K}$. We need to do one more step to introduce the so-called successive **C**-inradii of **K** as follows.

Let r be the **C**-inradius of **K**. For any $0 < \rho \leq r$ let the ρ**C**-*rounded body of* **K** be denoted by $\mathbf{K}^{\rho\mathbf{C}}$ and be defined as the union of all translates of $\rho\mathbf{C}$ that are covered by $\mathbf{K}$.

Now, take a fixed integer $m \geq 1$. On the one hand, if $\rho > 0$ is sufficiently small, then $w_{\mathbf{C}}(\mathbf{K}^{\rho\mathbf{C}}) > m\rho$. On the other hand, $w_{\mathbf{C}}(\mathbf{K}^{r\mathbf{C}}) = r \leq mr$. As $w_{\mathbf{C}}(\mathbf{K}^{\rho\mathbf{C}})$ is a decreasing continuous function of $\rho > 0$ and $m\rho$ is a strictly increasing continuous function of ρ, there exists a uniquely determined $\rho > 0$ such that

$$w_{\mathbf{C}}(\mathbf{K}^{\rho\mathbf{C}}) = m\rho.$$

This uniquely determined ρ is called the *mth successive* **C**-*inradius of* **K** and is denoted by $r_{\mathbf{C}}(\mathbf{K}, m)$.

Now, the two equivalent versions of Conjecture 1.2 and Conjecture 1.4 introduced in [**7**] can be phrased as follows.

CONJECTURE 1.5. *If a convex body* $\mathbf{K}$ *in* $\mathbb{E}^d, d \geq 2$ *is covered by the planks* $\mathbf{P}_1, \mathbf{P}_2, \ldots, \mathbf{P}_n$, *then* $\sum_{i=1}^{n} w_{\mathbf{C}}(\mathbf{P}_i) \geq w_{\mathbf{C}}(\mathbf{K})$ *for any convex body* $\mathbf{C}$ *in* $\mathbb{E}^d$.

CONJECTURE 1.6. *Let* $\mathbf{K}$ *and* $\mathbf{C}$ *be convex bodies in* $\mathbb{E}^d, d \geq 2$. *If* $\mathbf{K}$ *is sliced by* $n-1$ *hyperplanes, then the minimum of the greatest* **C**-*inradius of the pieces is equal to the nth successive* **C**-*inradius of* **K**, *i.e., it is* $r_{\mathbf{C}}(\mathbf{K}, n)$.

Recall that Theorem 1.3 gives a proof of (Conjecture 1.5 as well as) Conjecture 1.6 for centrally symmetric convex bodies $\mathbf{K}$ in $\mathbb{E}^d, d \geq 2$ (with $\mathbf{C}$ being an arbitrary convex body in $\mathbb{E}^d, d \geq 2$). Another approach that leads to a partial solution of Conjecture 1.6 was published in [**7**]. Namely, in that paper A. Bezdek and the author proved the following theorem that (under the condition that $\mathbf{C}$ is a ball) answers a question raised by Conway ([**6**]) as well as proves Conjecture 1.6 for successive hyperplane cuts.

THEOREM 1.7. *Let* $\mathbf{K}$ *and* $\mathbf{C}$ *be convex bodies in* $\mathbb{E}^d$, $d \geq 2$. *If* $\mathbf{K}$ *is sliced into* $n \geq 1$ *pieces by* $n-1$ *successive hyperplane cuts (i.e., when each cut divides one piece), then the minimum of the greatest* $\mathbf{C}$*-inradius of the pieces is the* n*th successive* $\mathbf{C}$*-inradius of* $\mathbf{K}$ *(i.e.,* $r_{\mathbf{C}}(\mathbf{K}, n)$*). An optimal partition is achieved by* $n-1$ *parallel hyperplane cuts equally spaced along the minimal* $\mathbf{C}$*-width of the* $r_{\mathbf{C}}(\mathbf{K}, n)\mathbf{C}$*-rounded body of* $\mathbf{K}$.

Akopyan and Karasev ([**1**]) just very recently have proved a related partial result on Conjecture 1.5. Their theorem is based on a nice generalization of successive hyperplane cuts. The more exact details are as follows. Under the *convex partition* $\mathbf{V}_1 \cup \mathbf{V}_2 \cup \cdots \cup \mathbf{V}_n$ of $\mathbb{E}^d$ we understand the family $\mathbf{V}_1, \mathbf{V}_2, \ldots, \mathbf{V}_n$ of closed convex sets having pairwise disjoint non-empty interiors in $\mathbb{E}^d$ with $\mathbf{V}_1 \cup \mathbf{V}_2 \cup \cdots \cup \mathbf{V}_n = \mathbb{E}^d$. Then we say that the convex partition $\mathbf{V}_1 \cup \mathbf{V}_2 \cup \cdots \cup \mathbf{V}_n$ of $\mathbb{E}^d$ is an *inductive partition* of $\mathbb{E}^d$ if for any $1 \leq i \leq n$, there exists an inductive partition $\mathbf{W}_1 \cup \cdots \cup \mathbf{W}_{i-1} \cup \mathbf{W}_{i+1} \cup \cdots \cup \mathbf{W}_n$ of $\mathbb{E}^d$ such that $\mathbf{V}_j \subset \mathbf{W}_j$ for all $j \neq i$. A partition into one part $\mathbf{V}_1 = \mathbb{E}^d$ is assumed to be inductive. We note that if $\mathbb{E}^d$ is sliced into n pieces by $n-1$ successive hyperplane cuts (i.e., when each cut divides one piece), then the pieces generate an inductive partition of $\mathbb{E}^d$. Also, the Voronoi cells of finitely many points of $\mathbb{E}^d$ generate an inductive partition of $\mathbb{E}^d$. Now, the main theorem of [**1**] can be phrased as follows.

THEOREM 1.8. *Let* $\mathbf{K}$ *and* $\mathbf{C}$ *be convex bodies in* $\mathbb{E}^d, d \geq 2$ *and let* $\mathbf{V}_1 \cup \mathbf{V}_2 \cup \cdots \cup \mathbf{V}_n$ *be an inductive partition of* $\mathbb{E}^d$ *such that* $\operatorname{int}(\mathbf{V}_i \cap \mathbf{K}) \neq \emptyset$ *for all* $1 \leq i \leq n$. *Then* $\sum_{i=1}^{n} r_{\mathbf{C}}(\mathbf{V}_i \cap \mathbf{K}, 1) \geq r_{\mathbf{C}}(\mathbf{K}, 1)$.

2. Extensions to Successive Inradii

First, we state the following stronger version of Theorem 1.7. Its proof is an extension of the proof of Theorem 1.7 published in [**7**].

THEOREM 2.1. *Let* $\mathbf{K}$ *and* $\mathbf{C}$ *be convex bodies in* $\mathbb{E}^d$, $d \geq 2$ *and let* m *be a positive integer. If* $\mathbf{K}$ *is sliced into* $n \geq 1$ *pieces by* $n-1$ *successive hyperplane cuts (i.e., when each cut divides one piece), then the minimum of the greatest* m*th successive* $\mathbf{C}$*-inradius of the pieces is the* (mn)*th successive* $\mathbf{C}$*-inradius of* $\mathbf{K}$ *(i.e.,* $r_{\mathbf{C}}(\mathbf{K}, mn)$*). An optimal partition is achieved by* $n-1$ *parallel hyperplane cuts equally spaced along the minimal* $\mathbf{C}$*-width of the* $r_{\mathbf{C}}(\mathbf{K}, mn)\mathbf{C}$*-rounded body of* $\mathbf{K}$.

Second, the method of Akopyan and Karasev ([**1**]) can be extended to prove the following stronger version of Theorem 1.8. In fact, that approach extends also the relavant additional theorems of Akopyan and Karasev stated in [**1**] and used in their proof of Theorem 1.8. However, in this paper following the recommendation of the referee, we derive the next theorem directly from Theorem 1.8.

THEOREM 2.2. *Let* $\mathbf{K}$ *and* $\mathbf{C}$ *be convex bodies in* $\mathbb{E}^d, d \geq 2$ *and let* m *be a positive integer. If* $\mathbf{V}_1 \cup \mathbf{V}_2 \cup \cdots \cup \mathbf{V}_n$ *is an inductive partition of* $\mathbb{E}^d$ *such that* $\mathrm{int}(\mathbf{V}_i \cap \mathbf{K}) \neq \emptyset$ *for all* $1 \leq i \leq n$, *then* $\sum_{i=1}^{n} r_{\mathbf{C}}(\mathbf{V}_i \cap \mathbf{K}, m) \geq r_{\mathbf{C}}(\mathbf{K}, m)$.

COROLLARY 2.3. *Let* $\mathbf{K}$ *and* $\mathbf{C}$ *be convex bodies in* $\mathbb{E}^d, d \geq 2$. *If* $\mathbf{V}_1 \cup \mathbf{V}_2 \cup \cdots \cup \mathbf{V}_n$ *is an inductive partition of* $\mathbb{E}^d$ *such that* $\mathrm{int}(\mathbf{V}_i \cap \mathbf{K}) \neq \emptyset$ *for all* $1 \leq i \leq n$, *then* $\sum_{i=1}^{n} w_{\mathbf{C}}(\mathbf{V}_i \cap \mathbf{K}) \geq w_{\mathbf{C}}(\mathbf{K})$.

For the sake of completeness we mention that in two dimensions one can state a bit more. Namely, recall that Akopyan and Karasev ([**1**]) proved the following: Let $\mathbf{K}$ and $\mathbf{C}$ be convex bodies in $\mathbb{E}^2$ and let $\mathbf{V}_1 \cup \mathbf{V}_2 \cup \cdots \cup \mathbf{V}_n = \mathbf{K}$ be a partition of $\mathbf{K}$ into convex bodies $\mathbf{V}_i$, $1 \leq i \leq n$. Then $\sum_{i=1}^{n} r_{\mathbf{C}}(\mathbf{V}_i, 1) \geq r_{\mathbf{C}}(\mathbf{K}, 1)$. Now, exactly the same way as Theorem 2.2 is derived from Theorem 1.8, it follows that $\sum_{i=1}^{n} r_{\mathbf{C}}(\mathbf{V}_i, m) \geq r_{\mathbf{C}}(\mathbf{K}, m)$ holds for any positive integer m.

Finally, we close this section stating that Conjectures 1.2, 1.4, 1.5, and 1.6 are all equivalent to the following two conjectures:

CONJECTURE 2.4. *Let* $\mathbf{K}$ *and* $\mathbf{C}$ *be convex bodies in* $\mathbb{E}^d, d \geq 2$ *and let* m *be a positive integer. If* $\mathbf{K}$ *is covered by the planks* $\mathbf{P}_1, \mathbf{P}_2, \ldots, \mathbf{P}_n$ *in* $\mathbb{E}^d$, *then* $\sum_{i=1}^{n} r_{\mathbf{C}}(\mathbf{P}_i, m) \geq r_{\mathbf{C}}(\mathbf{K}, m)$ *or equivalently,* $\sum_{i=1}^{n} w_{\mathbf{C}}(\mathbf{P}_i) \geq m r_{\mathbf{C}}(\mathbf{K}, m)$.

CONJECTURE 2.5. *Let* $\mathbf{K}$ *and* $\mathbf{C}$ *be convex bodies in* $\mathbb{E}^d, d \geq 2$ *and let the positive integer* m *be given. If* $\mathbf{K}$ *is sliced by* $n-1$ *hyperplanes, then the minimum of the greatest* m*th successive* $\mathbf{C}$*-inradius of the pieces is the* (mn)*th successive* $\mathbf{C}$*-inradius of* $\mathbf{K}$, *i.e., it is* $r_{\mathbf{C}}(\mathbf{K}, mn)$.

In the rest of the paper we prove the claims of this section.

3. Proof of Theorem 2.1

3.1. On Coverings of Convex Bodies by Two Planks. On the one hand, the following statement is an extension to higher dimensions of Theorem 4 in [**2**]. On the other hand, the proof presented below is based on Theorem 4 of [**2**].

LEMMA 3.1. *If a convex body* $\mathbf{K}$ *in* $\mathbb{E}^d, d \geq 2$ *is covered by the planks* $\mathbf{P}_1$ *and* $\mathbf{P}_2$, *then* $w_{\mathbf{C}}(\mathbf{P}_1) + w_{\mathbf{C}}(\mathbf{P}_2) \geq w_{\mathbf{C}}(\mathbf{K})$ *for any convex body* $\mathbf{C}$ *in* $\mathbb{E}^d$.

PROOF. Let H_1 (resp., H_2) be one of the two hyperplanes which bound the plank $\mathbf{P}_1$ (resp., $\mathbf{P}_2$). If H_1 and H_2 are translates of each other, then the claim is obviously true. Thus, without loss of generality we may assume that $L := H_1 \cap H_2$ is a $(d-2)$-dimensional affine subspace of $\mathbb{E}^d$. Let $\mathbb{E}^2$ be the 2-dimensional linear subspace of $\mathbb{E}^d$ that is orthogonal to L. If $(\cdot)'$ denotes the (orthogonal) projection of $\mathbb{E}^d$ parallel to L onto $\mathbb{E}^2$, then obviously, $w_{\mathbf{C}'}(\mathbf{P}_1') = w_{\mathbf{C}}(\mathbf{P}_1)$, $w_{\mathbf{C}'}(\mathbf{P}_2') = w_{\mathbf{C}}(\mathbf{P}_2)$ and $w_{\mathbf{C}'}(\mathbf{K}') \geq w_{\mathbf{C}}(\mathbf{K})$. Thus, it is sufficient to prove that

$$w_{\mathbf{C}'}(\mathbf{P}_1') + w_{\mathbf{C}'}(\mathbf{P}_2') \geq w_{\mathbf{C}'}(\mathbf{K}').$$

In other words, it is sufficient to prove Lemma 3.1 for $d = 2$. Hence, in the rest of the proof, $\mathbf{K}, \mathbf{C}, \mathbf{P}_1, \mathbf{P}_2, H_1$, and H_2 mean the sets introduced and defined above, however, for $d = 2$. Now, we can make the following easy observation

$$w_{\mathbf{C}}(\mathbf{P}_1) + w_{\mathbf{C}}(\mathbf{P}_2) = \frac{w(\mathbf{P}_1)}{w(\mathbf{C}, H_1)} + \frac{w(\mathbf{P}_2)}{w(\mathbf{C}, H_2)}$$

$$= \frac{w(\mathbf{P}_1)}{w(\mathbf{K}, H_1)} \frac{w(\mathbf{K}, H_1)}{w(\mathbf{C}, H_1)} + \frac{w(\mathbf{P}_2)}{w(\mathbf{K}, H_2)} \frac{w(\mathbf{K}, H_2)}{w(\mathbf{C}, H_2)}$$

$$\geq \left(\frac{w(\mathbf{P}_1)}{w(\mathbf{K}, H_1)} + \frac{w(\mathbf{P}_2)}{w(\mathbf{K}, H_2)} \right) w_{\mathbf{C}}(\mathbf{K})$$
$$= (w_{\mathbf{K}}(\mathbf{P}_1) + w_{\mathbf{K}}(\mathbf{P}_2))\, w_{\mathbf{C}}(\mathbf{K}).$$

Then recall that Theorem 4 in [**2**] states that if a convex set in the plane is covered by two planks, then the sum of their relative widths is at least 1. Thus, using our terminology, we have that $w_{\mathbf{K}}(\mathbf{P}_1) + w_{\mathbf{K}}(\mathbf{P}_2) \geq 1$, finishing the proof of Lemma 3.1. □

3.2. Minimizing the Greatest mth Successive C-Inradius. Let $\mathbf{K}$ and $\mathbf{C}$ be convex bodies in $\mathbb{E}^d$, $d \geq 2$. We prove Theorem 2.1 by induction on n. It is trivial to check the claim for $n = 1$. So, let $n \geq 2$ be given and assume that Theorem 2.1 holds for at most $n-2$ successive hyperplane cuts and based on that we show that it holds for $n-1$ successive hyperplane cuts as well. The details are as follows.

Let $H_1, \ldots, H_{n-1}$ denote the hyperplanes of the $n-1$ successive hyperplane cuts that slice $\mathbf{K}$ into n pieces such that the greatest mth successive $\mathbf{C}$-inradius of the pieces is the smallest possible say, ρ. Then take the first cut H_1 that slices $\mathbf{K}$ into the pieces $\mathbf{K}_1$ and $\mathbf{K}_2$ such that $\mathbf{K}_1$ (resp., $\mathbf{K}_2$) is sliced into n_1 (resp., n_2) pieces by the successive hyperplane cuts $H_2, \ldots, H_{n-1}$, where $n = n_1 + n_2$. The induction hypothesis implies that $\rho \geq r_{\mathbf{C}}(\mathbf{K}_1, mn_1) =: \rho_1$ and $\rho \geq r_{\mathbf{C}}(\mathbf{K}_2, mn_2) =: \rho_2$ and therefore

$$w_{\mathbf{C}}({\mathbf{K}_1}^{\rho\mathbf{C}}) \leq w_{\mathbf{C}}({\mathbf{K}_1}^{\rho_1\mathbf{C}}) = mn_1\rho_1 \leq mn_1\rho; \tag{3.1}$$

moreover,

$$w_{\mathbf{C}}({\mathbf{K}_2}^{\rho\mathbf{C}}) \leq w_{\mathbf{C}}({\mathbf{K}_2}^{\rho_2\mathbf{C}}) = mn_2\rho_2 \leq mn_2\rho. \tag{3.2}$$

Now, we need to define the following set.

Definition 3.2. Assume that the origin $\mathbf{o}$ of $\mathbb{E}^d$ belongs to the interior of the convex body $\mathbf{C} \subset \mathbb{E}^d$. Consider all translates of $\rho\mathbf{C}$ which are contained in the convex body $\mathbf{K} \subset \mathbb{E}^d$. The set of points in the translates of $\rho\mathbf{C}$ that correspond to $\mathbf{o}$ form a convex set called the inner $\rho\mathbf{C}$-parallel body of $\mathbf{K}$ denoted by $\mathbf{K}_{-\rho\mathbf{C}}$.

Clearly,

$$(\mathbf{K}_1)_{-\rho\mathbf{C}} \cup (\mathbf{K}_2)_{-\rho\mathbf{C}} \subset \mathbf{K}_{-\rho\mathbf{C}} \text{ with } (\mathbf{K}_1)_{-\rho\mathbf{C}} \cap (\mathbf{K}_2)_{-\rho\mathbf{C}} = \emptyset.$$

Also, it is easy to see that there is a plank $\mathbf{P}$ with $w_{\mathbf{C}}(\mathbf{P}) = \rho$ such that it is parallel to H_1 and contains H_1 in its interior; moreover,

$$\mathbf{K}_{-\rho\mathbf{C}} \subset (\mathbf{K}_1)_{-\rho\mathbf{C}} \cup (\mathbf{K}_2)_{-\rho\mathbf{C}} \cup \mathbf{P}.$$

Now, let H_1^+ (resp., H_1^-) be the closed halfspace of $\mathbb{E}^d$ bounded by H_1 and containing $\mathbf{K}_1$ (resp., $\mathbf{K}_2$) and let $\mathbf{P}^+ := \mathbf{P} \cap H_1^+$ (resp., $\mathbf{P}^- := \mathbf{P} \cap H_1^-$). Moreover, let $\mathbf{K}^+_{-\rho\mathbf{C}} := \mathbf{K}_{-\rho\mathbf{C}} \cap H_1^+$ (resp., $\mathbf{K}^-_{-\rho\mathbf{C}} := \mathbf{K}_{-\rho\mathbf{C}} \cap H_1^-$). Hence, applying Lemma 3.1 to $\mathbf{K}_{-\rho\mathbf{C}}$ partitioned into $\mathbf{K}^+_{-\rho\mathbf{C}} \cup \mathbf{K}^-_{-\rho\mathbf{C}}$ and to $\mathbf{K}^+_{-\rho\mathbf{C}}$ covered by the plank $\mathbf{P}^+$ and the plank generated by the minimal $\mathbf{C}$-width of $(\mathbf{K}_1)_{-\rho\mathbf{C}}$ as well as to $\mathbf{K}^-_{-\rho\mathbf{C}}$ covered by the plank $\mathbf{P}^-$ and the plank generated by the minimal $\mathbf{C}$-width of $(\mathbf{K}_2)_{-\rho\mathbf{C}}$ we get that

$$w_{\mathbf{C}}(\mathbf{K}_{-\rho\mathbf{C}}) \leq w_{\mathbf{C}}\left(\mathbf{K}^+_{-\rho\mathbf{C}}\right) + w_{\mathbf{C}}\left(\mathbf{K}^-_{-\rho\mathbf{C}}\right) \leq w_{\mathbf{C}}((\mathbf{K}_1)_{-\rho\mathbf{C}}) + \rho + w_{\mathbf{C}}((\mathbf{K}_2)_{-\rho\mathbf{C}}). \tag{3.3}$$

By definition $w_{\mathbf{C}}\left((\mathbf{K}_1)_{-\rho\mathbf{C}}\right) = w_{\mathbf{C}}({\mathbf{K}_1}^{\rho\mathbf{C}}) - \rho$, $w_{\mathbf{C}}\left((\mathbf{K}_2)_{-\rho\mathbf{C}}\right) = w_{\mathbf{C}}({\mathbf{K}_2}^{\rho\mathbf{C}}) - \rho$ and $w_{\mathbf{C}}\left(\mathbf{K}_{-\rho\mathbf{C}}\right) = w_{\mathbf{C}}(\mathbf{K}^{\rho\mathbf{C}}) - \rho$. Hence, (3.3) is equivalent to

$$w_{\mathbf{C}}(\mathbf{K}^{\rho\mathbf{C}}) \leq w_{\mathbf{C}}({\mathbf{K}_1}^{\rho\mathbf{C}}) + w_{\mathbf{C}}({\mathbf{K}_2}^{\rho\mathbf{C}}). \tag{3.4}$$

Finally, (3.1),(3.2), and (3.4) yield that

$$w_{\mathbf{C}}(\mathbf{K}^{\rho\mathbf{C}}) \leq mn_1\rho + mn_2\rho = mn\rho. \tag{3.5}$$

Thus, (3.5) clearly implies that $r_{\mathbf{C}}(\mathbf{K}, mn) \leq \rho$. As the case, when the optimal partition is achieved, follows directly from the definition of the mnth successive $\mathbf{C}$-inradius of $\mathbf{K}$, the proof of Theorem 2.1 is complete.

4. Proof of Theorem 2.2

Let $\mathbf{K}$ and $\mathbf{C}$ be convex bodies in $\mathbb{E}^d, d \geq 2$ and let m be a positive integer. It follows from the definition of $r_{\mathbf{C}}(\mathbf{K}, m)$ that $r_{\mathbf{C}}(\mathbf{K}, m)$ is a translation invariant, positively 1-homogeneous, inclusion-monotone functional over the family of convex bodies $\mathbf{K}$ in $\mathbb{E}^d$ for any fixed $\mathbf{C}$ and m. On the other hand, if $\mathbf{V}_1 \cup \mathbf{V}_2 \cup \cdots \cup \mathbf{V}_n$ is an inductive partition of $\mathbb{E}^d$ such that $\mathrm{int}(\mathbf{V}_i \cap \mathbf{K}) \neq \emptyset$ for all $1 \leq i \leq n$, then Theorem 1.8 applied to $\mathbf{C} = \mathbf{K}$ yields the existence of translation vectors $\mathbf{t}_1, \mathbf{t}_2, \ldots, \mathbf{t}_n$ and positive reals $\mu_1, \mu_2, \ldots, \mu_n$ such that $\mathbf{t}_i + \mu_i\mathbf{K} \subset \mathbf{V}_i \cap \mathbf{K}$ for all $1 \leq i \leq n$ satisfying $\sum_{i=1}^{n} \mu_i \geq 1$. Therefore,

$$r_{\mathbf{C}}(\mathbf{V}_i \cap \mathbf{K}, m) \geq r_{\mathbf{C}}(\mathbf{t}_i + \mu_i\mathbf{K}, m) = r_{\mathbf{C}}(\mu_i\mathbf{K}, m) = \mu_i r_{\mathbf{C}}(\mathbf{K}, m)$$

holds for all $1 \leq i \leq n$, finishing the proof of Theorem 2.2.

5. Proof of Corollary 2.3

Let $1 \leq m_1 \leq m_2$ be positive integers. Recall that if ρ_1 (resp., ρ_2) denotes the m_1th (resp., m_2th) successive $\mathbf{C}$-inradius of $\mathbf{K}$, then by definition $w_{\mathbf{C}}(\mathbf{K}^{\rho_1\mathbf{C}}) = m_1\rho_1$ (resp., $w_{\mathbf{C}}(\mathbf{K}^{\rho_2\mathbf{C}}) = m_2\rho_2$). As $w_{\mathbf{C}}(\mathbf{K}^{\rho\mathbf{C}})$ is a decreasing continuous function of $\rho > 0$, it follows that

$$m_1 r_{\mathbf{C}}(\mathbf{K}, m_1) = m_1\rho_1 \leq m_2\rho_2 = m_2 r_{\mathbf{C}}(\mathbf{K}, m_2)\ .$$

Thus, the sequence $mr_{\mathbf{C}}(\mathbf{K}, m), m = 1, 2, \ldots$ is an increasing one with

$$\lim_{m\to+\infty} mr_{\mathbf{C}}(\mathbf{K}, m) = w_{\mathbf{C}}(\mathbf{K})\ .$$

Hence, Corollary 2.3 follows from Theorem 2.2.

6. The equivalence of Conjectures 1.2, 1.4, 1.5, 1.6, 2.4, and 2.5

Recall that according to [7] Conjectures 1.2, 1.4, 1.5, and 1.6 are equivalent to each other. So, it is sufficent to show that Conjecture 1.5 implies Conjecture 2.4 and Conjecture 2.4 implies Conjecture 2.5 moreover, Conjecture 2.5 implies Conjecture 1.6.

As according to the previous section the sequence $mr_{\mathbf{C}}(\mathbf{K}, m), m = 1, 2, \ldots$ is an increasing one with $\lim_{m\to+\infty} mr_{\mathbf{C}}(\mathbf{K}, m) = w_{\mathbf{C}}(\mathbf{K})$ therefore Conjecture 1.5 implies Conjecture 2.4. Next, it is obvious that Conjecture 2.5 implies Conjecture 1.6. So, we are left to show that Conjecture 2.4 implies Conjecture 2.5. In order to do so we introduce the following equivalent description for $r_{\mathbf{C}}(\mathbf{K}, m)$. If $\mathbf{C}$ is a convex body in $\mathbb{E}^d$, then

$$\mathbf{t} + \mathbf{C}, \mathbf{t} + \lambda_2\mathbf{v} + \mathbf{C}, \ldots, \mathbf{t} + \lambda_m\mathbf{v} + \mathbf{C}$$

is called a *linear packing* of m translates of $\mathbf{C}$ positioned parallel to the line $\{\lambda\mathbf{v} \mid \lambda \in \mathbb{R}\}$ with direction vector $\mathbf{v} \neq \mathbf{o}$ if the m translates of $\mathbf{C}$ are pairwise non-overlapping, i.e., if

$$(\mathbf{t} + \lambda_i\mathbf{v} + \mathrm{int}\mathbf{C}) \cap (\mathbf{t} + \lambda_j\mathbf{v} + \mathrm{int}\mathbf{C}) = \emptyset$$

holds for all $1 \leq i \neq j \leq m$ (with $\lambda_1 = 0$). Furthermore, the line $l \subset \mathbb{E}^d$ passing through the origin $\mathbf{o}$ of $\mathbb{E}^d$ is called a *separating direction* for the linear packing

$$\mathbf{t} + \mathbf{C}, \mathbf{t} + \lambda_2\mathbf{v} + \mathbf{C}, \ldots, \mathbf{t} + \lambda_m\mathbf{v} + \mathbf{C}$$

if

$$\mathrm{Pr}_l(\mathbf{t} + \mathbf{C}), \mathrm{Pr}_l(\mathbf{t} + \lambda_2\mathbf{v} + \mathbf{C}), \ldots, \mathrm{Pr}_l(\mathbf{t} + \lambda_m\mathbf{v} + \mathbf{C})$$

are pairwise non-overlapping intervals on l, where $\mathrm{Pr}_l : \mathbb{E}^d \to l$ denotes the orthogonal projection of $\mathbb{E}^d$ onto l. It is easy to see that every linear packing

$$\mathbf{t} + \mathbf{C}, \mathbf{t} + \lambda_2\mathbf{v} + \mathbf{C}, \ldots, \mathbf{t} + \lambda_m\mathbf{v} + \mathbf{C}$$

possesses at least one separating direction in $\mathbb{E}^d$. Finally, let $\mathbf{K}$ be a convex body in $\mathbb{E}^d$ and let $m \geq 1$ be a positive integer. Then let $\overline{\rho} > 0$ be the largest positive real with the following property: for every line l passing through the origin $\mathbf{o}$ in $\mathbb{E}^d$ there exists a linear packing of m translates of $\overline{\rho}\mathbf{C}$ lying in $\mathbf{K}$ and having l as a separating direction. It is straightforward to show that

$$\overline{\rho} = r_{\mathbf{C}}(\mathbf{K}, m).$$

Now, let $\mathbf{K}$ and $\mathbf{C}$ be convex bodies in $\mathbb{E}^d, d \geq 2$ and let the positive integer m be given. Assume that the origin $\mathbf{o}$ of $\mathbb{E}^d$ lies in the interior of $\mathbf{C}$. Furthermore, assume that $\mathbf{K}$ is sliced by $n-1$ hyperplanes say, $H_1, H_2, \ldots, H_{n-1}$ and let ρ be the greatest mth successive $\mathbf{C}$-inradius of the pieces of $\mathbf{K}$ obtained in this way. Then let $\mathbf{P}_i := \bigcup_{\mathbf{p} \in H_i} (\mathbf{p} + (-m\rho)\mathbf{C})$, $1 \leq i \leq n-1$. Based on the above description of mth successive $\mathbf{C}$-inradii, it is easy to see that $\mathbf{K}_{-m\rho\mathbf{C}} \subset \bigcup_{i=1}^{n-1} \mathbf{P}_i$ with $w_{\mathbf{C}}(\mathbf{P}_i) = m\rho$ for all $1 \leq i \leq n-1$. Thus, Conjecture 2.4 implies that $(n-1)m\rho = \sum_{i=1}^{n-1} w_{\mathbf{C}}(\mathbf{P}_i) \geq m r_{\mathbf{C}}(\mathbf{K}_{-m\rho\mathbf{C}}, m) = m\left(r_{\mathbf{C}}(\mathbf{K}^{\rho\mathbf{C}}, m) - \rho\right)$ and so, $mn\rho \geq w_{\mathbf{C}}(\mathbf{K}^{\rho\mathbf{C}})$. Hence, $\rho \geq r_{\mathbf{C}}(\mathbf{K}, mn)$ finishing the proof of Conjecture 2.5.

7. Conclusion

Theorems 1.8 and 2.2 have covering analogues. Namely recall that Akopyan and Karasev ([**1**]) introduced the following definition. Under the *convex covering* $\mathbf{V}_1 \cup \mathbf{V}_2 \cup \cdots \cup \mathbf{V}_n$ of $\mathbb{E}^d$ we understand the family $\mathbf{V}_1, \mathbf{V}_2, \ldots, \mathbf{V}_n$ of closed convex sets in $\mathbb{E}^d$ with $\mathbf{V}_1 \cup \mathbf{V}_2 \cup \cdots \cup \mathbf{V}_n = \mathbb{E}^d$. Then we say that the convex covering $\mathbf{V}_1 \cup \mathbf{V}_2 \cup \cdots \cup \mathbf{V}_n$ of $\mathbb{E}^d$ is an *inductive covering* of $\mathbb{E}^d$ if for any $1 \leq i \leq n$, there exists an inductive covering $\mathbf{W}_1 \cup \cdots \cup \mathbf{W}_{i-1} \cup \mathbf{W}_{i+1} \cup \cdots \cup \mathbf{W}_n$ of $\mathbb{E}^d$ such that $\mathbf{W}_j \subset \mathbf{V}_j \cup \mathbf{V}_i$ for all $j \neq i$. A covering by one set $\mathbf{V}_1 = \mathbb{E}^d$ is assumed to be inductive. [**1**] proves that if $\mathbf{K}$ and $\mathbf{C}$ are convex bodies in $\mathbb{E}^d, d \geq 2$ and $\mathbf{V}_1 \cup \mathbf{V}_2 \cup \cdots \cup \mathbf{V}_n$ is an inductive covering of $\mathbb{E}^d$ such that $\mathrm{int}(\mathbf{V}_i \cap \mathbf{K}) \neq \emptyset$ for all $1 \leq i \leq n$, then $\sum_{i=1}^{n} r_{\mathbf{C}}(\mathbf{V}_i \cap \mathbf{K}, 1) \geq r_{\mathbf{C}}(\mathbf{K}, 1)$. Now, exactly the same way as Theorem 2.2 is derived from Theorem 1.8, it follows that

$$\sum_{i=1}^{n} r_{\mathbf{C}}(\mathbf{V}_i \cap \mathbf{K}, m) \geq r_{\mathbf{C}}(\mathbf{K}, m) \tag{7.1}$$

holds for any positive integer m. This raises the following rather natural question (see also Conjecture 2.4).

PROBLEM 7.1. *Let* $\mathbf{K}$ *and* $\mathbf{C}$ *be convex bodies in* $\mathbb{E}^d$, $d \geq 2$ *and let* m *be a positive integer. Prove or disprove that if* $\mathbf{V}_1 \cup \mathbf{V}_2 \cup \ldots \cup \mathbf{V}_n$ *is a convex partition (resp., covering) of* $\mathbb{E}^d$ *such that* $\mathrm{int}(\mathbf{V}_i \cap \mathbf{K}) \neq \emptyset$ *for all* $1 \leq i \leq n$, *then* $\sum_{i=1}^n r_{\mathbf{C}}(\mathbf{V}_i \cap \mathbf{K}, m) \geq r_{\mathbf{C}}(\mathbf{K}, m)$.

Next observe that (7.1) implies in a straightforward way that if $\mathbf{K}$ and $\mathbf{C}$ are convex bodies in $\mathbb{E}^d$ and $\mathbf{V}_1 \cup \mathbf{V}_2 \cup \ldots \cup \mathbf{V}_n$ is an inductive covering of $\mathbb{E}^d$ such that $\mathrm{int}(\mathbf{V}_i \cap \mathbf{K}) \neq \emptyset$ for all $1 \leq i \leq n$, then the greatest mth successive $\mathbf{C}$-inradius of the pieces $\mathbf{V}_i \cap \mathbf{K}, i = 1, 2, \ldots, n$ is at least $\frac{1}{n} r_{\mathbf{C}}(\mathbf{K}, m)$. As the sequence $m r_{\mathbf{C}}(\mathbf{K}, m), m = 1, 2, \ldots$ is an increasing one, therefore $\frac{1}{n} r_{\mathbf{C}}(\mathbf{K}, m) \leq r_{\mathbf{C}}(\mathbf{K}, mn)$ raising the following question (see also Conjecture 2.5).

PROBLEM 7.2. *Let* $\mathbf{K}$ *and* $\mathbf{C}$ *be convex bodies in* $\mathbb{E}^d$, $d \geq 2$ *and let* m *be a positive integer. Prove or disprove that if* $\mathbf{V}_1 \cup \mathbf{V}_2 \cup \ldots \cup \mathbf{V}_n$ *is a convex partition (resp., covering) of* $\mathbb{E}^d$ *such that* $\mathrm{int}(\mathbf{V}_i \cap \mathbf{K}) \neq \emptyset$ *for all* $1 \leq i \leq n$, *then the greatest* m*th successive* $\mathbf{C}$*-inradius of the pieces* $\mathbf{V}_i \cap \mathbf{K}, i = 1, 2, \ldots, n$ *is at least* $r_{\mathbf{C}}(\mathbf{K}, mn)$.

References

[1] Arseniy Akopyan and Roman Karasev, *Kadets-type theorems for partitions of a convex body*, Discrete Comput. Geom. **48** (2012), no. 3, 766–776, DOI 10.1007/s00454-012-9437-1. MR2957644

[2] Ralph Alexander, *A problem about lines and ovals*, Amer. Math. Monthly **75** (1968), 482–487. MR0234351 (38 #2668)

[3] Keith Ball, *The plank problem for symmetric bodies*, Invent. Math. **104** (1991), no. 3, 535–543, DOI 10.1007/BF01245089. MR1106748 (92c:52003)

[4] Thøger Bang, *On covering by parallel-strips*, Mat. Tidsskr. B. **1950** (1950), 49–53. MR0038085 (12,352d)

[5] Thøger Bang, *A solution of the "plank problem."*, Proc. Amer. Math. Soc. **2** (1951), 990–993. MR0046672 (13,769a)

[6] A. Bezdek and K. Bezdek, *A solution of Conway's fried potato problem*, Bull. London Math. Soc. **27** (1995), no. 5, 492–496, DOI 10.1112/blms/27.5.492. MR1338694 (96e:52016)

[7] A. Bezdek and K. Bezdek, *Conway's fried potato problem revisited*, Arch. Math. (Basel) **66** (1996), no. 6, 522–528, DOI 10.1007/BF01268872. MR1388103 (97b:52006)

DEPARTMENT OF MATHEMATICS AND STATISTICS, UNIVERSITY OF CALGARY, CALGARY, ALBERTA, CANADA T2N 1N4

E-mail address: bezdek@math.ucalgary.ca

Contemporary Mathematics
Volume **625**, 2014
http://dx.doi.org/10.1090/conm/625/12490

Minimal fillings of finite metric spaces: The state of the art

Alexandr Ivanov and Alexey Tuzhilin

ABSTRACT. We present a review on a new branch of one-dimensional geometrical optimization problem, the minimal fillings theory. This theory is connected closely with generalized Steiner problem and gives an opportunity to look at many classical questions appearing in optimal connection theory from new point of view.

1. Introduction: Length-Minimizing Connections

Problems related to length minimization form a very popular and important class of geometric optimization problems. We start with a discussion of possible approaches to the problem (Section 1) and then focus on minimal fillings of finite metric spaces (Sections 2–12).

In general terms, the problem of length minimization is stated as follows. Let $M = \{A_1, \dots, A_n\}$ be a finite set of points in a metric space (X, ρ). We would like to find a minimum-length connection of the points in X in terms of the total length of the connecting curves. We assume that we know how to connect pairs of points in X using lines or curves, so our goal is limited to organizing the set of shortest curves between the points of M in an optimal way. There are several natural statements of the problem, and Minimal Fillings Problem is just one of them. Below we list the most popular statements and discuss natural relations between them (Sections 1.1–1.3).

1.1. No Additional Forks Case: Spanning Trees. In this problem the paths between the points branch only at the points themselves, in other words, no forks between the points of M are allowed. As a result, we obtain a particular case of Graph Theory problem about minimal spanning trees in a connected weighted graph. We recall only necessary concepts of Graph Theory, the details can be found, for example in [**4**].

A (*simple*) *graph* can be considered as a pair $G = (V, E)$, consisting of a finite set $V = \{v_1, \dots, v_n\}$ of *vertices* and a finite set $E = \{e_1, \dots, e_m\}$ of *edges*, where each edge e_i is a two-element subset of V. If $e = \{v, v'\}$, then we say that v and v' are *neighboring*, edge e *joins* or *connects* them, the edge e and each of the vertices

---

2010 *Mathematics Subject Classification.* Primary 58E15; Secondary 51K99.
Key words and phrases. Minimal networks, Steiner problem, minimal spanning trees, minimal fillings of finite metric spaces.

v and v' are *incident.* The number of vertices neighboring to a vertex v is called the *degree of* v and is denoted by $\deg v$. A graph $H = (V_H, E_H)$ is said to be a *subgraph* of a graph $G = (V_G, E_G)$, if $V_H \subset V_G$ and $E_H \subset E_G$. The subgraph H is called *spanning,* if $V_H = V_G$.

A *route* γ in a graph G is a sequence $v_{i_1}, e_{i_1}, v_{i_2} \ldots, e_{i_k} v_{i_{k+1}}$ of its vertices and edges such that each edge e_{i_s} connects vertices v_{i_s} and $v_{i_{s+1}}$. We also say that the route γ *connects* the vertices v_{i_1} and $v_{i_{k+1}}$ which are said to be *ending vertices* of the route. A route is said to be *cyclic,* if its ending vertices coincide with each other. A route with pairwise distinct edges is referred as a *path.* A cyclic path is referred as a *cycle.* A graph without cycles is said to be *acyclic.* A graph is said to be *connected,* if any two its vertices can be connected by a route. An acyclic connected graph is called a *tree.*

If we are given with a function $\omega\colon E \to \mathbb{R}$ on the edge set of a graph G, then the pair (G, ω) is referred as a *weighted graph.* For any subgraph $H = (V_H, E_H)$ of a weighted graph $\big(G, \omega\big)$ the value $\omega(H) = \sum_{e \in E_H} \omega(e)$ is called the *weight of* H. Similarly, for any route $\gamma = v_{i_1}, e_{i_1}, v_{i_2} \ldots, e_{i_k} v_{i_{k+1}}$ the value $\omega(\gamma) = \sum_{s=1}^{k} \omega(e_{i_s})$ is called the *weight of* γ.

For a weighted connected graph $\big(G, \omega\big)$ with positive weight function ω, a spanning connected subgraph of minimal possible weight is called by a *minimal spanning tree.* The positivity of ω implies that such subgraph is *acyclic*; i.e., it is a tree indeed. The weight of any minimal spanning tree for (G, ω) is denoted by $\operatorname{mst}(G, \omega)$. The following result is well-known in Graph Theory.

ASSERTION 1.1. *For any weighted connected graph $\big(G, \omega\big)$ there exists a minimal spanning tree. A minimal spanning tree can be constructed in a polynomial time with respect to the number of graph's G vertices.*

Optimal connection problem without additional forks can be considered as minimal spanning tree problem for a special graph. Let $M = \{A_1, \ldots, A_n\}$ be a finite set of points in a metric space (X, ρ) as above. Consider the complete graph $K(M)$ with vertex set M and edge set consisting of all two-element subsets of M. In other words, any two vertices A_i and A_j are connected by an edge in $K(M)$. By A_iA_j we denote the corresponding edge. The number of edges in $K(M)$ is, evidently, $n(n-1)/2$. We define the positive weight function $\omega_\rho(A_iA_j) = \rho(A_i, A_j)$. Then any minimal spanning tree T in $K(M)$ can be considered as a set of shortest curves in (X, ρ) joining corresponding points and forming a network in X connecting M without additional forks in an optimal way; i.e., with the least possible length. Such a network is called a *minimal spanning tree for M in (X, ρ)*. Its total weight $\omega_\rho(T)$ is called *length* and is denoted by $\operatorname{mst}_X(M)$. Assertion 1.1 implies the following general result.

COROLLARY 1.2. *For any finite subset M of any metric space the optimal connection problem without additional forks can be solved in polynomial time with respect to the number of points in M.*

But it is well-known that in general case the length of minimal spanning tree can be decreased using *additional forks.* This idea is discussed in the next Subsection.

1.2. Shortest tree: Fermat–Steiner Problem. Already P. Fermat and C. F. Gauss understood that additional forks can be profitable; i.e., can give an opportunity to decrease the length of optimal connection. For example, see Figure 1, if we consider the vertex set $M = \{A_1, A_2, A_3\}$ of a regular triangle with side 1 in the Euclidean plane, then the corresponding graph $K(M)$ consists of three edges of the same weight 1 and each minimal spanning tree consists of two edges, so $\mathrm{mst}_{\mathbb{R}^2}(M) = 2$. But if we add the center T of the triangle and consider the network consisting of three straight segments A_1T, A_2T, A_3T, then its length is equal to $3\frac{2}{3}\frac{\sqrt{3}}{2} = \sqrt{3} < 2$, so it is shorter than the minimal spanning tree.

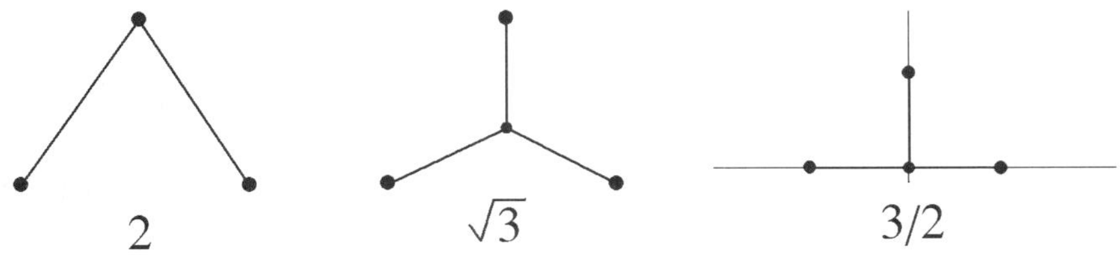

FIGURE 1. Minimal spanning tree (left), shortest tree (center), and minimal filling, connecting the vertex set of regular triangle in Euclidean plane.

This example leads to the following general definition. Let $M = \{A_1, \ldots, A_n\}$ be a finite set of points in a metric space (X, ρ) as above. Consider a larger finite set N, $M \subset N \subset X$, and a minimal spanning tree for N in X. Then this tree contains M as a subset of its vertex set N, but also may contain some other additional vertices-forks. Such additional vertices are referred as *Steiner points*. Further, we define a value $\mathrm{smt}_X(M) = \inf_{N: M \subset N \subset X} \mathrm{mst}_X(N)$ and call it by the *length of shortest tree connecting* M or of *Steiner minimal tree for* M. If this infimum attains at some set N, then each minimal spanning tree G for this N is called a *shortest tree* or a *Steiner minimal tree* connecting M, and the set M is referred as *boundary* of G. The celebrated *Steiner problem* is the problem of finding a shortest tree for a given finite subset of a metric space. The shortest tree for the vertex set of a regular triangle in the Euclidean plane is depicted in Figure 1.

Remark. There are no analogues of Corollary1.2 for the Steiner problem. Namely, there are examples of boundary subsets M of complete metric spaces (even Banach spaces), such that shortest tree does not exist, see for example [**24**]. Moreover, even for metric spaces where the existence of a shortest tree is proved for an arbitrary finite boundary (for example, Euclidean spaces), there is no a polynomial algorithm constructing a shortest tree for a given boundary. A review and details on shortest trees an be found in [**1**].

It turns out, that the length of shortest tree can be decreased in some sense also. This opportunity appears due to minimal fillings which are the main subject of the paper.

1.3. Minimizing over Different Ambient Spaces: Minimal Fillings. Shortest trees give the least possible length of connecting network for a given finite set in a fixed ambient space. But sometimes it is possible to decrease the length of connection by choosing another ambient space. Let $M = \{A_1, \ldots, A_n\}$ be a finite set of points in a metric space (X, ρ) as above, and consider M as a finite metric space with the distance function ρ_M obtained as the restriction of the distance function ρ. Consider an isometric embedding $\varphi\colon (M, \rho_M) \to (Y, \rho_Y)$ of this finite metric space (M, ρ_M) into a (compact) metric space (Y, ρ_Y) and consider the value $\mathrm{smt}_Y\big(\varphi(M)\big)$. It could be less than $\mathrm{smt}_X(M)$. For example, the vertex set of the regular triangle with side 1 can be embedded into Manhattan plane as the set $\big\{(-1/2, 0), (0, 1/2), (1/2, 0)\big\}$, see Figure 1. Than the unique additional vertex of the shortest tree is the origin, and the length of the tree is $3/2 < \sqrt{3}$.

Thus, for a finite metric space $\mathcal{M} = (M, \rho_M)$, consider the value $\mathrm{mf}(\mathcal{M}) = \inf_\varphi \mathrm{smt}_Y\big(\varphi(M)\big)$ which is referred as *weight of minimal filling* of the finite metric space $\mathcal{M}$. Minimal fillings for finite metric spaces were defined in [**3**] and turned out to be connected closely with shortest trees geometry. On the other hand, minimal fillings of finite metric spaces can be considered as a generalization of Gromov's concept of minimal fillings for Riemannian manifolds. The present paper gives the state of the art of this modern rapidly developing branch of discrete geometrical optimization.

In Section 2 we discuss a combinatorial definition of a filling of a finite metric space M as a weighted connected graph connecting M and having special metric properties, and the relation of this definition with classical Gromov minimal fillings concept [**22**]. Section 3 is devoted to so-called parametric minimal fillings; i.e., minimal fillings with prescribed graph structure (referred as topology). General minimal fillings problem can be considered as minimization of parametric minimal fillings with respect to the topologies of the fillings. It turns out, that for a given metric space the set of the topologies under consideration can be made finite. In Section 4 we show that any minimal filling can be realized as a shortest network in an appropriate ambient space. This result implies relations between minimal fillings and shortest networks properties. It turns out, that metric spaces where each shortest tree is a minimal filling for its boundary have many interesting properties. In Section 5 we reduce parametric minimal filling problem to a linear programming and prove the existence results for minimal fillings. Sections 6 and 7 are devoted to general formula calculating the weight of a minimal filling in terms of so-called multi-tours of a metric space. It turns out that to obtain this formula one needs to generalize a concept of a filling permitting negative weights of the edges. In Section 8 we discuss non-uniqueness results. It is shown that as uniqueness, so as non-uniqueness of a minimal filling can hold on open families of merit spaces. Sections 9 and 10 are devoted to special examples of metric spaces, which the minimal fillings can be constructed for. In particular, we consider additive and pseudo-additive metric spaces (the spaces with four-points rule and generalized four-points rule). Section 11 is devoted to Steiner type ratios. The classical Steiner ratio is a well-known characteristic of a metric space measuring the relative error of shortest tree approximation by a minimal spanning tree. We consider some natural generalization of this concept by means of minimal fillings. Section 12 devoted to generalization of the minimal filling concept to infinite metric spaces permitting connection by a tree of finite length. This object is closely related to metric geometry and classical Gromov minimal filling definition.

2. Combinatorial Definition of Minimal Filling

The concept of a minimal filling for Riemannian manifolds appeared in papers of Gromov, see [**22**]. Let M be a manifold endowed with a distance function ρ. Consider all possible films X spanning M; i.e., compact manifolds with the boundary M. Consider on X a distance function d that does not decrease the

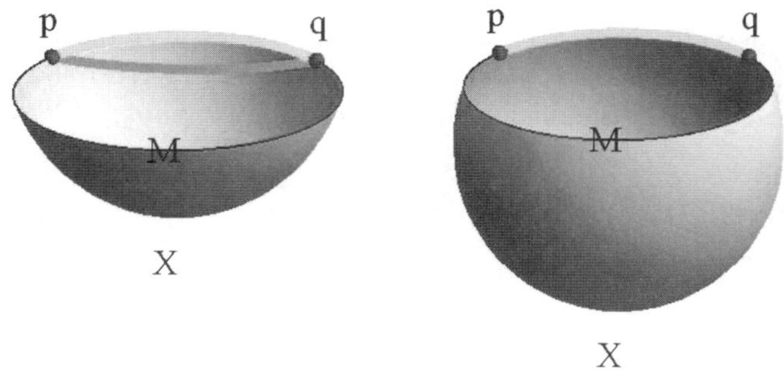

FIGURE 2. The space M is the circle S^1 with arc-metric. The films X in the both Figures are parts of the standard sphere containing M as a parallel. The left film X is not a filling since the distance between the points p and q in X is less than in M (the shortest path is shown in orange color). The right film X is a filling of M.

distances between points in M. Such a metric space $\mathcal{X} = (X, d)$ is cal,led a *filling* of the metric space $\mathcal{M} = (M, \rho)$, see example in Figure 2. The Gromov Problem consists in calculating the infimum of the volumes of the fillings and describing the spaces $\mathcal{X}$ which this infimum is achieved at (such spaces are called *minimal fillings*).

In the scope of Steiner problem, it is natural to expand the concept of fillings to finite metric spaces. Then the possible fillings are metric spaces having the structure of one-dimensional stratified manifolds which can be considered as graphs whose edges have nonnegative weights. This leads to the following particular case of generalized Gromov problem.

Let M be an arbitrary finite set, and $G = (V, E)$ be a connected graph. We say, that G *connects* M or *joins* M, if $M \subset V$. In this case M is referred as a *boundary* of G and is denoted by ∂G. Now, let $\mathcal{M} = (M, \rho)$ be a finite metric space, $G = (V, E)$ be a connected graph joining M, and $\omega \colon E \to \mathbb{R}_+$ is a mapping into non-negative numbers, which is usually referred as a *weight function* and which generates the *weighted graph* $\mathcal{G} = (G, \omega)$. The function ω generates on V the pseudo-metric d_ω (recall that some distances in a pseudo-metric can be equal to zero), namely, the d_ω-distance between two vertices of the graph $\mathcal{G}$ is defined as the least possible weight of the paths in $\mathcal{G}$ joining these vertices. If for any two points p and q from M the inequality $\rho(p, q) \le d_\omega(p, q)$ holds, then the weighted graph $\mathcal{G}$ is called a *filling* of the space $\mathcal{M}$, and the graph G is referred as the *type* of this filing. The value $\mathrm{mf}(\mathcal{M}) = \inf \omega(\mathcal{G})$, where the infimum is taken over all fillings $\mathcal{G}$ of the space $\mathcal{M}$ is the *weight of minimal filling*, and each filling $\mathcal{G}$ such that $\omega(\mathcal{G}) = \mathrm{mf}(\mathcal{M})$ is called a *minimal filling*.

3. Parametric Minimal Fillings

Let $\mathcal{M} = (M, \rho)$ be a finite metric space and $G = (V, E)$ be an arbitrary connected graph connecting M. By $\Omega(\mathcal{M}, G)$ we denote the set of all weight functions $\omega \colon E \to \mathbb{R}$ such that (G, ω) is a filling of the space $\mathcal{M}$. We put

$$\mathrm{mpf}(\mathcal{M}, G) = \inf_{\omega \in \Omega(\mathcal{M}, G)} \omega(G)$$

and we call this value the *weight of minimal parametric filling of the type G for the space $\mathcal{M}$*. If there exists a weight function $\omega \in \Omega(\mathcal{M}, G)$ such that $\omega(G) = \mathrm{mpf}(\mathcal{M}, G)$, then (G, ω) is called a *minimal parametric filling of the type G for the space $\mathcal{M}$*.

ASSERTION 3.1. *Let $\mathcal{M} = (M, \rho)$ be a finite metric space. Then*

$$\mathrm{mf}(\mathcal{M}) = \inf\big\{\mathrm{mpf}(\mathcal{M}, G)\big\},$$

where the infimum is taken over all connected graphs G joining M.

It is not difficult to show that to investigate minimal fillings one can restrict the consideration to trees such that all their vertices of degree 1 and 2 belong to their boundaries. **In what follows, we always assume that this assumption holds, providing the opposite is not declared.**

To be more precise, we recall the following definition. We say that a tree is a *binary* one if the degrees of its vertices can be 1 or 3 only, and the boundary ∂G consists just of all the vertices of degree 1. Then each finite metric space has a minimal filling whose type is a binary tree (possibly, with some degenerate edges; i.e., the edges of weight zero), and a minimal filling whose type is a tree and all weights are positive (and all whose vertices of degree 1 and 2 belong to its boundary in accordance with the above agreement), see [**3**].

4. Realization of Minimal Filling as a Minimal Network

It turns out that the problem on minimal filling can be reduced to Steiner problem in special metric spaces and for special boundaries.

We recall a useful concept of a network in a metric space. Let $G = (V, E)$ be a connected graph, and $\mathcal{X} = (X, \rho)$ be a metric space. A mapping $\Gamma\colon V \to X$ is called by a *network of the type G in the space $\mathcal{X}$*. By a *vertex* or an *edge* of the network Γ we mean the restriction of Γ on a vertex or an edge of the graph G, respectively. If some finite subset $M \subset X$ is contained in $\Gamma(V)$, then we say that the network Γ *connects* or *joins* the set M. Usually we suppose that a boundary $\partial G \subset V$ is fixed. Then the restriction of Γ on to ∂G is called by the *boundary of the network* and is denoted by $\partial\Gamma$. In this case, we say that the network Γ connects $M = \Gamma(\partial G)$ by the boundary mapping $\partial\Gamma$. Each network $\Gamma\colon V \to X$ generates a weight function ω_Γ on the edges of the graph G that is defined as follows: $\omega_\Gamma(uv) = \rho\big(\Gamma(u), \Gamma(v)\big)$. The weight $\omega_\Gamma(G)$ of this graph is referred as the *length of the network* Γ.

Consider a finite set $M = \{p_1, \ldots, p_n\}$, and let $\mathcal{M} = (M, \rho)$ be a metric space. We put $\rho_{ij} = \rho(p_i, p_j)$. By $\mathbb{R}^n_\infty$ we denote the n-dimensional arithmetic space with the norm

$$\big\|(v^1, \ldots, v^n)\big\|_\infty = \max\big\{|v^1|, \ldots, |v^n|\big\},$$

and by ρ_∞ the metric on $\mathbb{R}^n_\infty$ generated by $\|\cdot\|_\infty$; i.e., $\rho_\infty(v, w) = \|w - v\|_\infty$. Let us define a mapping $\varphi_\mathcal{M}\colon M \to \mathbb{R}^n_\infty$ as follows:

$$\varphi_\mathcal{M}(p_i) = \bar{p}_i = (\rho_{i1}, \ldots, \rho_{in}).$$

ASSERTION 4.1. *The mapping $\varphi_\mathcal{M}$ is an isometry of $\mathcal{M}$ onto its image.*

PROOF. This easily follows from the triangle inequality. Indeed,

$$\big\|\bar{p}_i - \bar{p}_j\big\| = \max_k |\rho_{ik} - \rho_{jk}| \ge \rho_{ij},$$

because the value ρ_{ij} stands at the ith and jth places of the vector $\bar{p}_i - \bar{p}_j$. On the other hand, $\rho_{ij} \geq \rho_{ik} - \rho_{jk}$ for any k, due to the triangle inequality, hence $\|\bar{p}_i - \bar{p}_j\| \leq \rho_{ij}$, and Assertion is proved. □

The mapping $\varphi_{\mathcal{M}}$ is called the *Kuratowski isometry.*

Let $\mathcal{G} = (G, \omega)$ be a filling of a space $\mathcal{M} = (M, \rho)$, where $G = (V, E)$, and d_ω be the pseudo-metric on V generated by the weight function ω. By E_M we denote the edges set of the complete graph on M and put $\bar{G} = (V, \bar{E} = E \cup E_M)$. Let $\bar{\omega}$ be the weight function on $\bar{E}$ coinciding with metric ρ on E_M and with ω on $\bar{E} \setminus E_M$. Recall that $d_{\bar{\omega}}$ denotes the pseudo-metric on V generated by $\bar{\omega}$.

We define the network $\Gamma_{\mathcal{G}} \colon V \to \mathbb{R}^n_\infty$ of the type G as follows:

$$\Gamma_{\mathcal{G}}(v) = \big(d_{\bar{\omega}}(v, p_1), \dots, d_{\bar{\omega}}(v, p_n)\big).$$

This network is called the *Kuratowski network for the filling $\mathcal{G}$.*

ASSERTION 4.2. *We have $\partial\Gamma_{\mathcal{G}} = \varphi_{\mathcal{M}}$.*

PROOF. This easily follows from the filling definition. Indeed, the mapping $\partial\Gamma_{\mathcal{G}}$ is defined on the set M only. By definition,

$$\Gamma_{\mathcal{G}}(p_i) = \big(d_{\bar{\omega}}(p_i, p_1), \dots, d_{\bar{\omega}}(p_i, p_n)\big),$$

hence it suffices to show that $d_{\bar{\omega}}(p_i, p_k) = \rho_{ik}$ for any k. The vertices p_i and p_k are joined by the edge p_ip_k of the weight ρ_{ik} in the graph $\bar{G}$, and the weight of any other path in G connecting p_i and p_k is more than or equal to ρ_{ik}, because G is a filling. Assertion is proved. □

COROLLARY 4.3. *Let $\mathcal{G} = (G, \omega)$ be a minimal parametric filling of a metric space (M, ρ) and $\Gamma = \Gamma_{\mathcal{G}}$ be the corresponding Kuratowski network. Then $\omega = \omega_\Gamma$.*

Let Γ be a network in a metric space $\mathcal{X}$, let G be its parameterizing graph, and $\mathcal{H} = (H, \omega)$ be a weighted graph. We say that Γ *and $\mathcal{H}$ are isometric*, if there exists an isomorphism of the weighted graphs $\mathcal{H}$ and $\mathcal{G} = (G, \omega_\Gamma)$.

Corollary 4.3 and the existence of a shortest trees in a finite-dimensional normed space [**1**] imply the following result.

COROLLARY 4.4. *Let $\mathcal{M} = (M, \rho)$ be a metric space consisting of n points, and $\varphi_{\mathcal{M}} \colon M \to \mathbb{R}^n_\infty$ be the Kuratowski isometry. Then there exists a minimal filling $\mathcal{G}$ for $\mathcal{M}$, and the corresponding Kuratowski network $\Gamma_{\mathcal{G}}$ is a shortest network in the space $\mathbb{R}^n_\infty$ joining the set $\varphi_{\mathcal{M}}(M)$. Conversely, each shortest network on $\varphi_{\mathcal{M}}(M)$ is isometric to some minimal filling of the space $\mathcal{M}$.*

Thus, shortest networks on some special boundaries in $\mathbb{R}^n_\infty$ give minimal fillings for these boundaries considered as finite metric spaces. Z. N. Ovsyannikov has proved that the same is true for arbitrary boundaries in $\mathbb{R}^n_\infty$.

ASSERTION 4.5 (Z. Ovsyannikov). *For any finite subset M of the normed space $\mathbb{R}^n_\infty$ any shortest tree connecting M is a realization of a minimal filling of M. In particular,* $\mathrm{smt}(M) = \mathrm{mf}(M)$.

Recently, B. B. Bednov and P. A. Borodin [**25**] investigate realization problem in general Banach spaces. Let us say, that a metric space $\mathcal{X} = (X, \rho)$ *realizes minimal filling* for its finite subset $M \subset X$, if (1) a shortest tree connecting M does exist, and (2) the equality $\mathrm{smt}(M) = \mathrm{mf}(M)$ holds.

THEOREM 4.6 (B. Bednov and P. Borodin). *A Banach space realizes minimal filling for all it finite subsets, if and only if it is a Lindenstrauss space.*

Notice that a finite dimensional normed space is Lindenstrauss, if and only if it is $\mathbb{R}^n_\infty$.

B. B. Bednov and P. A. Borodin [**25**] proved that a Banach space realizes minimal fillings for all its finite subsets, if and only if it realizes minimal fillings for all its subsets consisting of at most four points. The class of Banach spaces realizing minimal fillings for all three points subsets only is wider and coincide with another well-known class of spaces satisfying so-called 3.2.I.P property. The latter class contains L_1-spaces that can be characterized by uniqueness of minimal filling for any triplet of points.

5. Minimal Parametric Fillings and Linear Programming

Let $\mathcal{M} = (M, \rho)$ be a finite metric space connected by a (connected) graph $G = (V, E)$. As above, by $\Omega(\mathcal{M}, G)$ we denote the set consisting of all the weight functions $\omega\colon E \to \mathbb{R}_+$ such that $\mathcal{G} = (G, \omega)$ is a filling of $\mathcal{M}$, and by $\Omega_m(\mathcal{M}, G)$ we denote its subset consisting of the weight functions such that $\mathcal{G}$ is a minimal parametric filling of $\mathcal{M}$.

ASSERTION 5.1. *The set $\Omega(\mathcal{M}, G)$ is closed and convex in the linear space $\mathbb{R}^E$ of all the functions on E, and $\Omega_m(\mathcal{M}, G) \subset \Omega(\mathcal{M}, G)$ is a nonempty convex compact.*

PROOF. It is easy to see, that the set $\Omega(\mathcal{M}, G) \subset \mathbb{R}^E$ is determined by the linear inequalities of two types: $\omega(e) \ge 0$, $e \in E$, and $\sum_{e\in\gamma_{pq}} \omega(e) \ge \rho(p, q)$, where γ_{pq} stands for the unique path in the tree G connecting the boundary vertices p and q. Therefore, $\Omega(\mathcal{M}, G)$ is a convex closed polyhedral subset of $\mathbb{R}^E$ that is equal to the intersection of the corresponding closed half-spaces. The weight functions of minimal parametric fillings correspond to minima points of the linear function $\sum_{e\in E} \omega(e)$ restricted to the set $\Omega(\mathcal{M}, G)$. Thus, the problem of minimal parametric filling finding is a linear programming problem, and the set $\Omega_m(\mathcal{M}, G)$ of all minima points is a nonempty convex compact polyhedron (the boundedness and, hence, compactness of this set follows from increasing of the objective function with respect to each its variable). □

6. Generalized Fillings

Investigating the fillings of finite metric spaces, it turns out to be convenient to expand the class of weighted trees under consideration permitting arbitrary weights of the edges (not only non-negative). The corresponding objects are called *generalized fillings*, *minimal generalized fillings* and *minimal parametric generalized fillings*. Their weights for a metric space $\mathcal{M}$ and a tree G are denoted by $\mathrm{mf}_-(\mathcal{M})$ and $\mathrm{mpf}_-(\mathcal{M}, G)$, respectively.

For any finite metric space $\mathcal{M} = (M, \rho)$ and a tree G connecting M, the next evident inequality is valid: $\mathrm{mpf}_-(\mathcal{M}, G) \le \mathrm{mpf}(\mathcal{M}, G)$. And it is not difficult to construct an example, when this inequality becomes strict, see Figure 3. However, for minimal generalized fillings the following result holds, see [**26**].

THEOREM 6.1 (Ivanov, Ovsyannikov, Strelkova, Tuzhilin). *For an arbitrary finite metric space $\mathcal{M}$, the set of all its minimal generalized fillings contains its minimal filling; i.e., a generalized minimal filling with nonnegative weight function. Hence,* $\mathrm{mf}_-(\mathcal{M}) = \mathrm{mf}(\mathcal{M})$.

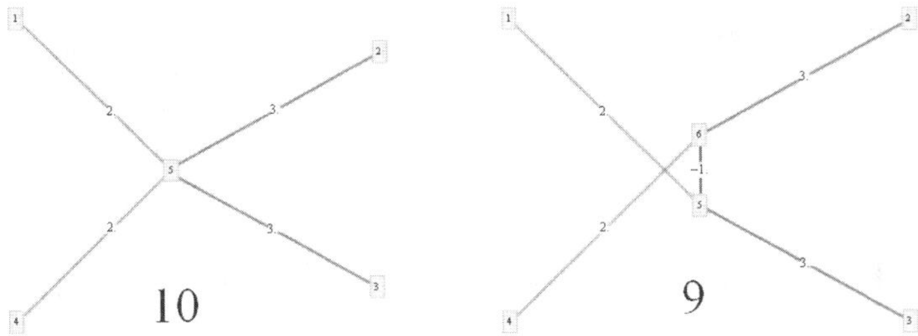

FIGURE 3. Minimal parametric filling (left) and minimal generalized parametric filling (right) of the vertex set of the plane rectangle with sides 3 and 4. The type is the same: the moustaches connects the diagonal pairs of the vertices. The interior edge has to be zero in the case of the filling and can be negative in the case of the generalized filling. Here $9 = \mathrm{mpf}_-(\mathcal{M}, G) < \mathrm{mpf}(\mathcal{M}, G) = 10$.

7. Formula for the Weight of Minimal Filling

It turns out, that the concept of generalized filling gives an opportunity to derive a formula for the weight of minimal filling in some geometrical terms.

7.1. Tours and Perimeters. Let $\mathcal{M} = (M, \rho)$ be a finite metric space, and G be a tree connecting M. Choose an arbitrary embedding G' of the tree G into the plane. Consider a walk around the tree G' in the plane. We draw the points of M consecutive with respect to this walk as a consecutive points of a circle S^1. Notice that each vertex p from M appears $\deg p$ times. For each vertex $p \in M$ of degree more than 1, we choose just one arbitrary point from the corresponding points of the circle. So, we construct an injection $\nu\colon M \to S^1$. Define a cyclic permutation π as follows: $\pi(p) = q$, where $\nu(q)$ follows after $\nu(p)$ on the circle S^1. We say that *π is generated by the embedding G'* (this procedure is not unique due to different possible choices of ν). Each π generated in this manner is called a *tour of M with respect to G*. The set of all tours on M with respect to G is denoted by $\mathcal{O}(M, G)$. For each tour $\pi \in \mathcal{O}(M, G)$ we put

$$p(\mathcal{M}, G, \pi) = \frac{1}{2} \sum_{x \in M} \rho\big(x, \pi(x)\big)$$

and we call this value by the *half-perimeter of the space $\mathcal{M}$ with respect to the tour π*. The minimal value of $p(\mathcal{M}, G, \pi)$ over all $\pi \in \mathcal{O}(M, G)$ for all possible trees G (in fact, over all possible cyclic permutations π on M) is called the *half-perimeter of the space $\mathcal{M}$*.

A. Ivanov and A. Tuzhilin conjectured that the weight of minimal parametric filling of a given type G, where G is a binary tree, can be found as maximal half-perimeter over all permutations $\pi \in \mathcal{O}(M, G)$. A. Yu. Eremin [**27**] showed that to make this conjecture valid it is necessary to generalize the concept of tour.

7.2. Mutitours and $\min\max$ **Formula.** To start with, consider the doubling of a tree G connecting M; i.e., the graph with the same vertex set, but containing each edge of G with multiplicity 2, see Figure 4. The resulting graph is Euler's, and each Euler cycle in it can be decomposed into the union of consecutive *irreducible*

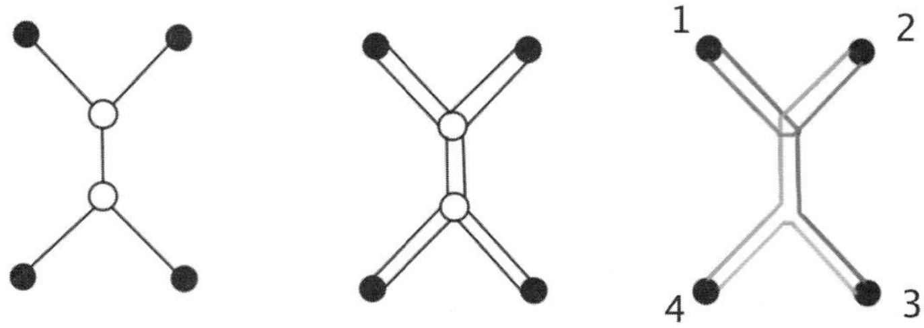

FIGURE 4. Binary tree (left), its doubling (middle) and an Euler cycle in the doubling decomposed into the union of irreducible boundary paths. The corresponding tour has the form $\pi = (1, 3, 4, 2)$.

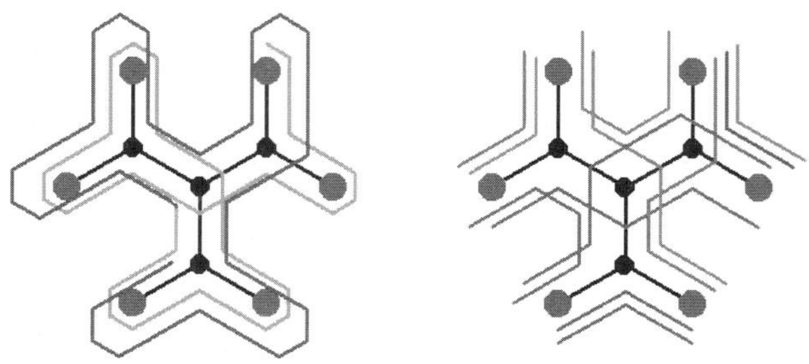

FIGURE 5. A part of a moultitour with multiplicity 2 (left), and the irreducible boundary paths forming this multitour (right). The multitour starts as a green polygonal line and becomes blue when multiplicity of edges becomes more than 2.

boundary paths; i.e., the paths connecting boundary vertices and do not containing other boundary vertices. The corresponding permutation π maping the beginning vertex of each irreducible boundary path onto its ending one is a tour of M with respect to G.

Now let us consider the graph in which every edge of G is taken with the multiplicity $2k$, $k \geq 1$. The resulting graph possesses an Euler cycle consisting of irreducible boundary paths, see an example in Figure 5. This Euler cycle generates a bijection $\pi\colon X \to X$, where $X = \sqcup_{i=1}^{k} M$, which is called *multitour of M with respect to G*. As above, this bijection maps the beginning vertices of irreducible boundary paths onto the ending ones. The set of all multitours on M with respect to G is denoted by $\mathcal{O}_\mu(M, G)$.

Let $\mathcal{M} = (M, \rho)$ be a finite metric space, and G be a tree connecting M. As in the case of tours, for each multitour $\pi \in \mathcal{O}_\mu(M, G)$ we put

$$p(\mathcal{M}, G, \pi) = \frac{1}{2k} \sum_{x \in X} \rho\big(x, \pi(x)\big).$$

THEOREM 7.1 (A. Yu. Eremin). *For an arbitrary finite metric space $\mathcal{M} = (M, \rho)$ and an arbitrary binary tree G joining M, the weight of minimal parametric* *generalized filing can be calculated as follows:*

$$\mathrm{mpf}_-(\mathcal{M},G)=\max\{p(\mathcal{M},G,\pi)\mid \pi\in\mathcal{O}_\mu(M,G)\}.$$

The weight of minimal filling can be calculated as follows:

$$\mathrm{mf}(\mathcal{M})=\mathrm{mf}_-(\mathcal{M})=\min_G\max\{p(\mathcal{M},G,\pi)\mid \pi\in\mathcal{O}_\mu(M,G)\},$$

where minimum is taken over all binary trees G connecting M.

Remark. A. Eremin proves Theorem 7.1 for so-called ρ-spaces (a generalization of the concept of metric space where "distance" function ρ is just symmetric).

Remark. Formally, the maximum in the formula for the weight of minimal parametric generalized filling in Theorem 7.1 is taken over the infinite set of all its multitours. In [**27**] it is proved that the maximum can be taken over so-called *irreducible* multitours, namely, the multitours that can not be presented as a union of consecutive multitours. The set of irreducible multitours is finite, since the multiplicity of each irreducible multitour does not exceed $\binom{n}{2}!$, see details in [**27**].

Now we list several corollaries of Theorem 7.1.

7.3. Minimal Fillings for Generic Metric Spaces. The set of all metric spaces consisting of n points can be naturally identified with a convex cone in $\mathbb{R}^{n(n-1)/2}$ (it suffices to enumerate the set of all two-elements subsets of these spaces and assign to each such space the vector of the distances between the pairs of points). This representation gives us an opportunity to speak about topological properties of families of metric spaces consisting of a fixed number of points.

We say, that some property holds for a *generic metric space*, if for any n this property is valid for an everywhere dense set of n-point metric spaces.

COROLLARY 7.2 (A. Yu. Eremin). *Any generalized minimal filling of a generic finite metric space is a binary tree with a positive weight function.*

7.4. Exact and Maximal Multitours. Let $\mathcal{M}=(M,\rho)$ be a metric space, and G be a binary tree connecting M. A multitour π of a binary tree G is said to be *maximal*, if $p(\mathcal{M},G,\pi)=\max_{\sigma\in\mathcal{O}_\mu(M,G)}p(\mathcal{M},G,\sigma)$. A multitour of a filling (G,ω) of the space $\mathcal{M}$ is said to be *exact*, if any irreducible boundary path γ forming the corresponding Euler cycle is exact in the sense of [**3**]; i.e., if γ connects boundary vertices u and v, than $\omega(\gamma)=\rho(u,v)$. The details concerning exact paths can be found in [**3**].

COROLLARY 7.3. *Let G be a binary tree connecting M, and $\mathcal{M}=(M,\rho)$ be a metric space. A multitour π of the tree G is maximal, if and only if it is exact for any minimal parametric generalized filling of type G of the space $\mathcal{M}$.*

Any minimal parametric generalized filling of type G of the space $\mathcal{M}$ has at least one exact irreducible multitour.

8. Uniqueness Problem

Uniqueness "in general position" seems to be a very natural property for boundary optimization problems. In the case of classical Steiner problem A. Ivanov and A. Tuzhilin [**36**] proved that the shortest tree is unique for generic finite subsets of the Euclidean plane. But the example of the Manhattan plane shows that even the set of shortest segments connecting a pair of points can be infinite for generic pairs. Since each minimal filling can be realized as a shortest tree in $\mathbb{R}^n_\infty$, then non-uniqueness of minimal filling even in "general position" becomes an expected result. But this intuitively expected non-uniqueness refers primary to non-uniqueness of weight function for a fixed topology.

But it turns out that there is no uniqueness of minimal filling for the metric spaces "in general position" even in the sense of topology. This observation and example belong to Z. Ovsyannikov [**29**] and is based on Theorem 7.1.

Let us consider the convex cone C_n of all $n \times n$ pseudo-distance matrices as a subset of $\mathbb{R}^{n(n-1)/2}$. Let us fix a set M and by ρ_m denote the pseudo-distance function generated by the matrix $m \in C_n$ on M. So, the cone C_n is a configuration space of all n-element metric spaces.

Let G be a binary tree whose boundary ∂G is equal to M (recall that in accordance with our Agreement ∂G coincides with the set of all the vertices of degree 1). Each multitour π of G generates the linear function $p(\mathcal{M}, G, \pi)$ on C_n, where $\mathcal{M} = (M, \rho_m)$ (the value of this function at point $m \in C_n$ is equal just to the sum of the appropriate elements of the matrix m). Hence, in accordance with Theorem 7.1, the weight of minimal parametric filling can be considered as a piecewise linear function $f_G\colon C_n \to \mathbb{R}$, namely, $f_G(m) = \max_\pi p(\mathcal{M}, G, \pi)$, — the maximum of a set of liner function. Moreover f_G is a convex function on C_n.

LEMMA 8.1. *The weight of a minimal parametric filling considered as a function on the cone C_n of the pseudo-distance matrices is piecewise linear convex function.*

Finally, the weight of minimal filling can be written as $h(m) = \min_G f_G(m)$, the minimum of piecewise linear functions which is a piecewise linear function itself. Thus, the following simple Lemma holds.

LEMMA 8.2. *The weight of minimal filling considered as a function on the cone C_n of the pseudo-distance matrices is piecewise linear.*

In accordance with Lemma 8.2, the cone C_n is partitioned into full dimensional "polyhedra–chambers" Δ_i in such a way that function h is linear on each the chamber. And, due to the construction, $h|_{\Delta_i}$ coincides with some of the functions $p(\mathcal{M}, G, \pi)$ for some G and π. Notice that the form of the linear function $h|_{\Delta_i}$ is completely determined by the multitour π, and do not depend on G. So, if π is a maximal moultitour for several topologies, say $G_1, \ldots, G_k$, then $\mathrm{mf}(\mathcal{M}) = p(\mathcal{M}, G_i, \pi)$ for all $i = 1, \ldots, n$. And if the perimeters of all the other multitours are strictly less than $\mathrm{mf}(\mathcal{M})$, then these relations remains the same for arbitrary small perturbations of the distance matrix m, and hence, we get an open subset U of $C_n \subset \mathbb{R}^{n(n-1)/2}$ such that any space (M, ρ_m) with $m \in U$ has minimal fillings of all the topologies G_i.

Z. Ovsyannikov constructed the 6 points metric space $\mathcal{M}_z = (M, \rho_{m_z})$ with the following distance matrix:

$$m_z = \begin{pmatrix} 0 & 13783 & 13200 & 17254 & 12758 & 13170 \\ 13783 & 0 & 19324 & 15679 & 12512 & 18790 \\ 13200 & 19324 & 0 & 12174 & 12307 & 13797 \\ 17254 & 15679 & 12174 & 0 & 14446 & 12616 \\ 12758 & 12512 & 12307 & 14446 & 0 & 17870 \\ 13170 & 18790 & 13797 & 12616 & 17870 & 0 \end{pmatrix}.$$

The permutation $\pi = (1,4,3,6,5,2)$ is a tour for many binary trees, three of them are depicted in Figure 6. Using a computer program written in *Mathematica* on the base of the direct verification of all possible topologies of binary trees and linear programming, the minimal fillings of the space $\mathcal{M}_z$ are found. It turns out that $\mathcal{M}_z$ has two topologically distinct minimal fillings, shown in Figure 6, and the tour π is a maximal tour for the both trees, say G_1 and G_2. Therefore, $\mathrm{mf}(\mathcal{M}_z) = p(\mathcal{M}_z, G_1, \pi) = p(\mathcal{M}_z, G_2, \pi)$.

It remains to verify that the perimeters of all the other multitours are strictly less than $\mathrm{mf}(\mathcal{M}_z)$. To do that we use the following trick. We increase by a small number a all the distances in m_z which are not contained in π, and apply our program to the resulting space again. It is clear that after this operation the perimeters of all the multitours except π increase. Therefore, if we had some multitour with the same perimeter, then it becomes more than $\mathrm{mf}(\mathcal{M}_z)$, and we get another minimal filling. But it is not the case, q.e.d.

ASSERTION 8.3. *All 6 points metric spaces (M, ρ_m), where m belongs to a sufficiently small open neighborhood $U \subset C_6$ of the matrix m_z has two minimal filling with distinct topologies G_1 and G_2.*

Due to minimal realization results, see Corollary 4.4, Assertion 8.3 implies non-uniqueness example of an open set of 6 points boundaries in $\mathbb{R}^6_\infty$ having two Steiner minimal trees of distinct topologies.

COROLLARY 8.4. *Uniqueness of the shortest tree structure for boundaries in general position does not hold in $\mathbb{R}^n_\infty$.*

9. Minimal Fillings of Additive and Pseudo-Additive Spaces

The additive spaces are very popular in bioinformatics, playing an important role in evolution theory and, more general, in an hierarchy modeling. Recall that a finite metric space $\mathcal{M} = (M, \rho)$ is called *additive* or *tree-like*, if M can be joined by a weighted tree $\mathcal{G} = (G, \omega)$ such that ρ coincides with the restriction of d_ω onto M. The tree $\mathcal{G}$ in this case is called a *generating tree* for the space $\mathcal{M}$.

Not any metric space is additive. An additivity criterion can be stated in terms of so-called *four points rule*: for any four points p_i, p_j, p_k, p_l, the values $\rho(p_i,p_j)+\rho(p_k,p_l)$, $\rho(p_i,p_k)+\rho(p_j,p_l)$, $\rho(p_i,p_l)+\rho(p_j,p_k)$ are the lengths of sides of an isosceles triangle whose base does not exceed its other sides. Equivalently, the *four points inequality*

$$\rho(p_i,p_j)+\rho(p_k,p_l) \le \max\{\rho(p_i,p_k)+\rho(p_j,p_l), \rho(p_i,p_l)+\rho(p_j,p_k)\}$$

is valid four any 4-tuple of points.

THEOREM 9.1 ([**30**], [**31**], [**32**], [**33**]). *A metric space is additive, if and only if it satisfies the four points rule. In the class of non-degenerate weighted trees, the generating tree of an additive metric space is unique.*

The next result solves completely the minimal filling problem for additive metric spaces.

THEOREM 9.2. *Minimal fillings of an additive metric space are exactly its generating trees.*

The next additivity criterion is obtained by O. V. Rubleva, see [**34**].

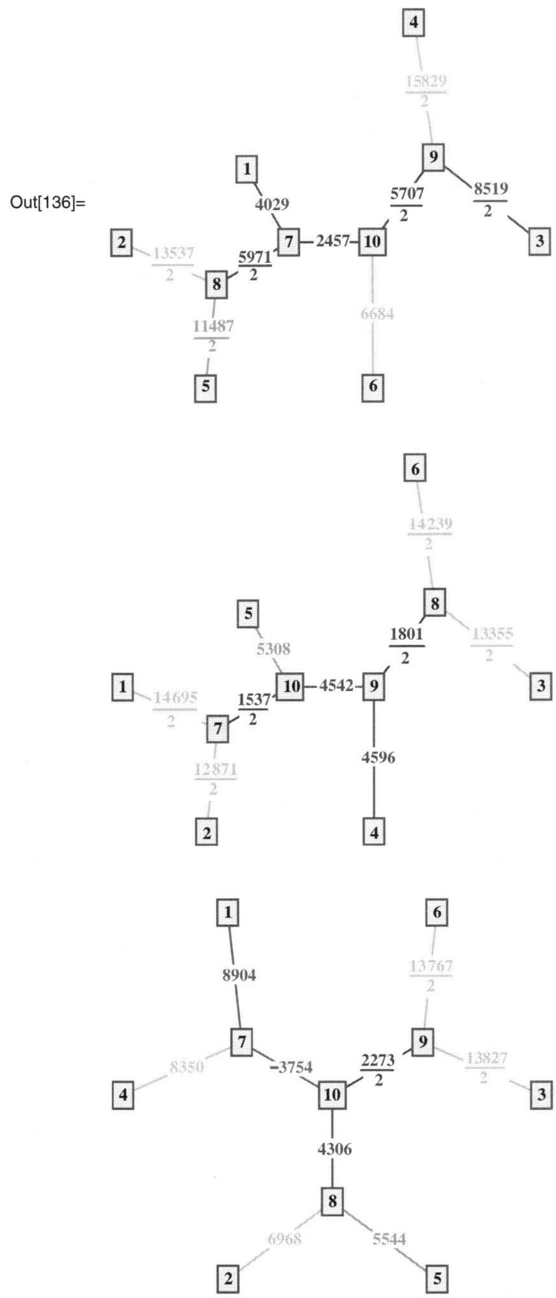

FIGURE 6. Minimal fillings of $\mathcal{M}_z$ with distinct topologies having an exact tour π (top and middle), and parametric generalized minimal filling of $\mathcal{M}_z$ with non-exact tour π (top).

THEOREM 9.3 (O. V. Rubleva). *The weight of a minimal filling of a finite metric space is equal to the half-perimeter of this space, if and only if this space is additive.*

In the scope of Theorem 9.3, it is natural to try to describe all the finite metric spaces permitting a connecting tree such that all the corresponding half-perimeters are equal to each other. It turns out that this class is wider than the class of additive spaces. Z. N. Ovsyannikov suggested to call such spaces pseudo-additive and obtained their description, see [**28**].

A finite metric space $\mathcal{M} = (M, \rho)$ is said to be *pseudo-additive*, if the metric ρ coincides with d_ω for a generalized weighted tree (G, ω) (which is also called *generating*), where the weight function ω can take arbitrary (not necessary non-negative) values. Z. N. Ovsyannikov shows that these spaces can be described in terms of so-called *weak four points rule*: for any four points p_i, p_j, p_k, p_l, the values $\rho(p_i, p_j) + \rho(p_k, p_l)$, $\rho(p_i, p_k) + \rho(p_j, p_l)$, $\rho(p_i, p_l) + \rho(p_j, p_k)$ are the lengths of sides of an isosceles triangle.[1] The generating tree is also unique in the class of non-degenerate trees. Moreover, the following result is valid, see [**28**].

THEOREM 9.4 (Z. N. Ovsyannikov). *Let $\mathcal{M} = (M, \rho)$ be a finite metric space. Then the following statements are equivalent.*

- *There exist a tree G such that M coincides with the set of degree 1 vertices of G and all the half-perimeters $p(M, G, \pi)$ of M corresponding to the tours around G are equal to each other.*
- *The space $\mathcal{M}$ is pseudo-additive.*

Moreover, the three G in this case is a generating tree for the space $\mathcal{M}$.

10. Examples of Minimal Fillings

Here we give several examples of minimal filling and demonstrate how to use the technique elaborated above.

10.1. Triangle. Let $\mathcal{M} = (M, \rho)$ consist of three points p_1, p_2, and p_3. Put $\rho_{ij} = \rho(p_i, p_j)$. Consider the tree $G = (V, E)$ with $V = M \cup \{v\}$ and $E = \{vp_i\}_{i=1}^3$. Define the weight function ω on E by the following formula:

$$\omega(vp_i) = \frac{\rho_{ij} + \rho_{ik} - \rho_{jk}}{2},$$

where $\{i, j, k\} = \{1, 2, 3\}$. Notice that d_ω restricted onto M coincides with ρ. Therefore, $\mathcal{M}$ is an additive space, $\mathcal{G} = (G, \omega)$ is a generating tree for $\mathcal{M}$, and, due to Theorem 9.2, $\mathcal{G}$ is a minimal filling of $\mathcal{M}$.

Recall that the value $(\rho_{ij} + \rho_{ik} - \rho_{jk})/2$ is called by the *Gromov product* $(p_j, p_k)_{p_i}$ of the points p_j and p_k of the space $\mathcal{M}$ with respect to the point p_i, see [**35**].

10.2. Regular Simplex. Let all the distances in the metric space $\mathcal{M}$ are the same and are equal to d; i.e., $\mathcal{M}$ is a regular simplex. Then the weighted tree $\mathcal{G} = (G, \omega)$, $G = (V, E)$, with the vertex set $V = M \cup \{v\}$ and edges vm, $m \in M$, of the weight $d/2$ is generating for $\mathcal{M}$. Therefore, the space $\mathcal{M}$ is additive, and, due to Theorem 9.2, $\mathcal{G}$ is its unique nondegenerate minimal filling. If n is the number of points in M, then the weight of the minimal filling is equal to $dn/2$.

---

[1] In the literature these spaces are also referred as *relaxed tree-like spaces*, and the rule is also known as *relaxed four points rule*, see [**40**], but we save here the terminology from [**28**].

10.3. Star. If a minimal filling $\mathcal{G} = (G, \omega)$ of a space $\mathcal{M} = (M, \rho)$ is a star whose single interior vertex v is joined with each point $p_i \in M$, $1 \le i \le n$, $n \ge 3$, then the metric space $\mathcal{M}$ is additive [**3**]. In this case the weights of edges can be calculated easily. Indeed, put $e_i = vp_i$. Since a subspace of an additive space is additive itself, then we can use the results for three-points additive space, see above. So, we have $\omega(e_i) = (p_j, p_k)_{p_i}$, where p_i, p_j, and p_k are arbitrary distinct boundary vertices. The weight of the minimal filling in this case can be calculated as

$$\frac{1}{2}\sum_{i=1}^{n} \rho(p_i, p_{i+1})$$

for an arbitrary enumeration $\{p_1, \dots, p_n\}$ of the set M, where p_{n+1} stands for p_1, see also Theorem 9.3.

10.4. Parametric Star. B. Bednov find out an interesting formula for the weight of parametric minimal filling of the star type. Notice that Theorem 7.1 works for parametric minimal fillings that are binary trees only. A particular case of this general formula appeared in [**25**].

ASSERTION 10.1 (B. Bednov, [**41**]). *Let G be a star connecting the set M of its vertices of degree 1. Then the weight of parametric minimal filling of type G of a metric space $\mathcal{M} = (M, \rho)$ can be calculated as follows:*

$$\frac{1}{2}\max_{\{c_i\}}\sum_{i=1}^{k} \rho(c_i),$$

where the maximum is taken over all finite coverings $\{c_i\}$ of the set M by pairwise non-intersecting simple cycles and edges c_i of the complete graph on the vertex set M.

10.5. Mustaches of Degree more than 2. Let $G = (V, E)$ be an arbitrary tree, and $v \in V$ be an interior vertex of degree $(k+1) \ge 3$ adjacent with k vertices $w_1, \dots, w_k$ from ∂G. Then the set of the vertices $\{w_1, \dots, w_k\}$, and also the set of the edges $\{vw_1, \dots, vw_k\}$, are referred as *mustaches*. The number k is called by the *degree*, and the vertex v is called by the *common vertex of the mustaches*. An edge incident to v and not belonging to $\{vw_1, \dots, vw_k\}$ is called the *root edge* of the mustaches under consideration.

As it is shown in [**3**], any mustaches of a minimal filling of a metric space form an additive subspace. If the degree of such mustaches is more than 2, then we can calculate the weights of all the edges contained in the mustaches just in the same way as we did in the case of a star.

10.6. Four-Points Spaces. Here we give a complete description of minimal fillings for four-points spaces, see details in [**3**].

PROPOSITION 10.2. *Let $M = \{p_1, p_2, p_3, p_4\}$, and ρ be an arbitrary metric on M. Put $\rho_{ij} = \rho(p_i, p_j)$. Then the weight of a minimal filling $\mathcal{G} = (G, \omega)$ of the space $\mathcal{M} = (M, \rho)$ is given by the following formula*

$$\frac{1}{2}\big(\min\{\rho_{12}+\rho_{34},\, \rho_{13}+\rho_{24},\, \rho_{14}+\rho_{23}\} + \max\{\rho_{12}+\rho_{34},\, \rho_{13}+\rho_{24},\, \rho_{14}+\rho_{23}\}\big).$$

If the minimum in this formula is equal to $\rho_{ij}+\rho_{rs}$, then the type of minimal filling is the binary tree with the mustaches $\{p_i, p_j\}$ and $\{p_r, p_s\}$.

We apply the obtained result to the vertex set of a planar convex quadrangle.

COROLLARY 10.3. *Let M be the vertex set of a convex quadrangle $p_1p_2p_3p_4 \subset \mathbb{R}^2$ and $\rho(p_i, p_j) = \|p_i - p_j\|$. The weight of a minimal filling of the space (M, ρ) is equal to*

$$\frac{1}{2}\min\big(\rho_{12} + \rho_{34},\, \rho_{14} + \rho_{23}\big) + \frac{\rho_{13} + \rho_{24}}{2}.$$

The topology of minimal filling is a binary tree with mustaches corresponding to opposite sides of the less total length.

11. Ratios

As we have already discussed above, the problem of finding a shortest tree or a minimal filling for a given boundary is exponential. On the other hand, in practice it is necessary to solve optimal connection problems of this kind for several thousands of boundary points many times a day. Therefore, in practice some heuristic algorithms have to be used. One of the most popular heuristics for a shortest tree is corresponding minimal spanning tree. But using such approximate solutions instead of exact one it is important to know the value of possible error appearing under the approximation. The classical *Steiner ratio of a metric space* is just the measure of maximal possible relative error for the approximation of a shortest tree by the corresponding minimal spanning tree. In this Section we remind what is known concerning the Steiner ratio and introduce two similar ratios dealing with minimal fillings. These three ratios seem to be related closely, so investigating one we can obtain some information on the others.

11.1. Steiner Ratio. Let M be a finite subset of a metric space (X, ρ), and assume that $|M| \ge 2$. We put $\operatorname{sr} M = \operatorname{smt}(M)/\operatorname{mst}(M)$. Evidently, $\operatorname{sr} M \le 1$. The next statement is also easy to prove considering any tour of a shortest tree and using the triangle inequality.

ASSERTION 11.1. *For any metric space (X, ρ) and any its finite subset $M \subset X$, $|M| \ge 2$, the inequality $\operatorname{sr} M > 1/2$ is valid.*

The value $\operatorname{sr}(M)$ is the relative error appearing under approximation of the length of a shortest tree for a given set M by the length of a minimal spanning tree. The *Steiner ratio of a metric space* (X, ρ) is defined as the value $\operatorname{sr}(X) = \inf_{M \subset X} \operatorname{sr}(M)$, where the infimum is taken over all finite subsets M, $|M| \ge 2$, of the metric space X. So, the Steiner ratio of X is the value of the relative error in the worse possible case.

ASSERTION 11.2. *For arbitrary metric space (X, ρ) the inequality $1/2 \le \operatorname{sr}(X) \le 1$ is valid. Moreover, for any $r \in [1/2, 1]$ there exists a metric space (X, ρ) with $\operatorname{sr}(X) = r$.*

Sometimes, it is convenient to consider so-called *Steiner ratios* $\operatorname{sr}_n(X)$ *of degree* n, where $n \ge 2$ is an integer, which are defined as: $\operatorname{sr}_n(X) = \inf_{M \subset X, |M| \le n} \operatorname{sr}(M)$. Evidently, $\operatorname{sr}_2(X) = 1$. It is also clear that $\operatorname{sr}(X) = \inf_n \operatorname{sr}_n(X)$.

Steiner ratio was firstly defined for the Euclidean plane in [**11**], and during the subsequent years the problem of Steiner ratio calculation is one of the most attractive, interesting and difficult problems in geometrical optimization. A short review can be found in [**2**] and in [**12**]. One of the most famous stories here is connected with several attempts to prove so-called *Gilbert–Pollack Conjecture*, see [**11**], saying that $\mathrm{sr}(\mathbb{R}^2, \rho_2) = \sqrt{3}/2$, and hence $\mathrm{sr}(\mathbb{R}^2, \rho_2)$ is attained at the vertex set of a regular triangle, see Figure 1 (here ρ_2 stands for the Euclidean metric). In 1990s D. Z. Du and F. K. Hwang announced that they proved the Steiner Ratio Gilbert–Pollak Conjecture [**13**], and their proof was published in Algorithmica [**14**]. In spite of the appealing ideas of the paper, the questions concerning the proof appeared just after the publication, because the text did not appear formal. In about 2003–2005 it became clear that the gaps in the D. Z. Du and F. K. Hwang work are too deep and can not be repaired, see details in [**15**] and [**49**].

In [**11**] Gilbert and Pollack proved that $\mathrm{sr}_3(\mathbb{R}^2, \rho_2) = \sqrt{3}/2$. Now the equality $\mathrm{sr}_n(\mathbb{R}^2, \rho_2) = \sqrt{3}/2$ is proved for $n \le 7$ (the recent progress is obtained by O. de Wet, see [**16**]). The proof of de Wet is based on the analysis of Du and Hwand's method from [**14**] and understanding that it works for boundary sets with $n \le 7$ points. Also in 60s several lower bounds for $\mathrm{sr}(\mathbb{R}^2, \rho_2)$ were obtained, and the best of them is worse than $\sqrt{3}/2$ less than by $1/100$.

PROBLEM 11.3. *Very attractive problem is to prove that* $\mathrm{sr}(\mathbb{R}^2, \rho_2) = \sqrt{3}/2$*; i.e., to prove Gilbert–Pollack Conjecture. The attempts to repair the proof of Du and Hwang have remained unsuccessful, so some fresh ideas might be necessary.*

We have no aim to give here a detailed review on the Steiner ratio problem referring to [**2**] and [**12**]. Here we mention just several results important for what follows.

The following statement is evident.

ASSERTION 11.4. *If Y is a subspace of a metric space X; i.e., the distance function on Y is the restriction of the distance function of X, then* $\mathrm{sr}(Y) \ge \mathrm{sr}(X)$.

This implies, that $\mathrm{sr}(\mathbb{R}^n, \rho_2) \le \mathrm{sr}(\mathbb{R}^2, \rho_2) \le \sqrt{3}/2$. Recall that Gilbert–Pollack conjecture implies that the Steiner ratio of Euclidean plane attains at the vertex set of a regular triangle. In multidimensional case the situation is more complicated. The following result was obtained by Du and Smith [**17**]

ASSERTION 11.5. *If $M \subset \mathbb{R}^n$ is the vertex set of a regular n-dimensional simplex, then* $\mathrm{sr}(M) > \mathrm{sr}(\mathbb{R}^n, \rho_2)$ *for* $n \ge 3$.

PROOF. Consider the boundary set P in $\mathbb{R}^{n+1}$, consisting of the following $1 + n(n+1)$ points: one point $(0, \dots, 0)$ and $n(n+1)$ points all whose coordinates except two are zero, one is equal to 1, and the remaining one is -1. It is clear that P is a subset of n-dimensional plane defined by the next linear condition: sum of all coordinates is equal to zero. Represent P as the union of the subsets $P^i = \{x \in P \mid x^i = 1\} \cup \{(0, \dots, 0)\}$. Notice that each set P^i, $i = 1, \dots, n+1$, consists of $n+1$ points and forms the vertex set of a regular n-dimensional simplex (to see that it suffices to verify that all the distances between the pairs of points from P^i are the same and are equal to $\sqrt{2}$). The configuration of 7 points in $\mathbb{R}^3$ is shown in Figure 7 (this case is not important for us, but it is easy to draw). Now, $\mathrm{mst}(P) = (n+1)\,\mathrm{mst}(P^i)$, but for $n \ge 3$ we conclude that $\mathrm{smt}(P) < (n+1)\,\mathrm{smt}(P^i)$, because the degree of the vertex $(0, \dots, 0)$ in the corresponding network which is the union of the shortest networks for P^i is equal to $n + 1 \ge 4$ that is impossible in the shortest network due to the Local Structure Theorem, see, for example, [**1**].

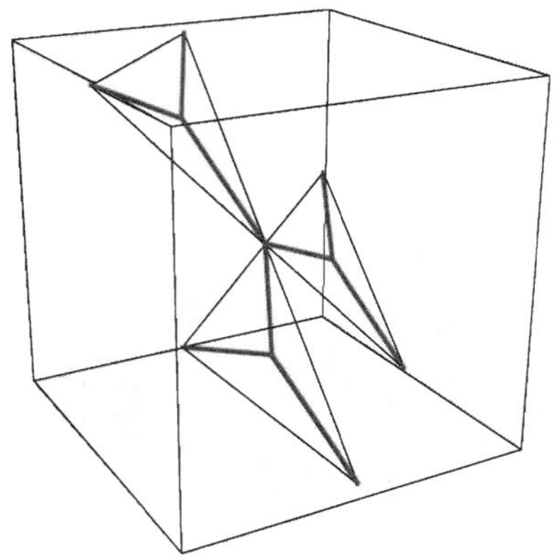

FIGURE 7. Construction of the set P in $\mathbb{R}^3$ (non-interesting but visual case $n = 2$).

So,

$$\operatorname{sr}(P) = \operatorname{smt}(P)/\operatorname{mst}(P) < \frac{(n+1)\operatorname{smt}(P^i)}{(n+1)\operatorname{mst}(P^i)} = \operatorname{sr}(P^i).$$

□

Notice that the idea of the proof is also used in paper [**17**] to estimate from below the number of points in a subset M of $\mathbb{R}^n$ such that $\operatorname{sr}(M) = \operatorname{sr}(\mathbb{R}^n, \rho_2)$ by a function $f(n)$ increasing on n rapidly. But recently Z. Ovsyannikov and B. Bednov find out a gap in the proof of these theorem (a counter example to the key lemma is constructed), so the existence of such an estimate is still an interesting open question. But, anyway, it lead to the best known upper estimate for $\mathbb{R}^3$ obtained by Smith and Smith [**18**]. It is attained at an infinite boundary set which is known as "Smith sausage" and depicted in Figure 8. The corresponding value, obtained as the limit of the ratios for finite fragments, is as follows:

$$\sqrt{\frac{283}{700} - \frac{3\sqrt{21}}{700} + \frac{9\sqrt{22 - 2\sqrt{21}}}{140}} = \frac{17 + 3\sqrt{21}}{50}$$

(we include the first expression also since it is given in the original publication [**18**]).

Recently, the Steiner ratio of the Lobachevskii plane, and more general, of any multidimensional Lobachevskii space has bin calculated by Innami and Kim, see [**20**].

THEOREM 11.6. *Steiner ratio of Lobachevskii space L^n for any $n \geq 2$ is equal to* 1/2.

For general Riemannian manifold Ivanov, Cieslik and Tuzhilin, see [**21**], obtained the following general result.

THEOREM 11.7. *The Steiner ratio of n-dimensional Riemannian manifold is less than or equal to the Steiner ratio of the Euclidean space $\mathbb{R}^n$.*

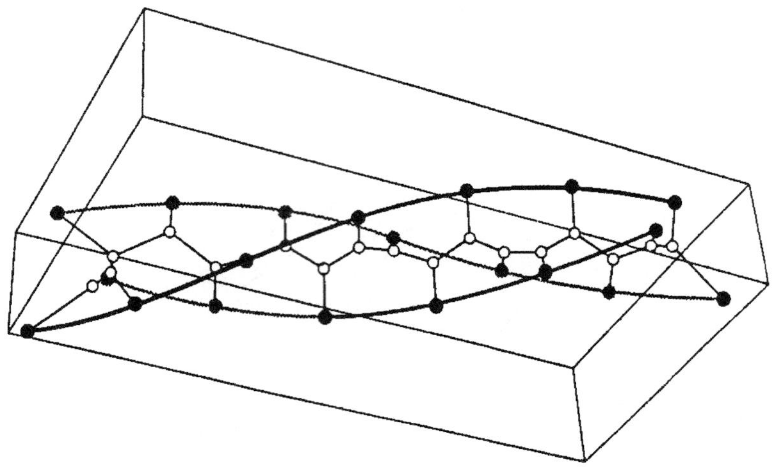

FIGURE 8. A finite fragment of infinite "Smith sausage".

11.2. Steiner–Gromov Ratio. For convenience, the sets consisting of more than a single point are referred as *nontrivial.* Let $\mathcal{X} = (X, \rho)$ be an arbitrary metric space, and let $M \subset X$ be some finite subset. For nontrivial M, we define the value

$$\operatorname{sgr}(M) = \operatorname{mf}(M, \rho)/\operatorname{mst}(M, \rho)$$

and call it the *Steiner–Gromov ratio* of the subset M. The value $\inf \operatorname{sgr}(M)$, where the infimum is taken over all nontrivial finite subsets of $\mathcal{X}$, consisting of at most n vertices is denoted by $\operatorname{sgr}_n(\mathcal{X})$ and is called the *degree n Steiner–Gromov ratio of the space $\mathcal{X}$*. At last, the value $\inf \operatorname{sgr}_n(\mathcal{X})$, where the infimum is taken over all positive integers $n > 1$ is called the *Steiner–Gromov ratio of the space $\mathcal{X}$* and is denoted by $\operatorname{sgr}(\mathcal{X})$, or by $\operatorname{sgr}(X)$, if it is clear what particular metric on X is considered. Notice that $\operatorname{sgr}_n(\mathcal{X})$ is a non-increasing function on n.

As in the case of the classical Steiner ratio, it is not difficult to show that the Steiner–Gromov ratio of an arbitrary metric space is not less than $1/2$ and is not greater than 1. More precise result is obtained by A. S. Pakhomova [**38**].

THEOREM 11.8 (A. Pakhomova). *For any metric space $\mathcal{X}$ the estimate*

$$\operatorname{sgr}_n(\mathcal{X}) \ge \frac{n}{2n-2}$$

is valid. Moreover, this estimate is exact; i.e., for any $n \ge 2$ there exists a metric space $\mathcal{X}_n$ such that $\operatorname{sgr}_n(\mathcal{X}_n) = n/(2n-2)$.

For any metric space $\mathcal{X}$ the estimate $1/2 \le \operatorname{sgr} \mathcal{X} \le 1$ is valid. Moreover, for any $s \in [1/2, 1]$ there exists a metric space $\mathcal{X}$ such that $\operatorname{sgr} \mathcal{X} = s$.

Let us also list several particular answers and estimates. It is easy to see that $\operatorname{sgr}_2(\mathcal{X}) = 1$, and $\operatorname{sgr}_3(\mathcal{X}) \ge 3/4$ for any nontrivial metric space $\mathcal{X}$. It is also easy to see, that for the vertex set M of a regular triangle the equality $\operatorname{sgr}(M) = 3/4$ holds. Therefore, if a space $\mathcal{X}$ contains a regular triangle, then $\operatorname{sgr}_3 \mathcal{X} = 3/4$. This is a particular case of the following general Lemma.

LEMMA 11.9. *If a metric space $\mathcal{X}$ contains a regular simplex consisting of n points, then* $\operatorname{sgr}_n \mathcal{X} = n/(2n-2)$.

For example, $\operatorname{sgr}_3 \mathbb{R}^m = 3/4$ for $m \geq 2$, and $\operatorname{sgr}_4(\mathbb{R}^m) = 2/3$ for $m \geq 3$. I. Laut and E. Stepanova showed that $\operatorname{sgr}_4 \mathbb{R}^2$ is also equal to $3/4$. But recently, Z. Ovsyannikov proved that $\operatorname{sgr}_5 \mathbb{R}^2 < 0.742 < 3/4$.

It is clear from definitions that $\operatorname{sgr} \mathcal{X} \leq \operatorname{sr} \mathcal{X}$ for any metric space $\mathcal{X}$. Together with Theorem 11.8 that implies the following result.

ASSERTION 11.10. *If* $\operatorname{sr} \mathcal{X} = 1/2$, *then* $\operatorname{sgr} \mathcal{X} = 1/2$.

Hence, Theorem 11.6 implies the following answer for Lobachevski space.

COROLLARY 11.11. *The Steiner–Gromov ratio of Lobachevski space L^n, $n \geq 2$, is equal to* $1/2$.

Lemma 11.9 can be used to calculate the Steiner–Gromov ratio for the space ℓ_p, $1 \leq p \leq \infty$, of all real sequences $x = \{x_i\}$ having finite ℓ_p-norm

$$\|x\|_p = \Big(\sum_i |x_i|^p\Big)^{1/p} < \infty.$$

The same arguments can be also applied to phylogenetic space (the space of words with so-called edit or Levenshtein distance) under the assumption that the alphabet consists of at least two letters. Namely, the following result is valid.

ASSERTION 11.12 (A. Pakhomova). *Let $\mathcal{X}$ be either the space ℓ_p, $1 \leq p \leq \infty$, or the space of words over an alphabet $A = \{a_1, \ldots, a_k\}$, $k \geq 2$, endowed with the Levenshtein metric. Then for all $n \geq 2$ the following relations hold:*

$$\operatorname{sgr}_n \mathcal{X} = \frac{n}{2n-2}, \qquad \operatorname{sgr} \mathcal{X} = \frac{1}{2}.$$

Also recently, Z. Ovsyannikov [**39**] investigated the metric space of all compact subsets of Euclidean plane endowed with Hausdorff metric.

ASSERTION 11.13 (Z. Ovsyannikov). *The Steiner ratio and the Steiner–Gromov ratio of the metric space of all compact subsets of Euclidean plane endowed with Hausdorff metric are equal to* $1/2$.

For Riemannian manifolds V. Mishchenko [**47**] obtained the direct an analogue of Theorem 11.7.

ASSERTION 11.14 (V. Mishchenko). *The Steiner–Gromov ratio of an arbitrary n-dimensional Riemannian manifold is less than or equal to the Steiner–Gromov ratio of the Euclidean space $\mathbb{R}^n$.*

11.3. Steiner Subratio. Let $\mathcal{X} = (X, \rho)$ be an arbitrary metric space, and let $M \subset \mathcal{X}$ be some its finite subset. Recall that by $\operatorname{smt}(M, \rho)$ we denote the length of Steiner minimal tree joining M. Further, for nontrivial subsets M, we define the value

$$\operatorname{ssr}(M) = \operatorname{mf}(M, \rho) / \operatorname{smt}(M, \rho)$$

and call it by the *Steiner subratio* of the set M. The value $\inf \operatorname{ssr}(M)$, where infimum is taken over all nontrivial finite subsets of $\mathcal{X}$ consisting of at most $n > 1$ points, is denoted by $\operatorname{ssr}_n(\mathcal{X})$ and is called the *degree n Steiner subratio of the space $\mathcal{X}$*. At last, the value $\inf \operatorname{ssr}_n(\mathcal{X})$, where the infimum is taken over all positive integers

$n > 1$, is called the *Steiner subratio of the space* $\mathcal{X}$ and is denoted by $\mathrm{ssr}(\mathcal{X})$, or by $\mathrm{ssr}(X)$, if it is clear what particular metric on X is considered. Notice that $\mathrm{ssr}_n(\mathcal{X})$ is a nonincreasing function on n.

As above, it is not difficult to show that the Steiner–Gromov ratio of an arbitrary metric space is not less than $1/2$ and is not greater than 1. More precise result is obtained by A. S. Pakhomova [**38**].

THEOREM 11.15 (A. Pakhomova). *For any metric space* $\mathcal{X}$ *the estimate*

$$\mathrm{ssr}_n(\mathcal{X}) \geq \frac{n}{2n-2}$$

is valid. Moreover, this estimate is exact; i.e., for any $n \geq 2$ *there exists a metric space* $\mathcal{X}$ *such that* $\mathrm{sgr}_n(\mathcal{X}) = n/(2n-2)$.

For any metric space $\mathcal{X}$ *the estimate* $1/2 \leq \mathrm{ssr}\,\mathcal{X} \leq 1$ *is valid. Moreover, for any* $s \in [1/2, 1]$ *there exists a metric space* $\mathcal{X}$ *such that* $\mathrm{ssr}\,\mathcal{X} = s$.

Let us again list several particular answers and estimates. It is easy to see that $\mathrm{ssr}_2(\mathcal{X}) = 1$ for an arbitrary metric space. A. Ivanov and A. Tuzhilin [**3**] calculated $\mathrm{ssr}_3\,\mathbb{R}^n$. It turns out that $\mathrm{ssr}_3(\mathbb{R}^n) = \sqrt{3}/2$. Than E. Stepanova, see [**37**], expanded this result to four points set and shown that $\mathrm{ssr}_4(\mathbb{R}^2) = \sqrt{3}/2$. But recently Z. Ovsyannikov [**48**] constracted the example demonstrating that $\mathrm{ssr}_5\,\mathbb{R}^2 < 0.8562 < \sqrt{3}/2$. He also found out that $\mathrm{ssr}_4\,\mathbb{R}^3 = (2\sqrt{3}+\sqrt{5})/7 < 0.82 < \sqrt{3}/2$.

B. Bednov and P. Borodin [**25**] suggested consider also the values

$$\mathrm{ssr}(d) = \inf\{\mathrm{ssr}\,V \mid V \text{ is a Banach space of dimension } d\}.$$

They proved the folioing estimates.

ASSERTION 11.16 (B. Bednov, P. Borodin). *The next inequalities are valid*:

$$\frac{3}{4} \leq \mathrm{ssr}(2) \leq \frac{5}{6}, \qquad \frac{2}{\sqrt{5}+1} \leq \mathrm{ssr}(3) \leq \frac{4}{5}.$$

A. Pakhomova calculated the Steiner sabratio for some phylogenetic spaces.

ASSERTION 11.17 (A. Pakhomova). *Let* $\mathcal{X}$ *be a phylogenetic space of words over an alphabet containing at least* $n-1$ *letters. Then* $\mathrm{ssr}_n\,\mathcal{X} = n/(2n-2)$. *Hence,* $\inf_{\mathcal{X}} \mathrm{ssr}\,\mathcal{X} = 1/2$, *where infimum is taken over all phylogenetic spaces.*

A. Erokhovets [**46**] investigated Steiner subratio and Steiner–Gromov ratio for the spaces of p-adic numbers that are important for applications as a useful example of an ultarmetric space, see for example [**40**]. Recall the construction of the p-adic norm on the set $\mathbb{Q}$ of rationals. Let p be a prime. We put

$$|0|_p = 0, \qquad |x|_p = p^{-\gamma(x)}, \quad \text{for non-zero } x \in \mathbb{Q},$$

where the integer $\gamma(x)$ is defined from the following representation:

$$x = p^{\gamma(x)}\frac{m}{n}, \quad m \text{ and } n \text{ are integers coprine to } p.$$

The norm $|\cdot|_p$ defines an ultrametric on $\mathbb{Q}$, and the completion of $\mathbb{Q}$ with respect to this metric is referred as the *space of p-adic numbers*.

THEOREM 11.18 (A. Erokhovets). *Steiner subratio and Steiner–Gromov ratio of the space* $\mathbb{Q}$ *are equal to* $1/2$ *for any prime* p.

Also recently, Z. Ovsyannikov [**39**] investigated the metric space of all compact subsets of Euclidean plane endowed with Hausdorff metric.

PROPOSITION 11.19 (Z. Ovsyannikov). *Let $\mathcal{C}$ be the metric space of all compact subsets of Euclidean plane endowed with Hausdorff metric. Then* $\operatorname{ssr}_3(\mathcal{C}) = 3/4$ *and* $\operatorname{ssr}_4(\mathcal{C}) = 2/3$.

11.4. Ratios Continuity and Discontinuity. Recently A. Pakhomova [**43**] investigated continuity properties of all three above ratios in the sense of Gromov–Hausdorff distance. Recall the necessary definitions.

Let $\mathcal{X} = (X, \rho)$ be a metric space. Then, for any $a \in X$ and positive ε we put $U_\varepsilon(a) = \{x \in X \mid \rho(a, x) < \varepsilon\}$, and for any $A \subset X$ by $U_\varepsilon(A)$ we denote the union of all $U_\varepsilon(a)$, $a \in A$, and call $U_\varepsilon(A)$ by *ε-neighborhood of the set A*. Further, for $A, B \subset X$, the value

$$d_H(A, B) = \inf\bigl\{\varepsilon \mid A \subset U_\varepsilon(B) \quad \text{and} \quad B \subset U_\varepsilon(A)\bigr\}$$

is referred as the *Hausdorff distance between A and B*. It is well-known that on the set of closed bounded subsets of $\mathcal{X}$ the function d_H is a distance function.

Further, let $\mathcal{X}_1$ and $\mathcal{X}_2$ be some metric spaces. By a *realization* of metric spaces $\mathcal{X}_1$ and $\mathcal{X}_2$ we call a metric space $\mathcal{Y}$ with two its subspaces $\mathcal{Z}_1$ and $\mathcal{Z}_2$, where $\mathcal{X}_i$ is isometric to $\mathcal{Z}_i$, $i = 1, 2$. The *Gromov–Hausdorff distance between the spaces $\mathcal{X}_1$ and $\mathcal{X}_2$* is defined as the value

$$d_{GH}(\mathcal{X}_1, \mathcal{X}_2) = \inf\bigl\{r \in \mathbb{R} \mid \exists \text{ a realization } \mathcal{Z}_i \subset \mathcal{Y} \text{ such that } d_H(\mathcal{Z}_1, \mathcal{Z}_2) \le r\bigr\}.$$

It is well-known that this function is a finite metric on the space of all compact metric spaces. The proof can be found, for example, in [**42**].

THEOREM 11.20 (A. Pakhomova). *Let $\mathcal{C}_{GH}$ be the space of all compacts with the Gromov–Hausdorff metric. By r_i we denote the functions on $\mathcal{C}_{GH}$ defined as* $r_1(\mathcal{X}) = \operatorname{sr}(\mathcal{X})$, $r_2(\mathcal{X}) = \operatorname{sgr}(\mathcal{X})$, *and* $r_3(\mathcal{X}) = \operatorname{ssr}(\mathcal{X})$. *Then the functions r_i are upper semi-continuous. Moreover, r_i is continuous at $\mathcal{X}$, if and only if $r_i(\mathcal{X}) = 1/2$.*

Similar result can be stated for the degree n ratios also, see details in [**43**].

12. Generalizations for Infinite Sets

Recently A. Eremin [**45**] made an attempt to expand the above technique to the case of infinite metric spaces. In this Section we briefly overview his results.

12.1. Infinite Graphs and Minimal Fillings. Defining a combinatorial graph as a pair $G = (V, E)$ we need not really to assume (as we did above) that the sets V, and hence E, are finite. We do not assume it in this Section. There are several approaches to the concept of a connected infinite graph. Here we proceed as follows. By a *route* in a graph G we understand a **finite** sequence $v_{i_1}, e_{i_1}, v_{i_2}, \dots, e_{i_k} v_{i_{k+1}}$ of its vertices and edges such that each edge e_{i_s} is incident to both v_{i_s} and $v_{i_{s+1}}$, $s = 1, \dots, k$. Then all the above concepts (connected graph, tree, graph connecting a set, weighted graph, a (generalized) filling of a metric space, a minimal filling, etc.) can be extended word-by-word to the case of infinite graphs. Notice that in this case the weight of a tree can be infinite.

If $G = (V, E)$ is a tree, and $L \subset V$ is a subset, then by $G|_L$ we denote the minimal subtree of G connecting L. This subtree is the union of all the paths connecting pairs of vertices from L in G. If ω is a weight function on the edges of G and $\mathcal{G} = (G, \omega)$ is the corresponding weighted tree, then by $\mathcal{G}|_L$ we denote the weighted tree (G_L, ω_L), where by ω_L we denote the restriction of the function ω

onto the edge set of the tree G_L. The weighted tree $\mathcal{G}|_L$ is referred as the *restriction* of $\mathcal{G}$ to L.

If $\mathcal{M} = (M, \rho)$ is a metric space, and $L \subset M$ is a subset, then by $\mathcal{M}|_L$ we denote the metric space $(L, \rho|_L)$, where by $\rho|_L$ we denote the restriction of the distance function ρ onto the set $L \times L$. The space $\mathcal{M}|_L$ is referred as the *restriction* of $\mathcal{M}$ onto L.

In [**44**] the possibility of connecting a subset of a metric space by a tree of finite length is studied. A subset of a metric space permitting such a connection is referred as *good*. It turns out that just the good metric spaces possess fillings of a finite weight.

ASSERTION 12.1 (A. Eremin). *A metric space $\mathcal{M}$ is good, if and only if* $\operatorname{mf} \mathcal{M} < \infty$.

A sequence $\{L_i\}$ of subsets of a set M is called by an *exhausting* of M if $L_i \subset L_{i+1}$ for any i, and $\cup_i L_i = M$. It tuns out that the weight of minimal filling of a good metric space can be calculated in term of the weights of minimal fillings of its exhausting by finite subsets. The notation $L \sqsubset M$ stands for a finite subset L of M.

THEOREM 12.2 (A. Eremin). *Let $\mathcal{M} = (M, \rho)$ be a good metric space, $\{L_i\}$ be an exhausting of M by its finite subsets, and $\mathcal{L}_i = \mathcal{M}|_{L_i}$. Then*

$$\operatorname{mf} \mathcal{M} = \sup_{L \sqsubset M} \operatorname{mf} \mathcal{M}|_L = \lim_{i \to \infty} \operatorname{mf} \mathcal{L}_i.$$

Example. The condition that $\mathcal{M}$ is good can not be omitted. For example, in [**44**] it is shown that $\mathcal{M} = \big([0,1] \cap \mathbb{Q}, \rho(x,y) = |x-y|\big)$ is not good. Therefore, $\operatorname{mf} \mathcal{M} = \infty$, due to Assertion 12.1. On the other hand, for any its finite subset L we have: $\operatorname{mf} \mathcal{M}|_L \le \operatorname{diam} \mathcal{M}|_L \le 1$, therefore $\operatorname{mf} \mathcal{M}$ is strictly greater than the supremum from Theorem 12.2.

Remark. Similar results can be obtained for parametric minimal fillings. In the case we need no assumption that the space is good. See details in [**45**].

Remark. For infinite minimal fillings the equality $\operatorname{mf}_- \mathcal{M} = \operatorname{mf} \mathcal{M}$ is also valid. See Theorem 7.1 and [**45**].

12.2. Minimal Metric Hulls. Return to the previous example $\mathcal{M} = \big([0,1] \cap \mathbb{Q}, \rho(x,y) = |x-y|\big)$. In some sense it seems natural to consider the segment $[0,1]$ as a filling of the space $\mathcal{M}$. This leads us to another generalization of the minimal filling concept appeared in [**45**], which is even more close to the initial concept of M. Gromov [**22**].

Recall that a mapping $f\colon X \to Y$ of metric spaces $\mathcal{X} = (X, \rho)$ and $\mathcal{Y} = (Y, d)$ is said to be *non-compressing*, if $d\big(f(a), f(b)\big) \ge \rho(a,b)$ for any points a and b from X.

By a *metric hull* of a metric space $\mathcal{M} = (M, \rho)$ we call a pair $(\mathcal{X}, f)$, where $\mathcal{X} = (X, d)$ is a pathwise connected pseudo metric space, and $f\colon M \to X$ is a non-compressing mapping. By the *weight of the metric hull* $(\mathcal{X}, f)$ we call the one-dimensional Hausdorff measure $\mathcal{H}^1(\mathcal{X})$ of the space $\mathcal{X}$. We put $\operatorname{hmf}(\mathcal{M}) = \inf \mathcal{H}^1(\mathcal{X})$, where the infimum is taken over all metric hulls $(\mathcal{X}, f)$ of the space $\mathcal{M}$. A metric hull $(\mathcal{X}, f)$ such that $\mathcal{H}^1(\mathcal{X}) = \operatorname{hmf}(\mathcal{M})$ is called by a *minimal metric hull of the space $\mathcal{M}$*. The next Theorem gives a relation between the concept of metric hull and above concept of minimal filling.

THEOREM 12.3 (A. Eremin). *For any pseudo metric space the following inequality is valid*: $\operatorname{hmf}\mathcal{M} \le \operatorname{mf}\mathcal{M}$, *where each part of the inequality could be infinite. Moreover, for good metric spaces* $\operatorname{hmf}\mathcal{M} = \operatorname{mf}\mathcal{M}$.

It turns out that for the metric hulls an analogue of Theorem 12.2 is valid.

THEOREM 12.4 (A. Eremin). *For an arbitrary metric space* $\mathcal{M} = (M, \rho)$ *the following equality is valid*:

$$\operatorname{hmf}\mathcal{M} = \sup_{L \sqsubset M} \operatorname{hmf}\mathcal{M}|_L = \sup_{L \sqsubset M} \operatorname{mf}\mathcal{M}|_L.$$

Moreover, there exists a metric hull $(\mathcal{X}, f)$ *of* $\mathcal{M}$ *such that* $\mathcal{H}^1(\mathcal{X}) = \operatorname{hmf}(\mathcal{M})$; *i.e., there exists a minimal metric hull.*

Remark. It is shown in [**45**] that a minimal metric hull can be found among so-called metric trees, a direct topological analogues of combinatorial weighted trees. Recall that a metric space with an intrinsic metric is called a *metric tree*, if any its two points are connected by a unique curve, and this curve is a geodesic; i.e., an isometric embedding of a standard segment.

Acknowledgments

We are grateful to Professor A. Fomenko for his kind attention to our work, and to the participants of the seminar "Optimal Networks" at Mechanics and Mathematics Department of Lomonosov Moscow State University for useful discussions. We are also grateful to Professor O. Musin for inviting us to submit this paper to the present volume. This work is partly supported by RFBR, Project 13–01–00664a, by the President of RF program for Leading Scientific Schools support, Project NSh–581.2014.1, and by a grant of the Government of RF, Project 11.G34.31.0053.

References

[1] A. O. Ivanov and A. A. Tuzhilin, *Branching solutions to one-dimensional variational problems*, World Scientific Publishing Co., Inc., River Edge, NJ, 2001. MR1825364 (2002a:90003)
[2] A. O. Ivanov, and A. A. Tuzhilin, *Extreme Networks Theory*, Inst. Komp. Issledov., Moscow, Izhevsk, 2003 [in Russian].
[3] A. O. Ivanov and A. A. Tuzhilin, *The one-dimensional Gromov minimal filling problem* (Russian, with Russian summary), Mat. Sb. **203** (2012), no. 5, 65–118, DOI 10.1070/SM2012v203n05ABEH004239; English transl., Sb. Math. **203** (2012), no. 5-6, 677–726. MR2977099
[4] V. A. Emelichev, at al., *Lections on Graph Theory*, Nauka, Moscow, 1990 [in Russian].
[5] Bernard Chazelle, *The soft heap: an approximate priority queue with optimal error rate*, J. ACM **47** (2000), no. 6, 1012–1027, DOI 10.1145/355541.355554. MR1866455 (2003a:68033)
[6] A. O. Ivanov and A. A. Tuzhilin, *The geometry of inner spanning trees for planar polygons* (Russian, with Russian summary), Izv. Ross. Akad. Nauk Ser. Mat. **76** (2012), no. 2, 3–36, DOI 10.1070/IM2012v076n02ABEH002581; English transl., Izv. Math. **76** (2012), no. 2, 215–244. MR2976276
[7] A. O. Ivanov and A. A. Tuzhilin, *The Steiner problem on a plane or planar minimal networks* (Russian), Mat. Sb. **182** (1991), no. 12, 1813–1844; English transl., Math. USSR-Sb. **74** (1993), no. 2, 555–582. MR1138637 (93d:05148)
[8] F. K. Hwang, *A linear time algorithm for full Steiner trees*, Oper. Res. Lett. **4** (1986), no. 5, 235–237, DOI 10.1016/0167-6377(86)90008-8. MR829887 (87c:05047)
[9] Z. A. Melzak, *On the problem of Steiner*, Canad. Math. Bull. **4** (1961), 143–148. MR0125466 (23 #A2767)

[10] A. O. Ivanov and A. A. Tuzhilin, *Solution of the Steiner problem for convex boundaries* (Russian), Uspekhi Mat. Nauk **45** (1990), no. 2(272), 207–208, DOI 10.1070/RM1990v045n02ABEH002345; English transl., Russian Math. Surveys **45** (1990), no. 2, 214–215. MR1069355 (92e:05056)
[11] E. N. Gilbert and H. O. Pollak, *Steiner minimal trees*, SIAM J. Appl. Math. **16** (1968), 1–29. MR0223269 (36 #6317)
[12] Dietmar Cieslik, *The Steiner ratio*, Combinatorial Optimization, vol. 10, Kluwer Academic Publishers, Dordrecht, 2001. MR1882260 (2003e:90001)
[13] D.-Z. Du and F. K. Hwang, *The Steiner ratio conjecture of Gilbert and Pollak is true*, Proc. Nat. Acad. Sci. U.S.A. **87** (1990), no. 23, 9464–9466, DOI 10.1073/pnas.87.23.9464. MR1080946 (91m:05061)
[14] D.-Z. Du and F. K. Hwang, *A proof of the Gilbert-Pollak conjecture on the Steiner ratio*, Algorithmica **7** (1992), no. 2-3, 121–135, DOI 10.1007/BF01758755. The Steiner problem. MR1146492 (92m:05060)
[15] A. O. Ivanov and A. A. Tuzhilin, *The Steiner ratio Gilbert-Pollak conjecture is still open*, Algorithmica **62** (2012), no. 1-2, 630–632, DOI 10.1007/s00453-011-9508-3. MR2886059 (2012m:05093)
[16] P. O. de Wet, *Geometric Steiner Minimal Trees*, PhD Thesis. UNISA, Pretoria, 2008.
[17] Ding-Zhu Du and Warren D. Smith, *Disproofs of generalized Gilbert-Pollak conjecture on the Steiner ratio in three or more dimensions*, J. Combin. Theory Ser. A **74** (1996), no. 1, 115–130, DOI 10.1006/jcta.1996.0040. MR1383508 (97h:05044)
[18] Warren D. Smith and J. MacGregor Smith, *On the Steiner ratio in 3-space*, J. Combin. Theory Ser. A **69** (1995), no. 2, 301–332, DOI 10.1016/0097-3165(95)90055-1. MR1313899 (95k:05053)
[19] R. L. Graham and F. K. Hwang, *A remark on Steiner minimal trees*, Bull. Inst. Math. Acad. Sinica **4** (1976), no. 1, 177–182. MR0437371 (55 #10302)
[20] Nobuhiro Innami and Byung Hak Kim, *Steiner ratio for hyperbolic surfaces*, Proc. Japan Acad. Ser. A Math. Sci. **82** (2006), no. 6, 77–79. MR2255998 (2007j:05044)
[21] A. O. Ivanov, A. A. Tuzhilin, and D. Tsislik, *The Steiner ratio for manifolds* (Russian, with Russian summary), Mat. Zametki **74** (2003), no. 3, 387–395, DOI 10.1023/A:1026106802540; English transl., Math. Notes **74** (2003), no. 3-4, 367–374. MR2022502 (2004i:52013)
[22] Mikhael Gromov, *Filling Riemannian manifolds*, J. Differential Geom. **18** (1983), no. 1, 1–147. MR697984 (85h:53029)
[23] A. O. Ivanov and A. A. Tuzhilin, *The Steiner ratio: the current state* (Russian), Mathematical problems in cybernetics. No. 11 (Russian), FizMatLit, Moscow, 2002, pp. 27–48. MR2269103 (2007m:05057)
[24] P. A. Borodin, *An example of the nonexistence of a Steiner point in a Banach space* (Russian, with Russian summary), Mat. Zametki **87** (2010), no. 4, 514–518, DOI 10.1134/S0001434610030260; English transl., Math. Notes **87** (2010), no. 3-4, 485–488. MR2762739 (2011m:41062)
[25] B. B. Bednov and P. A. Borodin, *Banach spaces realizing minimal fillings*, Matem. Sbornik **205** (2014); translation in Sbornik: Mathematics **205** (2014) (to appear).
[26] A. O. Ivanov, Z. N. Ovsyannikov, N. P. Strelkova, and A. A. Tuzhilin, *One-dimensional minimal fillings with edges of negative weight* (Russian, with English and Russian summaries), Vestnik Moskov. Univ. Ser. I Mat. Mekh. **5** (2012), 3–8, DOI 10.3103/S0027132212050014; English transl., Moscow Univ. Math. Bull. **67** (2012), no. 5-6, 189–194. MR3076491
[27] A. Yu. Eremin, *A formula for the weight of a minimal filling of a finite metric space* (Russian, with Russian summary), Mat. Sb. **204** (2013), no. 9, 51–72. MR3137134
[28] Z. N. Ovsyannikov, *Pseudo-additive metric spaces and minimal fillings*, Vestnik MGU, Sr. Mat., Mekh., (2014); translation in Moscow Univ. Math. Bull **69** (2014) (to appear).
[29] Z. N. Ovsyannikov, *An open family of sets that have several minimal fillings*, Fundamentalnayai prikladnaya matematika **18** (2013), no. 2, 153–156; translation in J. of Math. Sci. (New York) (to appear).
[30] K. A. Zareckiĭ, *Constructing a tree on the basis of a set of distances between the hanging vertices* (Russian), Uspehi Mat. Nauk **20** (1965), no. 6, 90–92. MR0199124 (33 #7274)
[31] J. M. S. Simões Pereira, *A note on the tree realizability of a distance matrix*, J. Combinatorial Theory **6** (1969), 303–310. MR0237362 (38 #5650)

[32] Ye. A. Smolenskii, *A method for the linear recording of graphs*, U.S.S.R. Comput. Math. and Math. Phys. **2** (1962), 396–397. MR0190037 (32 #7453)

[33] S. L. Hakimi and S. S. Yau, *Distance matrix of a graph and its realizability*, Quart. Appl. Math. **22** (1965), 305–317. MR0184873 (32 #2344)

[34] O. V. Rubleva, *An additivity criterion for a finite metric space and minimal fillings* (Russian, with English and Russian summaries), Vestnik Moskov. Univ. Ser. I Mat. Mekh. **2** (2012), 8–11, DOI 10.3103/S0027132212020027; English transl., Moscow Univ. Math. Bull. **67** (2012), no. 2, 52–54. MR2985887

[35] M. Gromov, *Hyperbolic groups*, Essays in group theory, Math. Sci. Res. Inst. Publ., vol. 8, Springer, New York, 1987, pp. 75–263, DOI 10.1007/978-1-4613-9586-7_3. MR919829 (89e:20070)

[36] A. O. Ivanov and A. A. Tuzhilin, *Uniqueness of the Steiner minimal tree for boundaries in general position* (Russian, with Russian summary), Mat. Sb. **197** (2006), no. 9, 55–90, DOI 10.1070/SM2006v197n09ABEH003800; English transl., Sb. Math. **197** (2006), no. 9-10, 1309–1340. MR2273168 (2008d:05042)

[37] E. I. Filonenko (Stepanova), *Degree* 4 *Steiner subratio of Euclidean plane*, Vestnik MGU, Ser. Matem., Mekh. (2014); translation in Moscow Univ. Math. Bull **69** (2014) (to appear).

[38] A. S. Pakhomova, *Estimates for Steiner–Gromov ratio and Steiner subratio*, Vestnik MGU, Ser. Matem., Mekh. (2014), no. 1; translation in Moscow Univ. Math. Bull **69** (2014), no. 1–2 (to appear).

[39] Z. N. Ovsyannikov, *Steiner ratio, Steiner–Gromov ratio and Steiner subratio for the metric space of all compact subsets of the Euclidean space with Hausdorff metric*, Fundam. i Prikladn. Matem., **18** (2013), no. 2, 157–165; translation in J. of Math. Sci. (New York) (to appear).

[40] Michel Marie Deza and Elena Deza, *Encyclopedia of distances*, Springer-Verlag, Berlin, 2009. With 1 CD-ROM (Windows, Macintosh and UNIX). MR2538177 (2011b:51001)

[41] B. Bednov, *Star type parametric minimal fillings*, In preparation.

[42] D. Burago, Yu. Burago, and S. Ivanov, *A Course in Metric Geometry*, Graduate Studies in Mathematics, vol. 33, A.M.S., Providence, RI, 2001.

[43] A. Palhomova, *Steiner Type Ratios Continuity Criteria in the Gromov–Hausdorff Space*, Mat. Zametki, (2014); translation in Math. Motes (2014) (to appear).

[44] A. O. Ivanov, I. M. Nikonov, and A. A. Tuzhilin, *Sets admitting connection by graphs of finite length* (Russian, with Russian summary), Mat. Sb. **196** (2005), no. 6, 71–110, DOI 10.1070/SM2005v196n06ABEH000903; English transl., Sb. Math. **196** (2005), no. 5-6, 845–884. MR2164552 (2006g:52015)

[45] A. Eremin, *Minimal Fillings of Infinite sets*, in preparation.

[46] A. Erokhovets, *Steiner sub-ratio for the space of p-adic numbers*, Vestnik MGU, Ser. Matem. i Mekhan., (2014); translation in Moscow Univ. Math. Bull **69** (2014) (to appear).

[47] V. A. Mishchenko, *Estimates for the Steiner-Gromov ratio of Riemannian manifolds*, Fundamentalnaya i prikladnaya matematika **18** (2013), no. 2, 119–124; translation in J. of Math. Sci. (New York) (to appear).

[48] Z. N. Ovsyannikov, *The Steiner subratio of five points on a plane and four points in three-dimensional space*, Fundamentalnaya i prikladnaya matematika **18** (2013), no. 2, 167–179; translation in J. of Math. Sci. (New York) (to appear).

[49] A. Ivanov, A. Tuzhilin, Du-Hwang characteristic area: Catch-22, // ArXiv e-prints, 1402.6079 [math.MG], 2014

Mech. Math. Faculty, Lomonosov Moscow State University, Leninskie Gory 1, Moscow, 119991 Russia, and Bauman Moscow State Technical University, Vtoraya Baumanskaya ul. 5, Moscow, 105005 Russia

E-mail address: `aoiva@mech.math.msu.su`

Mech. Math. Faculty, Lomonosov Moscow State University, Leninskie Gory 1, Moscow, 119991 Russia

E-mail address: `tuz@mech.math.msu.su`

Contemporary Mathematics
Volume **625**, 2014
http://dx.doi.org/10.1090/conm/625/12491

Combinatorics and geometry of transportation polytopes: An update

Jesús A. De Loera and Edward D. Kim

ABSTRACT. A transportation polytope consists of all multidimensional arrays or tables of non-negative real numbers that satisfy certain sum conditions on subsets of the entries. They arise naturally in optimization and statistics, and also have interest for discrete mathematics because permutation matrices, latin squares, and magic squares appear naturally as lattice points of these polytopes.

In this paper we survey advances on the understanding of the combinatorics and geometry of these polyhedra and include some recent unpublished results on the diameter of graphs of these polytopes. In particular, this is a thirty-year update on the status of a list of open questions last visited in the 1984 book by Yemelichev, Kovalev and Kravtsov and the 1986 survey paper of Vlach.

1. Introduction

Transportation polytopes are well-known objects in mathematical programming and statistics. In the operations research literature, classical transportation problems arise from the problem of transporting goods from a set of factories, each with given supply outcome, and a set of consumer centers, each with an amount of demand. Assuming the total supply equals the total demand and that costs are specified for each possible pair (factory, consumer center), one may wish to optimize the cost of transporting goods. Indeed this was the original motivation that led Kantorovich (see [**103**]), Hitchcock (see [**97**]), and T. C. Koopmans (see [**109**]) to look at these problems. They are indeed among the first linear programming problems investigated, and Koopmans received the Nobel Prize in Economics for his work in this area (see [**98**] for an interesting historical perspective). Not much later Birkhoff (see [**18**]), von Neumann (see [**148**]), and Motzkin (see, e.g., [**120**]) were key contributors to the topic. The success of combinatorial algorithms such as the Hungarian method (see [**8, 22, 75, 76, 105, 110, 111, 121, 141**]) depends on the rich combinatorial structure of the convex polyhedra that defined the possible solutions, the so called *transportation polytopes.*

In statistics, people have looked at the *integral* transportation tables, which are widely known as *contingency tables.* In statistics, a contingency table represents sample data arranged or tabulated by categories of combined properties. Several questions motivate the study of the geometry of contingency tables, for instance,

2010 *Mathematics Subject Classification.* Primary 37F20, 52B05, 90B06, 90C08.

in the table entry security problem: given a table T (multi-dimensional perhaps) with statistics on private data about individuals, we may wish to release aggregated marginals of such a table without disclosing information about the exact entries of the table. What can a data thief discover about T from the published marginals? When is T uniquely identifiable by its margins? This problem has been studied by many researchers (see [**35, 43, 48, 49, 69–71, 77, 101**] and the references therein). Another natural problem is whether a given table presents strong evidence of significant relations between the characteristics tabulated (e.g., is cancer related to smoking). There is a lot of interest among statisticians on testing significance of independence for variables. Some methods depend on counting all possible contingency tables with given margins (see e.g., [**68, 116**]). This in turn is an interesting combinatorial geometric problem on the lattice points of transportation polytopes.

In this article we survey the state of the art in the combinatorics and geometry of transportation polytopes and contingency tables. The survey [**145**] by Vlach, the 1984 monograph [**150**] by Yemelichev, Kovalev, and Kravtsov, and the paper [**106**] by Klee and Witzgall summarized the status of transportation polytopes up to the 1980s. Due to recent advances on the topic by the authors and others, we decided to write a new updated survey collecting remaining open problems and presenting recent solutions. We also included details on some unpublished new work on the diameter of the graphs of these polytopes.

In what follows we will denote by $[q] = \{1, 2, \ldots, q\}$. Similarly $\mathbb{R}^n_{\geq 0}$ denotes those vectors in $\mathbb{R}^n$ whose entries are non-negative. Our notation and terminology on polytopes follows [**88**] and [**151**].

2. Classical transportation polytopes (2-ways)

We begin by introducing the most well-known subfamily, the classical transportation polytopes in just two indices. We call them 2-*way* transportation polytopes and in general d-ways refers to the case of variables with d indices. Many of these facts are well-known and can be found in [**150**], but we repeat them here as we will use them in what follows.

Fix two integers $p, q \in \mathbb{Z}_{>0}$. The *transportation polytope* P of size $p \times q$ defined by the vectors $u \in \mathbb{R}^p$ and $v \in \mathbb{R}^q$ is the convex polytope defined in the pq variables $x_{i,j} \in \mathbb{R}_{\geq 0}$ ($i \in [p], j \in [q]$) satisfying the $p + q$ equations

$$\sum_{j=1}^{q} x_{i,j} = u_i \ (i \in [p]) \quad \text{and} \quad \sum_{i=1}^{p} x_{i,j} = v_j \ (j \in [q]). \tag{2.1}$$

Since the coordinates $x_{i,j}$ of P are non-negative, the conditions (2.1) imply P is bounded. The vectors u and v are called *marginals* or *margins*. These polytopes are called transportation polytopes because they model the transportation of goods from p supply locations (with the ith location supplying a quantity of u_i) to q demand locations (with the jth location demanding a quantity of v_j). The feasible points $x = (x_{i,j})_{i\in[p],j\in[q]}$ in a $p \times q$ transportation polytope P model the scenario where a quantity of $x_{i,j}$ of goods is transported from the ith supply location to the jth demand location. See Figure 1.

EXAMPLE 2.1. Let us consider the 3×3 transportation polytope $P_{3\times 3}$ defined by the marginals $u = (5, 5, 1)^T$ and $v = (2, 7, 2)^T$, which corresponds to the transportation problem shown in Figure 1. A point $x^* = (x^*_{i,j})$ in P is shown in

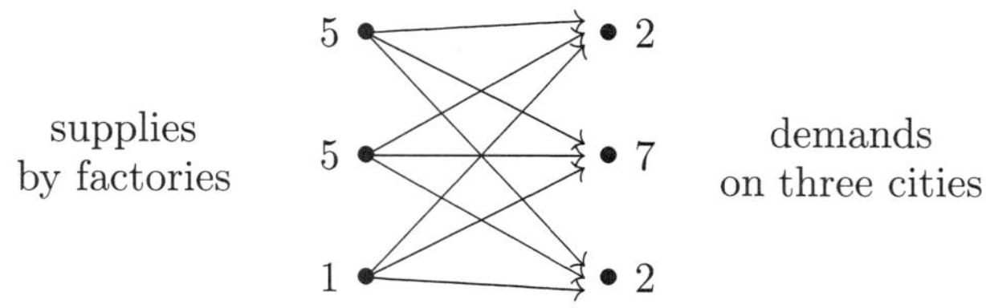

FIGURE 1. Three supplies and three demands

Figure 2. The equations in (2.1) are conditions on the row sums and column sums (respectively) of tables $x \in P$.

$$x^* = \begin{array}{|c|c|c|}\hline x^*_{1,1} & x^*_{1,2} & x^*_{1,3} \\ \hline x^*_{2,1} & x^*_{2,2} & x^*_{2,3} \\ \hline x^*_{3,1} & x^*_{3,2} & x^*_{3,3} \\ \hline \end{array} = \begin{array}{|c|c|c|}\hline 2 & 2 & 1 \\ \hline 0 & 5 & 0 \\ \hline 0 & 0 & 1 \\ \hline \end{array}$$

FIGURE 2. A point $x^* \in P_{3\times 3}$

2.1. Dimension and feasibility. Notice in Example 2.1 that $5+5+1 = 2+7+2$. The condition that the sum of the supply margins equals the sum of the demand margins is not only necessary but also sufficient for a classical transportation polytope to be non-empty:

LEMMA 2.2. *Let P be the $p \times q$ classical transportation polytope defined by the marginals $u \in \mathbb{R}^p_{\geq 0}$ and $v \in \mathbb{R}^q_{\geq 0}$. The polytope P is non-empty if and only if*

$$\sum_{i\in[p]} u_i = \sum_{j\in[q]} v_j. \tag{2.2}$$

The proof of this lemma uses the well-known *northwest corner rule algorithm* (see [**130**] or Exercise 17 in Chapter 6 of [**150**]).

The equations (2.1) and the inequalities $x_{i,j} \geq 0$ can be rewritten in the matrix form

$$P = \{x \in \mathbb{R}^{pq} \mid Ax = b, x \geq 0\}$$

with a 0-1 matrix A of size $(p+q) \times pq$ called the *constraint matrix* and a vector $b \in \mathbb{R}^{p+q}$. The constraint matrix for a $p \times q$ transportation polytope is the vertex-edge incidence matrix of the complete bipartite graph $K_{p,q}$.

LEMMA 2.3. *Let A be the constraint matrix of a $p \times q$ transportation polytope P. Then:*

(1) *Maximal rank submatrices of A correspond to spanning trees on $K_{p,q}$.*
(2) $\operatorname{rank}(A) = p + q - 1$.
(3) *Each subdeterminant of A is ± 1, thus A is totally unimodular.*
(4) *If $P \neq \emptyset$, its dimension is $pq - (p+q-1) = (p-1)(q-1)$.*

Part 4 follows from Part 2.

EXAMPLE 2.4. Continuing from Example 2.1, observe $P_{3\times 3} = \{x \in \mathbb{R}^9 \mid A_{3\times 3}x = b, x \geq 0\}$, where $A_{3\times 3}$ is the constraint matrix

$$A_{3\times 3} = \begin{bmatrix} 1 & 0 & 0 & 1 & 0 & 0 & 1 & 0 & 0 \\ 0 & 1 & 0 & 0 & 1 & 0 & 0 & 1 & 0 \\ 0 & 0 & 1 & 0 & 0 & 1 & 0 & 0 & 1 \\ 1 & 1 & 1 & 0 & 0 & 0 & 0 & 0 & 0 \\ 0 & 0 & 0 & 1 & 1 & 1 & 0 & 0 & 0 \\ 0 & 0 & 0 & 0 & 0 & 0 & 1 & 1 & 1 \end{bmatrix} \quad \text{and} \quad b = \begin{bmatrix} 2 \\ 7 \\ 2 \\ 5 \\ 5 \\ 1 \end{bmatrix}. \tag{2.3}$$

Up to permutation of rows and columns, the matrix $A_{3\times 3}$ is the unique constraint matrix for 3×3 classical transportation polytopes. It is a 6×9 matrix of rank five. Thus, $P_{3\times 3}$ is a four-dimensional polytope described in a nine-dimensional ambient space.

Birkhoff polytopes, first introduced by G. Birkhoff in [**18**], are an important subclass of transportation polytopes:

DEFINITION 2.5. The *pth Birkhoff polytope*, denoted by B_p, is the $p \times p$ classical transportation polytope with margins $u = v = (1, 1, \ldots, 1)^T$.

The Birkhoff polytope is also called the *assignment polytope* or the *polytope of doubly stochastic matrices* (see, e.g., [**7**]). It is the perfect matching polytope of the complete bipartite graph $K_{p,p}$. We can generalize the definition of the Birkhoff polytope to rectangular arrays:

DEFINITION 2.6. The *central transportation polytope* is the $p \times q$ classical transportation polytope with $u_1 = \cdots = u_p = q$ and $v_1 = \cdots = v_q = p$. This polytope is also called the *generalized Birkhoff polytope* of size $p \times q$.

2.2. Combinatorics of faces and graphs. The study of the faces of transportation polytopes is a nice combinatorial question (see, e.g., [**10**]). Unfortunately it is still incomplete, e.g., one does not know the number of i-dimensional faces of each dimension other than in a few cases. E.g., in [**126**], Pak presented an efficient algorithm for computing the f-vector of the generalized Birkhoff polytope of size $p \times (p + 1)$. Hartfiel (see [**92**]) and Dahl (see [**55**]) described the supports of certain feasible points in classical transportation polytopes. In this section, we fully describe the vertices and the edges of a 2-way transportation polytope P. The resulting graph has some interesting properties, but there are still open questions about it.

Let P be a $p\times q$ classical transportation polytope. For a point $x = (x_{i,j})_{i\in[p],j\in[q]}$, define the *support set* $\text{supp}(x) = \{(i, j) \in [p] \times [q] \mid x_{i,j} > 0\}$. We also define a bipartite graph $B(x)$, called the *support graph* of x. The graph $B(x)$ is the following subgraph of the complete bipartite graph $K_{p,q}$:

- **Vertices of** $B(x)$**.** The vertices of the graph $B(x)$ are the vertices of the complete bipartite graph $K_{p,q}$. We label the supply nodes $\sigma_1, \ldots, \sigma_p$ and the demand nodes $\delta_1, \ldots, \delta_q$.
- **Edges of** $B(x)$**.** There is an edge (σ_i, δ_j) if and only if $x_{i,j}$ is *strictly* positive. In other words, the edge set is indexed by $\text{supp}(x)$.

EXAMPLE 2.7. Let us consider the point $x^* \in P_{3\times 3}$ from Example 2.1. Here, $\text{supp}(x^*) = \{(\sigma_1, \delta_1), (\sigma_1, \delta_2), (\sigma_1, \delta_3), (\sigma_2, \delta_2), (\sigma_3, \delta_3)\}$. Figure 3 depicts the graph $B(x^*)$.

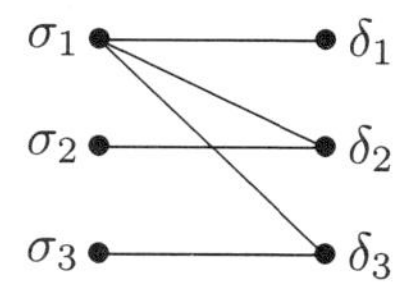

FIGURE 3. The support graph $B(x^*)$ of the point $x^* \in P_{3\times 3}$. The nodes of $B(x^*)$ on the left are the $p = 3$ supplies. The nodes on the right are the $q = 3$ demands.

An important subclass of transportation polytopes are those which are generic. Generic transportation polytopes are easiest to analyze in the proofs which follow and are the ones typically appearing in applications. Generic d-way transportation polytopes are those whose vertices have maximal possible non-zero entries. All generic transportation polytopes are simple, but not vice versa.

DEFINITION 2.8. A $p \times q$ classical transportation polytope P is *generic* if

$$\sum_{i\in Y} u_i \neq \sum_{j\in Z} v_j. \tag{2.4}$$

for every non-empty proper subset $Y \subsetneq [p]$ and non-empty proper subset $Z \subsetneq [q]$. (Of course, due to (2.2), we must disallow the case where $Y = [p]$ and $Z = [q]$.)

The graph properties of $B(x)$ provide a useful combinatorial characterization of the vertices of classical transportation polytopes:

LEMMA 2.9 (Klee, Witzgall [**106**]). *Let P be a $p \times q$ classical transportation polytope defined by the marginals $u \in \mathbb{R}^p_{>0}$ and $v \in \mathbb{R}^q_{>0}$, and let $x \in P$. Then the graph $B(x)$ is spanning. The point x is a vertex of P if and only if $B(x)$ is a spanning forest. Moreover, if P is generic, then x is a vertex of P if and only if $B(x)$ is a spanning tree.*

COROLLARY 2.10. *Let x be a point in a generic $p \times q$ classical transportation polytope P. Then x is a vertex of P if and only if $|\operatorname{supp}(x)| = p + q - 1$.*

A vertex of a $p\times q$ transportation polytope is *non-degenerate* if it has $p+q-1$ positive entries. Otherwise, the vertex is *degenerate.* A transportation polytope is *non-degenerate* if all its vertices are non-degenerate. Non-degenerate transportation polytopes are of particular interest, as they have the largest possible number of vertices and largest possible diameter among the graphs of all transportation polytopes of given type and parameters (e.g., p, q, and s). Indeed, if P is a degenerate transportation polytope, by carefully perturbing the marginals that define P we can get a non-degenerate polytope P'. (A careful explanation of how to do the perturbation is given in Lemma 4.6 of Chapter 6 in [**150**] on page 281.) The perturbed marginals are obtained by taking a feasible point x in P, perturbing the entries in the table and using the recomputed sums as the new marginals for P'. The graph of P can be obtained from that of P' by contracting certain edges, which cannot increase either the diameter nor the number of vertices.

Finally, note the following property on the vertices of a classical transportation polytope, which follows from part 3 of Lemma 2.3 and Cramer's rule:

COROLLARY 2.11. *Given integral marginals u, v, all vertices of the corresponding transportation polytope are integral.*

We now recall a classical characterization of the vertices of the Birkhoff polytope:

THEOREM 2.12 (Birkhoff-von Neumann Theorem). *The $p!$ vertices of the pth Birkhoff polytope B_p are the 0-1 permutation matrices of size $p \times p$.*

In other words, the vertices of the Birkhoff polytope are the permutation matrices, so every doubly stochastic matrix is a convex combination of permutation matrices. This theorem was proved by Birkhoff in [**18**] and proved independently by von Neumann (see [**148**]). Equivalent results were shown earlier in the thesis [**139**] of Steinitz, and the theorem also follows from [**107**] and [**108**] by Kőnig. For a more complete discussion, see the preface to [**114**]. See also the papers [**26–29**], where various various combinatorial and geometric properties of the Birkhoff polytope were studied such as its graph. Of course due to the above theorem, Birkhoff's polytopes play an important role in combinatorics and discrete optimization and the literature about their properties is rather large.

We also want to know how many vertices a transportation polytope can have. In particular there is a visible difference in behavior between generic and non-generic polytopes. How about maximum number of vertices? The exact formula is complicated but the following result of Bolker in [**19**] can serve as a reference:

LEMMA 2.13 (Bolker, [**19**]). *The maximum possible number of vertices among $p \times q$ transportation polytopes is achieved by the central transportation polytope whose marginals are $u = (q, q, \ldots, q)$ and $v = (p, p, \ldots, p)$.*

Indeed one can characterize which transportation polytopes reach the largest possible number of vertices. (See results by Yemelichev, Kravtsov and collaborators from the 1970's mentioned in [**150**].)

QUESTION 2.14. What are the possible values for the number of vertices of a generic $p \times q$ transportation polytope? Are there gaps or do all integer values on an interval occur?

A partial answer to this question is provided in Table 1, with more detail available at [**142**]. Another partial answer, given in [**60**], is:

THEOREM 2.15. *The number of vertices of a non-degenerate $p \times q$ classical transportation polytope is divisible by* $\gcd(p, q)$.

The support graph associated to a point of the transportation polytope also characterizes edges of classical transportation polytopes. (See Lemma 4.1 in Chapter 6 of [**150**].)

PROPOSITION 2.16. *Let x and x' be distinct vertices of a classical transportation polytope P. Then the vertices x and x' are adjacent if and only if the graph $B(x) \cup B(x')$ contains a unique cycle.*

This can be seen since the bases corresponding to the vertices x and x' differ in the addition and the removal of one element (see [**115, 133**]).

One can also characterize the facets of the $p \times q$ transportation polytope, which have dimension $(p-1)(q-1)-1$ by Lemma 2.3. The following lemma is Theorem 3.1 in Chapter 6 of [**150**].

sizes	Distribution of number of vertices in transportation polytopes
2×3	3 4 5 6
2×4	4 6 8 10 12
2×5	5 8 11 12 14 15 16 17 18 19 20 21 22 23 24 25 26 27 28 29 30
3×3	9 12 15 18
3×4	16 21 24 26 27 29 31 32 34 36 37 39 40 41 42 44 45 46 48 49 50 52 53 54 56 57 58 60 61 62 63 64 66 67 68 70 71 72 74 75 76 78 80 84 90 96
4×4	108 116 124 128 136 140 144 148 152 156 160 164 168 172 176 180 184 188 192 196 200 204 208 212 216 220 224 228 232 236 240 244 248 252 256 260 264 268 272 276 280 284 288 296 300 304 312 320 340 360

TABLE 1. Numbers of vertices of $p \times q$ transportation polytopes

LEMMA 2.17. *Let P be the $p \times q$ transportation polytope ($pq > 4$) defined by marginals u and v. Pick integers $1 \leq i^* \leq p$ and $1 \leq j^* \leq q$. The subset of points of P*

$$F_{i^*,j^*} = \{(x_{i,j}) \in P \mid x_{i^*,j^*} = 0\}$$

is a facet of P if and only if $u_{i^*} + v_{j^*} < \sum_{i=1}^{p} u_i$.

×			100
			6
		✓	6
38	37	37	

FIGURE 4. The equation $x_{3,3} = 0$ defines a facet, while the top-left corner entry corresponds to an equation $x_{1,1} = 0$ that does not.

See Figure 4 for an example. From this basic characterization we see:

COROLLARY 2.18. *For $2 \leq p \leq q$ and $q \geq 3$, the possible number of facets of a $p \times q$ transportation polytope is a number of the form $(p-1)q + k$ for $k = 0, \ldots, q$ and only such integers can occur.*

For example, 3×3 transportation polytopes can have 6, 7, 8, or 9 facets and only these values occur.

2.2.1. *Diameter of graphs of transportation polytopes.* Now we study a classical question about the graphs of transportation polytopes. Recall that the *distance* between two vertices x, y of a polytope P is the minimal number $\text{dist}_P(x, y)$ of edges needed to go from x to y in the graph of P. The *diameter* of a polytope is the maximum possible distance between pairs of vertices in the graph of the polytope. The Hirsch Conjecture asserts that the diameter of every polytope P is bounded above by the number of facets of P minus the dimension of P. Though the Hirsch Conjecture was finally shown to be false in general for polytopes (see [**131**]), the problem is still unsolved for transportation polytopes, and diameter bounds for this special class of polytopes are very interesting. Dyer and Frieze (see [**72**]) gave the first polynomial diameter bound for totally unimodular polytopes which applies to classical transportation polytopes (and more generally to network polytopes), but this was recently improved by Bonifas et al. in [**20**].

The diameters of classical transportation polytopes and their applications (see, e.g., [**51**]) have been studied extensively. In [**9**], Balinski proved that the Hirsch Conjecture holds and is tight for dual transportation polyhedra. For the specific case of transportation polytopes Yemelichev, Kovalev, and Kravtsov (see Theorem 4.6 in Chapter 6 of [**150**] and the references therein) and Stougie (see [**140**]) presented improved polynomial bounds. This was improved to a quadratic bound by van den Heuvel and Stougie in [**144**], and further improved to a linear bound:

THEOREM 2.19 (Brightwell, van den Heuvel, Stougie [**23**]). *The diameter of every $p \times q$ transportation polytope is at most* $8(p+q-2)$.

The bound follows from a crucial lemma which bounds the graph distance $\text{dist}_P(y, y')$ between any two vertices x and y of a $p \times q$ transportation polytope P, by constructing vertices x' and y' of P and nodes σ, δ of $K_{p,q}$ such that $\deg_{B(x')}(\delta) = \deg_{B(y')}(\delta) = 1$, $(\sigma, \delta) \in B(x') \cap B(y')$, and $\text{dist}_P(x, x') + \text{dist}_P(y, y') \leq 8$. In the arguments below, there is an important distinction between **vertices** of the polytope P (which we always denote by x or y) and **nodes** of the support graph $B(x) \subset K_{p,q}$ of a vertex x of P (which we always denote by σ or δ).

Theorem 2.19 was further improved by Cor Hurkens [**100**].

THEOREM 2.20 (Hurkens [**100**]). *The diameter of every $p \times q$ transportation polytope is at most* $4(p+q-2)$.

We present a brief sketch of Hurkens' proof. The result follows immediately from this lemma:

LEMMA 2.21 (Hurkens [**100**]). *For any two vertices x and y of a $p \times q$ transportation polytope P, there is an integer $r \geq 1$, a vertex y' of P, and nodes $\sigma, \delta_1, \ldots, \delta_r$ of $K_{p,q}$ such that:*

(1) $\deg_{B(x)}(\delta_k) = \deg_{B(y')}(\delta_k) = 1$ *for* $k = 1, \ldots, r$,
(2) $(\sigma, \delta_k) \in B(x), B(y')$ *for* $k = 1, \ldots, r$, *and*
(3) $\text{dist}_P(y, y') \leq 4r$.

The key idea that Hurkens showed is that four pivots are required (on average) to construct a common leaf node. More specifically, Hurkens proved this lemma by showing that for any two vertices x and y of a transportation polytope P, there is a node σ in $K_{p,q}$ (which can be assumed to be a supply) with r incident edges $(\sigma, \delta_1), \ldots, (\sigma, \delta_r)$ in $B(x)$ where $\delta_1, \ldots, \delta_r$ are all leaf nodes (which are necessarily demands) of $K_{p,q}$. Moreover, the nodes $\sigma, \delta_1, \ldots, \delta_r$ of $K_{p,q}$ identified in Hurkens' algorithm also satisfy the property that if

$$S := \{(\sigma, \delta_k) \mid (\sigma, \delta_k) \in B(y), \ k = 1, \ldots, r\},$$

then there is a vertex y' of P obtained after at most $4r$ pivots from the vertex y of P such that $B(x)$ and $B(y')$ have r common leaf nodes.

In the algorithm of Brightwell, van den Heuvel, and Stougie (see [**23**]), pivots are applied to vertices x and y of P, resulting in new vertices x' and y' of P. A key difference in Hurkens' algorithm in [**100**] is that pivots are only applied to *one* of the two vertices x and y of P. Without loss of generality, pivots are applied to the vertex y of P and not applied to the vertex x of P. Thus, we do not describe the vertex x further. Other than the property that the demand nodes $\delta_1, \ldots, \delta_r$ are leaf nodes in $B(x)$ adjacent to the node σ, the structure of $B(x)$ may be arbitrary.

We label the relevant supply and demands nodes participating in pivots. For each $k = 1, \ldots, r$ let $(\sigma_{k,n}, \delta_k)$ for $n = 1, \ldots, \ell_k$ be the edges in $B(y) \setminus S$ incident to δ_k, where $\ell_k = \deg_{B(y)\setminus S}(\delta_k)$. Let $(\sigma, \hat{\delta}_q)$ be the edges in $B(y) \setminus S$ incident to σ for $q = 1, \ldots, t$ where $t = \deg_{B(y)\setminus S}(\sigma)$. See Figure 5. Here we describe the successive

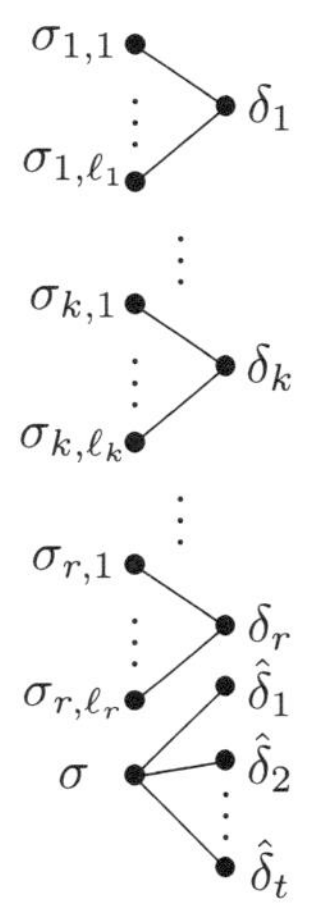

FIGURE 5. Example of supply nodes adjacent to demand nodes δ_k in $B(y)$ for $k = 1, \ldots, r$ and demand nodes adjacent to the supply node σ

pivots applied starting from the vertex y of P. For each $k = 1, \ldots, r$, we do the following:

(1) If (σ, δ_1) is not in the support graph, pivot to add (σ, δ_1). Then, pivot to add edges of the form $(\sigma_{1,n}, \hat{\delta}_q)$ for $n = 1, 2, \ldots$ until all edges of the form $(\sigma_{1,n}, \delta_1)$ are removed.
(2) If (σ, δ_2) is not in the support graph, pivot to add (σ, δ_2). Then, pivot to add edges of the form $(\sigma_{2,n}, \hat{\delta}_q)$ for $n = 1, 2, \ldots$ until all edges of the form $(\sigma_{2,n}, \delta_2)$ are removed.
(3) Continue in this way for $k = 3, \ldots, r$: If (σ, δ_k) is not in the support graph, pivot to add it. Then, pivot to add edges of the form $(\sigma_{k,n}, \hat{\delta}_q)$ for $n = 1, 2, \ldots$ until all edges of the form $(\sigma_{k,n}, \delta_k)$ are removed.

In the resulting vertex y' of P, the support graph $B(y')$ has $\delta_1, \ldots, \delta_r$ as leaf nodes adjacent to σ, which matches the support graph $B(x)$ of the vertex x of P. What remains to show (and we skip it) is that there is a choice of nodes $\sigma, \delta_1, \ldots, \delta_r$ where the number of pivots performed is at most $4r$. Instead, we illustrate the idea behind the sequence of prescribed pivots in an example:

EXAMPLE 2.22. Let y be a vertex of P where nodes $\sigma, \delta_1, \ldots, \delta_r$ in $B(y)$ are already identified. Figure 6 shows the support graph $B(y)$. (The vertex x and its associated support graph $B(x)$ can be arbitrary, thus we do not depict it in Figure 6.)

Since (σ, δ_1) is not in the support graph $B(y)$ of the vertex y of P, we insert it, and the pivot operation removes the edge $(\sigma_{1,3}, \delta_1)$. We now apply pivots to the resulting adjacent vertex of P as follows: After the pivot, only the edges $(\sigma_{1,1}, \delta_1)$ and $(\sigma_{1,2}, \delta_1)$ are incident to the demand node δ_1. These two edges are removed by pivoting to add the edges $(\sigma_{1,1}, \hat{\delta}_1)$ and $(\sigma_{1,2}, \hat{\delta}_1)$, respectively, which causes δ_1 to be a leaf node adjacent to σ.

After insertion of the edge (σ, δ_2) the remaining edge of the form $(\sigma_{2,n}, \delta_2)$ is removed the same way. Since δ_3 is already a leaf node, the insertion of (σ, δ_3) will cause it to be a leaf node adjacent to σ.

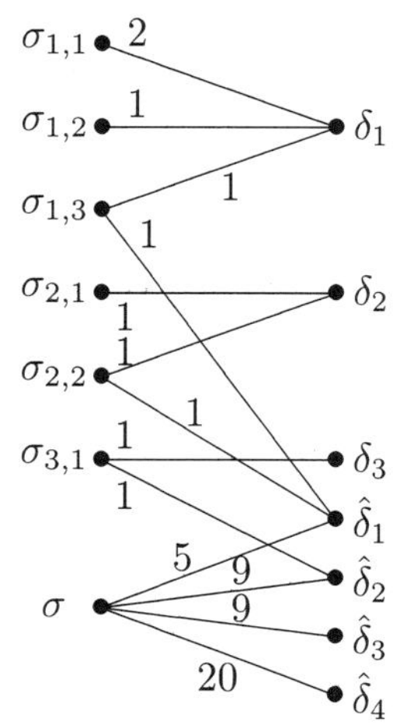

FIGURE 6. The support graph $B(y)$ of the vertex y of a transportation polytope P

To prove that the Hirsch Conjecture is true for transportation polytopes, one would hope that any pair of vertices that differ in k support elements has a pivot step that reduces the number of non-zero variables in which the vertices differ, but Brightwell et al. [**23**] noticed that this was not true. We show their counter-example in Figure 7.

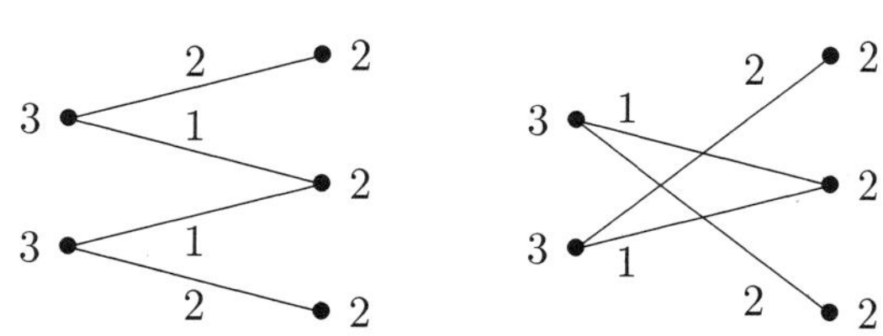

FIGURE 7. Support graphs of a pair of vertices where no pivot reduces the difference in support

OPEN PROBLEM 2.23. Prove or disprove the Hirsch Conjecture for 2-way transportation polytopes.

By Corollary 2.18, this would mean the diameter is less than or equal to $p+q-1$. This conjecture holds for many special cases that restrict the margins. For example the conjecture is true for Birkhoff's polytope and for some special right-hand sides (see e.g., [**21**]).

While transportation polytopes seem tame compared to other polytopes, it has been shown that they have some non-trivial topological structure: Diameter bounds for simple d-polyhedra can be studied via decomposition properties of related simplicial complexes. Each non-degenerate simple polytope has a polar *simplicial complex*, a simplicial polytope. Billera and Provan (see [**129**]) showed that polytopes whose dual simplicial polytope is *weakly vertex decomposable* have a linear diameter. But it has recently been shown (see [**61**]) that the infinite family of polars of $p \times 2$ transportation polytopes for $p \geq 5$ are not weakly vertex-decomposable, the first ever such examples. But despite the pathological behavior, one can prove the Hirsch Conjecture holds for $2 \times q$ transportation polytopes:

THEOREM 2.24. *Let $P \neq \emptyset$ be a generic classical transportation polytope of size $2 \times q$ with $n \leq 2q$ facets. Then, the diameter of P is at most $n-(q-1)$. In particular, the dimension of P is $d = q-1$, thus P satisfies the Hirsch Conjecture.*

To begin the proof we need to set up the properties of the polytope.

LEMMA 2.25. *Let P be a non-empty generic $2 \times q$ transportation polytope. If x is a vertex of P (a $2 \times q$ table), then there is exactly one index $j \in \{1, \ldots, q\}$ such that the number of non-zero entries in the jth column of x equals two, i.e., $|\{i \in \{1,2\} : x_{i,j} > 0\}| = 2$, but if $k \neq j$, then $|\{i \in \{1,2\} : x_{i,k} > 0\}| = 1$.*

PROOF. Because P is generic and non-empty we can assume the margins $u \in \mathbb{R}^2_{>0}$ and $v \in \mathbb{R}^q_{>0}$ of the table are strictly positive, because otherwise, if any of the values is zero, it forces all the entries of a row or a column to be zero and we are dealing with a degenerate polytope. By Lemma 2.9, if x is a vertex of a $2 \times q$ transportation polytope P, then there is at most one index $j \in \{1, \ldots, q\}$ such that the number of non-zero entries in the jth column of x equals two, i.e., $|\{i \in \{1,2\} : x_{i,j} > 0\}|$, otherwise there is a cycle of length four in the support graph of x, which would imply x is not a vertex. Similarly, by the characterization of vertices of generic transportation polytopes as spanning trees there must be at least one column $j \in \{1, \ldots, q\}$ with two non-zero entries, i.e., $|\{i \in \{1,2\} : x_{i,j} > 0\}| = 2$ (if all vertices have degree one there are not enough edges in the graph to be a spanning tree). Because we have only $q+1$ non-zero entries in the support of x, all remaining columns have exactly one support entry. □

Let X_1 denote the subset of columns $j \in \{1, \ldots, q\}$ where $x_{1,j} > 0$ in column j. Similarly, let X_2 denote the columns j where $x_{2,j}$ is positive. Note that from our discussion above $|X_1| + |X_2| = q+1$ and $|X_1 \cap X_2| = 1$. In parts of the proof we will need to perform *pivots* on a vertex x of P to obtain a vertex adjacent to x. We have seen these correspond to edge traversals in P. Earlier we had used the bipartite graph representation of the points of P (support graph), but we now mostly use the table representation. Suppose that r is the unique column index in $X_1 \cap X_2$. A pivot on x can be performed to insert an entry of the form $(2, s)$ for $s \in X_1 \setminus X_2$ or to insert an entry of the form $(1, t)$ for $t \in X_2 \setminus X_1$. If a pivot is performed to insert the $(2, s)$ entry for $s \in X_1 \setminus X_2$, then the values of $x_{2,s}$ and $x_{1,r}$ increase while the values of $x_{1,s}$ and $x_{2,r}$ decrease, all by the same amount. The amount of increase and decrease in these four coordinates equals $\min\{x_{1,s}, x_{2,r}\}$. Since P is generic, (2.4) implies that $x_{1,s}$ and $x_{2,r}$ are distinct, thus precisely one of $(1,s)$ or $(2,r)$ leave the support. A pivot to insert an entry of the form $(1,t)$ for $t \in X_2 \setminus X_1$ similarly removes either the $(2,t)$ or $(1,r)$ entry, but not both.

Let x and y be two arbitrary vertices of P. By simultaneously reordering columns, we can assume that the unique column of x where both $x_{1,j}$ and $x_{2,j}$ are strictly positive is the last column $j = q$. We can also assume by reordering that the columns $s = 1, \ldots, a$ have only the $x_{1,s}$ entry strictly positive (i.e., $x_{2,s} = 0$) and that columns $t = a+1, \ldots, q-1$ have $x_{2,t} > 0$ and $x_{1,t} = 0$. Note that in any other vertex y of P, either the unique column with two support entries is also indexed by q or by an index less than q. We will consider the two situations next:

LEMMA 2.26. *Let x and y be two arbitrary vertices of a generic classical transportation polytope of size $2 \times q$. Suppose the qth column is the unique column of x where $x_{1,q}$ and $x_{2,q}$ are positive, all columns $s = 1, \ldots, a$ have only the $x_{1,s}$ entry strictly positive and the columns $t = a+1, \ldots, q-1$ have $x_{2,t} > 0$. Suppose in addition $y_{1,q}$ and $y_{2,q}$ are positive entries and again the qth column is the unique column of y with two positive entries in y.*

Suppose at least one of the following two situations occurs:

1. *there is an index $s \in \{1, \ldots, a\}$ such that $0 < y_{2,s} < y_{1,q}$,*
2. *there is an index $t \in \{a+1, \ldots, q-1\}$ such that $0 < y_{1,t} < y_{2,q}$.*

Then there is a $2 \times (q-1)$ transportation polytope P' with precisely two fewer facets and $\operatorname{diam}(P) = 1 + \operatorname{diam}(P')$. *In particular, if P' satisfies the Hirsch Conjecture, then P satisfies the Hirsch Conjecture too.*

PROOF. In the first case, we apply a pivot on y to add the $(1,s)$ entry, which will remove the $(2,s)$ entry. Note that in the resulting new vertex and in x, the sth column contains exactly one support entry and this is the $(1,s)$ entry. Let $j = s$. In the second case, apply a pivot on y to add the $(2,t)$ entry, which will remove the $(1,t)$ entry. In the resulting new vertex as well as in x, the tth column contains exactly one support entry and this is the $(2,t)$ entry. Let $j = t$.

Remove the jth column to obtain two vertices of a $2 \times (q-1)$ transportation polytope P'. By Lemma 2.17, both removed coordinates $(1,j)$ and $(2,j)$ correspond to facets, so the number $n(P)$ of facets of P is two more than the number $n(P')$ of facets of P'. If P' satisfies the Hirsch Conjecture, since the dimension of P' is $q-2$ and the dimension of P is $q-1$, we have

$$\operatorname{diam}(P) = 1 + \operatorname{diam}(P') \leq 1 + n(P') - (q-2) = n(P) - (q-1),$$

and thus P also satisfies the Hirsch Conjecture. □

LEMMA 2.27. *Let x and y be two arbitrary vertices of a generic classical transportation polytope of size $2 \times q$. Suppose, as before, for vertex x the q-th column is the unique column with two positive entries (i.e., $x_{1,q}$ and $x_{2,q}$ are strictly positive), the columns indexed by $s = 1, \ldots, a$ have only the entry $x_{1,s} > 0$ and all the columns indexed by $t = a+1, \ldots, q-1$ have $x_{2,t} > 0$. Suppose also that y_{1,j^*} and y_{2,j^*} are strictly positive for some index $j^* < q$.*

Suppose at least one the following conditions occurs:

1. *there is an index $s \in \{1, \ldots, a\}$ with $s \neq j^*$ such that $0 < y_{2,s} < y_{1,j^*}$,*
2. *there is an index $t \in \{a+1, \ldots, q-1\}$ with $t \neq j^*$ such that $0 < y_{1,t} < y_{2,j^*}$.*

Then there is a $2 \times (q-1)$ *transportation polytope* P' *with precisely two fewer facets and* $\mathrm{diam}(P) = 1 + \mathrm{diam}(P')$. *In particular, if* P' *satisfies the Hirsch Conjecture, then* P *satisfies the Hirsch Conjecture too.*

The conditions and the proof are truly identical to the previous lemma, obtaining P' from P by doing a single pivot and then deleting a column that deletes two facets. We leave the details out.

PROOF OF THEOREM 2.24. The proof proceeds by induction on q by applying the lemmas. From the simultaneous reordering of columns, without loss of generality we can assume that there exists an index a, between 1 and q such that

- for the indices $s = 1, \dots, a$ only the entry $x_{1,s}$ is positive,
- for the indices $t = a+1, \dots, q-1$ only $x_{2,t} > 0$, and
- both $x_{1,q} > 0$ and $x_{2,q} > 0$.

Meanwhile applying the same reordering to the vertex y, either the unique column with two support entries is also the qth column as we have for vertex x, or it is a column $j^* < q$. We consider these two cases:

(1) **The unique column with two support entries for both vertices x, y is the qth column:** Note that in particular both $y_{1,q}$ and $y_{2,q}$ are strictly positive. After applying the previous lemmas until they can no longer be applied and removing any columns $j < q$ with identical support in x and y, then y satisfies the following: $y_{1,j} > 0$ if and only if $x_{2,j} > 0$ for all $j < q$. In summary, the complete list of entries in y which are strictly positive consists of $(2,1), \dots, (2,a), (1,a+1), \dots, (1,q-1), (1,q)$, and $(2,q)$. Moreover, since Lemma 2.26 does not apply anymore, we have

$$y_{2,s} > y_{1,q} \text{ for all } s = 1, \dots, a \tag{2.5}$$

and

$$y_{1,t} > y_{2,q} \text{ for all } t = a+1, \dots, q-1. \tag{2.6}$$

We will apply pivots to y and subsequent resulting vertices until we obtain x. First apply a pivot on y to add $(2, a+1)$, which can be performed since $y_{2,a+1} = 0$. This pivot increases the $(2, a+1)$ and $(1, q)$ entries while decreasing the $(2, q)$ and $(1, a+1)$ entries, and one of these decreasing entries will leave the support. The inequality (2.6) implies that the pivot removes the $(2, q)$ entry. The resulting vertex has $(2,1)$,…,$(2,a)$,$(1,a+1)$,…,$(1,q-1)$,$(2,a+1)$, and $(1,q)$ as support entries. Note that performing this pivot on y does not increase the number of entries having common support with x.

In the resulting vertex, we will perform pivots as prescribed below. Unlike the first pivot above which did not increase the number of entries having common support with y, each pivot performed below will increase the number of columns whose support is identical to the corresponding columns in vertex x. Our pivots will insert all entries of the form $(1, s)$ for $s \le a$ and all entries of the form $(2, t)$ for $a+2 \le t \le q-1$, starting with a pivot on $(1, 1)$, in the following manner:

- If a pivot inserting an entry of the form $(1, s)$ for $s < a$ replaces the corresponding $(2, s)$ entry, then the sth column of x and the sth column of the new vertex resulting from the pivot are identical, and the next pivot should be applied to insert $(1, s+1)$. No further pivot involves an entry from column s, thus in all subsequent vertices, the sth column will remain identical to the sth column of x.

If a pivot inserting an entry of the form $(1, s)$ for $s \leq a$ instead removes an entry of the form $(1, t)$ for $a+1 \leq t \leq q-1$, then the tth column of x and the tth column of the resulting vertex are identical, and the next pivot should be applied to insert $(2, t+1)$. No further pivot involves an entry from column t, thus in all subsequent vertices, the tth column will remain identical to the tth column of x.

- If a pivot inserting an entry of the form $(2, t)$ for $a+2 \leq t < q-1$ removes the corresponding $(1, t)$ entry, then the tth column of x and the tth column of the resulting vertex are identical, and the next pivot should be applied to insert $(2, t+1)$. No further pivot involves an entry from column t, thus in all subsequent vertices, the tth column will remain identical to the tth column of x.

  If a pivot inserting an entry of the form $(2, t)$ for $a+2 \leq t \leq q-1$ removes an entry of the form $(2, s)$ for $s \leq a$, then the sth column of x and the sth column of the resulting vertex are identical, and the next pivot should be applied to insert $(1, s+1)$. No further pivot involves an entry from column s, thus in all subsequent vertices, the sth column will remain identical to the sth column of x.

After the initial pivot, a total of $q-2$ additional pivots are applied, namely pivots to insert entries of the form $(1, s)$ for $s \leq a$ and pivots to insert entries of the form $(2, t)$ for $a+2 \leq t \leq q-1$, with each pivot causing an additional column to be identical to the corresponding column of x. After these $q-1$ pivots, by applying a single pivot to reintroduce the $(2, q)$ entry (namely, the pivot to insert the $(2, q)$ entry), the two remaining columns which were not identical are now identical, and the resulting vertex is x. A total of q pivots were performed from y to x and by Lemma 2.17 all $n = 2q$ coordinates correspond to facets, hence $q \leq n - (q-1)$ and thus P satisfies the Hirsch Conjecture.

(2) **The unique column with two support entries in x is the q-th column, while for y it is the j^*-th column for some $j^* < q$.** In this case y_{1,j^*} and y_{2,j^*} are strictly positive. After applying the previous lemmas until they can no longer be applied and removing any columns $j < q$ with identical support in x and y, then either $j^* \in \{1, \ldots, a\}$ or $j^* \in \{a+1, \ldots, q-1\}$. We assume that $j^* \in \{1, \ldots, a\}$, with the case $j^* \in \{a+1, \ldots, q-1\}$ being similar. Further, we may assume without loss of generality that $j^* = 1$.

Since columns with identical support in x and y were removed, the complete list of entries in y which are positive are $(1,1)$, $(2,1)$, $(2,2)$, $\ldots$, $(2,a)$, $(1,a+1)$, $\ldots$, $(1,q-1)$, and exactly one entry of the form (i^*, q) for i^* either 1 or 2. Moreover, since Lemma 2.27 does not apply anymore, we have

$$y_{2,s} > y_{1,1} \text{ for all } s = 2, \ldots, a \tag{2.7}$$

and

$$y_{1,t} > y_{2,1} \text{ for all } t = a+1, \ldots, q-1, \tag{2.8}$$

since $j^* = 1$.

We will apply pivots to y and subsequent resulting vertices until we obtain x: First apply a pivot on y to add $(2, a+1)$, which can be performed because $y_{2,a+1} = 0$. This pivot increases the $(2, a+1)$ and $(1,1)$ entries while decreasing the $(1, a+1)$ and $(2,1)$ entries, and one of these decreasing entries will leave the support. The inequality (2.8) implies that this pivot removes the $(2,1)$ entry. The resulting vertex has $(1,1), (2,2), (2,3), \ldots, (2,a), (1,a+1), \ldots, (1,q-1), (2,a+1)$, and (i^*, q) as support entries. Note that performing this pivot on y increases the number of entries having common support with x, however the $a+1$ column of the new vertex has both entries positive while the $a+1$ column of x has only the $(2, a+1)$ entry positive.

In the resulting vertex, we will perform pivots as prescribed below. Each pivot performed below will increase the number of columns whose support is identical to the vertex x. Our pivots will insert all entries of the form $(1, s)$ for $2 \leq s \leq a$ and pivots of the form $(2, t)$ for $a+2 \leq t \leq q-1$, starting with a pivot on $(1, 2)$, in the following manner:

- If a pivot inserting an entry of the form $(1, s)$ for $2 \leq s < a$ removes the corresponding $(2, s)$ entry, then the sth column of x and the sth column of the resulting vertex are identical, and the next pivot should be applied to insert $(1, s+1)$. No further pivot involves an entry from column s, thus in all subsequent vertices, the sth column will remain identical to the sth column of x.
  If a pivot inserting an entry of the form $(1, s)$ for $2 \leq s \leq a$ instead removes an entry of the form $(1, t)$ for $a+1 \leq t \leq q-1$, then the tth column of x and the tth column of the resulting vertex are identical, and the next pivot should be applied to insert $(2, t+1)$. No further pivot involves an entry from column t, thus in all subsequent vertices, the tth column will remain identical to the tth column of x.
- If a pivot inserting an entry of the form $(2, t)$ for $a+1 \leq t < q-1$ removes the corresponding $(1, t)$ entry, then the tth column of x and the tth column of the resulting vertex are identical, and the next pivot should be applied to insert $(2, t+1)$. No further pivot involves an entry from column t, thus in all subsequent vertices, the tth column will remain identical to the tth column of x.
  If a pivot inserting an entry of the form $(2, t)$ for $a+1 \leq t \leq q-1$ removes an entry of the form $(2, s)$ for $1 \leq s \leq a$, then the sth column of x and the sth column of the resulting vertex are identical, and the next pivot should be applied to insert $(1, s+1)$. No further pivot involves an entry from column s, thus in all subsequent vertices, the sth column will remain identical to the sth column of x.

A total of $q-2$ additional pivots are applied (and this total includes the initial pivot), namely to insert entries of the form $(1, s)$ for $2 \leq s \leq a$ and pivots of the form $(2, t)$ for $a+1 \leq t \leq q-1$, with each pivot causing an additional column to be identical to the corresponding column of x. After these $q-2$ pivots, by applying a final pivot to $(3-i^*, q)$, the two remaining columns are identical, and the resulting vertex is x. Thus a total of $q-1$ pivots were performed from y to x and by Lemma 2.17 at least $2q-1$ coordinates correspond to facets. Since $n \geq 2q-1$, we have $q-1 < (2q-1)-(q-1) \leq n-(q-1)$, thus P satisfies the Hirsch Conjecture.

Given arbitrary vertices x and y of P, in all cases, at most $n(P)-(q-1)$ pivots are performed to obtain the vertex x from y, and thus P satisfies the Hirsch Conjecture. □

EXAMPLE 2.28. Let us consider how the theorem applies to two crucial vertices x and y in the $2\times q$ generalized Birkhoff polytope P for odd q with margins $u=(q,q)$ and $v=(2,2,\dots,2)$. Let x be the vertex with $x_{1,1}=\dots=x_{1,\lfloor\frac{q}{2}\rfloor}=x_{2,\lceil\frac{q}{2}\rceil}=\dots=x_{2,q-1}=2$, $x_{1,q}=x_{2,q}=1$ and all other entries zero. Let y be the vertex with $y_{2,1}=\dots=y_{2,\lfloor\frac{q}{2}\rfloor}=y_{1,\lceil\frac{q}{2}\rceil}=\dots=y_{1,q-1}=2$, $y_{1,q}=y_{2,q}=1$ and all other entries zero. Then $a=\lfloor\frac{q}{2}\rfloor$ and since the lemmas cannot be applied, the inequalities (2.5) and (2.6) apply. The first pivot inserts $(2,a+1)$ and the $(2,q)$ entry goes from 1 to 0, removing it from the support. This pivot did not increase the number of entries having common support with x.

Then the pivot sequence in the proof of the theorem performs the following pivots: A pivot on $(1,1)$ removes the $(1,a+1)$ entry. A pivot on $(2,a+2)$ removes the $(2,1)$ entry. A pivot on $(1,2)$ removes the $(1,a+2)$ entry. A pivot on $(2,a+3)$ removes the $(2,2)$ entry. The pivots continue in this alternating fashion, with a pivot of the form $(1,j)$ removing the $(1,a+j)$ entry followed by a pivot of the form $(2,a+j+1)$ removing the $(2,j)$ entry, until a pivot to introduce the $(1,\lfloor\frac{q}{2}\rfloor)$ entry. A final pivot, the qth pivot, is performed on the $(2,q)$ entry. In particular, the case $q=3$ is the polytope of Brightwell et al. in [**23**] depicted in Figure 7.

Incidentally the Hirsch bound also holds for Birkhoff polytopes and can be seen with a very easy argument:

THEOREM 2.29. *Let B_p be the pth Birkhoff polytope. Then,*

(1) *the degree of each vertex of B_p is*

$$\sum_{k=0}^{p-2}\binom{p}{k}(p-k-1)!$$

(2) *If $p \geq 4$, the diameter of B_p is 2.*
(3) *(Billera-Sarangarajan [**17**]) Every pair of vertices x, y is contained in a cubical face. The dimension of this cubical face is the number of cycles in the union of $B(x)$ and $B(y)$.*

PROOF. For part 1, note that because the symmetric group acts transitively on the vertices (which are permutation matrices) the degree of all vertices is the same. It suffices to count how many vertices are adjacent to the vertex corresponding to the identity matrix. Any adjacent vertex y has k common edges with x for $k=0,\dots,p-2$. Now the k edges can be chosen in $\binom{p}{k}$ ways and for each choice we have a unique cycle being formed with the remaining $(p-k-1)$ pairs of vertices (i,i'). This can be done in $(p-k-1)!$ ways.

Now we prove part 2. Given two non-adjacent vertices x and y we have a third vertex z adjacent to both. Without loss of generality, the two graphs $B(x)$ and $B(y)$ have no common edges, otherwise apply induction. Thus they define p disjoint bipartite cycles, as shown in Figure 8. □

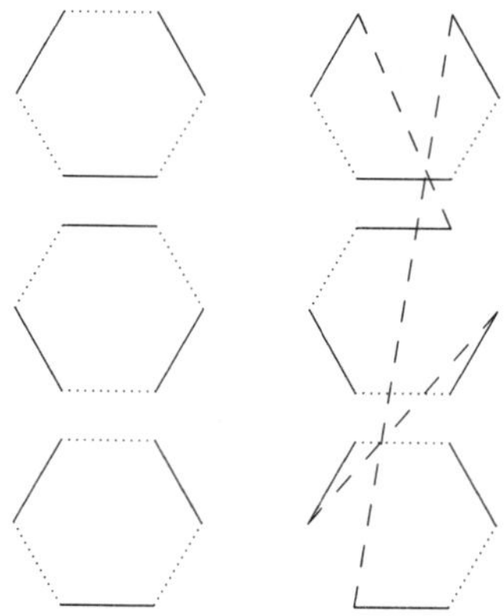

FIGURE 8. Diameter of Birkhoff polytope

It is worth noting that even if the Hirsch Conjecture for transportation polytopes is true, the simplex method may behave badly because there could be long decreasing pivot sequences:

THEOREM 2.30 (I. Pak [**125**]). *Consider the linear functional*

$$c_\alpha \cdot x = x_{1,1} + \alpha x_{1,2} + \cdots + \alpha^{p-1} x_{1,p} + \alpha^p x_{2,1} + \cdots + \alpha^{p^2-1} x_{p,p}.$$

For $1/p > \alpha > 0$, there exist a decreasing sequence of vertices of the $p \times p$ Birkhoff-von Neumann polytope of length $Kp!$ for a universal constant K.

However, Pak (see [**125**]) also showed the more encouraging result that the expected average running time of the simplex method on the Birkhoff polytope with cost vector c_α is $O(p \log p)$.

2.3. Integer points. Questions on the integer, or lattice, points of transportation polytopes are very popular in combinatorics. Objects such as magic squares, magic labelling of graphs and sudoku arrangements can be presented as lattice points of transportation polytopes. See, for instance, [**15**, **56**, **138**] and the references therein. How many ways are there to fill the entries of a $p \times q$ table with margins u and v using only non-negative *integer* entries $x_{i,j}$? E.g., see Figure 9. This counting problem is a $\#P$-complete problem, even for $2 \times q$ tables (see [**74**]).

68	119	26	7	220
20	84	17	94	215
15	54	14	10	93
5	29	14	16	64
108	286	71	127	

There are 1,225,914,276,768,514 such tables.

FIGURE 9. This transportation polytope has many lattice points!

The lattice points of dilations of the Birkhoff polytope are called *semi-magic squares*: that is to say, a semi-magic square is an integral lattice point in a transportation polytope where every row and column sum is the same, namely ζ. The number ζ is called the *magic number*. Counting these objects is a rather natural combinatorial problem that has been studied by many researchers. In [**62**], De Loera, Liu, and Yoshida presented a generating function for the number of semi-magic squares and formulas for the coefficients of the Ehrhart polynomial of the pth Birkhoff polytope B_p. In particular they also deduced a combinatorial formula for the volume of Birkhoff polytopes. The volume formula is a multivariate generating function for the lattice points of the Birkhoff polytope and all its dilations. Unfortunately the number of terms, which alternate in sign, is quite large. The summation runs over all the possible arborescences of a complete graph in p nodes (p^{p-2} of them) and the $p!$ permutations, thus the formula is quite large and not efficient to evaluate. The key elements of this formula come from understanding triangulations of the tangent cones of the Birkhoff polytope and the algorithmic theory of lattice points developed by Barvinok (see [**12, 13**]). More recently Liu (see [**113**]) described the same kind of generating functions for perturbations of the Birkhoff polytope into simple transportation polytopes (i.e., the margin sum conditions are not one but a small change in value). She obtained similar combinatorial formulas for the generalized Birkhoff $kp \times p$ polytope. She also recovered the formula for the maximum possible number of vertices of transportation polytopes of order $kp \times p$ that had been studied in the literature before. Prior work on enumeration includes [**37**], where Carlitz described lattice points of dilations of the Birkhoff polytope using exponential generating functions.

Counting magic squares and lattice points in (dilations of) Birkhoff polytopes is related to computing their volumes. The computation of volumes and triangulations of the Birkhoff polytope is related to the problem of generating a random doubly stochastic matrix (see [**38**]). The volume problem has been studied by many researchers (see [**1, 2, 15, 16, 38, 67, 91, 95, 125, 137, 138**], among others). The exact value of the volume of the pth Birkhoff polytope B_p is known (see [**128**]) only up to $p = 10$. Canfield and McKay (see [**36**]) presented an asymptotic formula for the volume of the pth Birkhoff polytope B_p. In [**11**], Barvinok also presented asymptotic upper and lower bounds for the volumes of $p \times q$ classical transportation polytopes and the number of $p \times q$ semi-magic rectangles.

The currently known exact values of $a(p)$ are summarized in Table 2.

p	$a(p)$
1	1
2	1
3	3
4	352
5	4718075
6	14666561365176
7	17832560768358341943028
8	12816077964079346687829905128694016
9	7658969897501574748537755050756794492337074203099
10	5091038988117504946842559205930853037841762820367901333706255223000

TABLE 2. Normalized volumes of Birkhoff polytopes

To compute more values it would be useful to know the answer to the following problem:

OPEN PROBLEM 2.31. Is there a short (polynomial time computable) formula for the normalized volume $a(p)$ of the $p \times p$ Birkhoff-von Neumann polytope?

Besides knowing the volumes and the number of vertices, we are interested in knowing the so-called *integer range* of a coordinate in a transportation polytope P. This asks the following: fixing i and j, do all integers in an interval appear as the value of the coordinate $x_{i,j}$ of among the set of lattice points of P? For classical transportation polytopes, the answer is yes:

LEMMA 2.32 (Diaconis and Gangolli [**68**], Integer range of a coordinate). *For an entry $x_{i,j}$ of the transportation polytope with marginals u and v, the set of all possible integral values are the integers on a segment.*

This gives a method of performing the so-called *sequential importance sampling* (see, e.g., [**41**, **149**]). Chen et al. (see [**41**]) use the interval property to justify correctness of their algorithm for the sequential sampling of entries in multi-way contingency tables with given constraints. This method of sampling contingency tables with given margins introduced in [**40**] is later extended by Chen in [**39**] to sample tables with fixed marginals and a given set of structural zeros.

For many applications, again including sampling and enumerating lattice points, we are interested in having a set of "local moves" or operations that connect the set of *all* integer contingency tables with fixed margins. E.g., such a set of moves is important in probability and statistics in the interest of running Markov chains on contingency tables (see [**63**]). As it turns out the set of moves necessary is quite simple:

LEMMA 2.33. *The set of "rectangular" vectors whose entries are* 0,-1, *and* 1 *(as in Figure* 10*) corresponding to 4-cycles in the complete bipartite graph $K_{p,q}$, with a* 1 *and a* -1 *in two rows and two columns are integer vectors in the kernel of the constraint matrix of* 2*-way transportation polytopes. They are simple moves that connect all lattice points of any* 2*-way transportation polytope.*

0	0	0	0	0	0	0	0	0
0	0	-1	0	0	0	1	0	0
0	0	0	0	0	0	0	0	0
0	0	0	0	0	0	0	0	0
0	0	1	0	0	0	-1	0	0
0	0	0	0	0	0	0	0	0
0	0	0	0	0	0	0	0	0

FIGURE 10. Typical monomial in Graver basis

Using these moves one can run a Markov chain on all the vertices of a transportation polytope, where we move from one vertex to another by adding one of the randomly generated moves that preserves non-negativity. Cryan et al. (see [**51**]) have shown that the associated Markov chain mixes rapidly when the number p or q of rows or columns is assumed fixed.

This set of vectors is an example of a *Graver basis* for the kernel of the matrix associated to the 2-way transportation polytope in question (see [**86**]). Formally, to define a Graver basis, we first describe a partial order $\sqsubseteq$ on $\mathbb{Z}^n$. Given two integer vectors $u, v \in \mathbb{Z}^n$, we say $u \sqsubseteq v$ if $|u_k| \leq |v_k|$ and $u_k v_k \geq 0$ for all $k = 1, \ldots, n$. Then the *Graver basis* of a matrix A is the set of all $\sqsubseteq$-minimal vectors in $\{x \in \mathbb{Z}^n \mid Ax = 0, x \neq 0\}$. Graver bases are quite important in optimization (see Chapters 3 and 4 of [**57**] and the nice book [**124**] for details).

In the next section, we discuss multi-way transportation polytopes. As we will see, their behavior is much more complicated.

3. Multi-way transportation polytopes

Classical transportation polytopes were called 2-*way transportation polytopes* because the coordinates $x_{i,j}$ have two indices. We can consider generalizations of 2-way transportation polytopes by having coordinates indexed by three or more integers (e.g., $x_{i,j,k}$ or $x_{i,j,k,l}$). As the number of indices grows the possible form and shape of constraints grows. There has been very active work on understanding the properties of the corresponding polyhedra (see e.g., [**89, 90, 118, 119, 132, 134–136, 145**]). As we will see here the case of 3-way transportation problems, i.e., three indices, is already so complicated that in a sense contains all polyhedral geometry and combinatorial optimization! Nevertheless one cannot forget that applications of problems with more than two indices are quite common and include not only problems in optimization and statistics (e.g., contingency tables [**48**]), but more recently the prior connections to combinatorics ([**112**]) and representation theory ([**5, 127, 143**]) have been reinforced by the new connection to quantum computation [**44**].

A *d-way table* of size $p_1 \times \cdots \times p_d$ is a $p_1 \times p_2 \times \cdots \times p_d$ array of non-negative real numbers $x = (x_{i_1,\ldots,i_d})$, $1 \leq i_\ell \leq p_\ell$. Given an integer m, with $0 \leq m < d$, an m**-margin** of the d-way table x is one of the $\binom{d}{m}$ possible m-tables obtained by summing the entries over all but m indices. For example, if $(x_{i,j,k})$ is a 3-way table then its 0-marginal is $x_{+,+,+} = \sum_{i=1}^{p_1} \sum_{j=1}^{p_2} \sum_{k=1}^{p_3} x_{i,j,k}$, it has three 1-margins, which are $x_{i,+,+}) = \sum_{j=1}^{p_2} \sum_{k=1}^{p_3} x_{i,j,k}$ and likewise $(x_{+,j,+})$, $(x_{+,+,k})$. Finally x has three 2-margins given by the sums $(x_{i,j,+}) = \sum_{k=1}^{p_3} x_{i,j,k}$ and likewise $(x_{i,+,k})$, $(x_{+,j,k})$.

A *d-way transportation polytope* of size $p_1 \times \cdots \times p_d$ defined by m-marginals is the set of all d-way tables of size $p_1 \times p_2 \times \cdots \times p_d$ with the specified marginals. When $d = 2$, we recover the classical transportation polytopes of the previous section. When $d \geq 3$, the transportation polytope is also called a *multi-way transportation polytope.* When $d = 3$, we will typically denote the size of the transportation polytope by $p \times q \times s$ instead of $p_1 \times p_2 \times p_3$.

In a well-defined sense the most important margins of a d-way transportation polytope are the $(d-1)$-margins:

THEOREM 3.1 (Junginger [**102**]). *There exists a polynomial time algorithm that, given a linear (integer) minimization problem over a d-way $p_1 \times \cdots \times p_d$ transportation polytope $T_{d,m}$ with fixed m-marginals and cost vector c, computes an associated linear functional $\hat{c}$ and a d-way $(p_1+1) \times \cdots \times (p_d+1)$ transportation polytope $T_{d,d-1}$ with fixed $(d-1)$-marginals such that if y is an optimal (integral) solution for $T_{d,d-1}$ its entries with indices with the original range also give an optimal (integral) solution of $T_{d,m}$.*

EXAMPLE 3.2. We illustrate Junginger's theorem in the 3-way case. Suppose we have a linear optimization problem over a 3-way $p \times q \times s$ transportation defined by 1-marginals:

$$\text{minimize} \quad \sum_{i_1=1}^{p} \sum_{i_2=1}^{q} \sum_{i_3=1}^{s} c_{i_1,i_2,i_3} x_{i_1,i_2,i_3}$$

$$\text{subject to} \quad \begin{cases} \displaystyle\sum_{i_2=1}^{q} \sum_{i_3=1}^{s} x_{i_1,i_2,i_3} = b_{i_1,+,+}, \\ \displaystyle\sum_{i_1=1}^{p} \sum_{i_3=1}^{s} x_{i_1,i_2,i_3} = b_{+,i_2,+}, \\ \displaystyle\sum_{i_1=1}^{p} \sum_{i_2=1}^{q} x_{i_1,i_2,i_3} = b_{+,+,i_3}, \\ x_{i_1,i_2,i_3} \geq 0. \end{cases}$$

Junginger showed this can be solved instead using a 3-way $(p+1)\times(q+1)\times(s+1)$ transportation polytope with fixed 2-marginals:

$$\text{minimize} \quad \sum_{i_1=1}^{p+1} \sum_{i_2=1}^{q+1} \sum_{i_3=1}^{s+1} \hat{c}_{i_1,i_2,i_3} y_{i_1,i_2,i_3}$$

$$\text{subject to} \quad \begin{cases} \displaystyle\sum_{i_1=1}^{p+1} y_{i_1,i_2,i_3} = a_{+,i_2,i_3}, \\ \displaystyle\sum_{i_2=1}^{q+1} y_{i_1,i_2,i_3} = a_{i_1,+,i_3}, \\ \displaystyle\sum_{i_3=1}^{s+1} y_{i_1,i_2,i_3} = a_{i_1,i_2,+}, \\ y_{i_1,i_2,i_3} \geq 0. \end{cases}$$

Here the cost coefficients $\hat{c}_{i_1,i_2,i_3}$ and the 2-marginals a are as follows:

$$\hat{c}_{i_1,i_2,i_3} = \begin{cases} c_{i_1,i_2,i_3}, & \text{if all 3 indices are within the original ranges} \\ M, & \text{if exactly 2 of the indices are within the original range} \\ 0, & \text{otherwise.} \end{cases}$$

Let $\beta = \max(b_{i_1,+,+}, b_{+,i_2,+}, b_{+,+,i_3})$.

When i_1, i_2, i_3 stay within the original ranges:

$$a_{+,i_2,i_3} = \beta; \quad a_{i_1,+,i_3} = \beta; \quad a_{i_1,i_2,+} = \beta.$$

When we go outside the ranges in exactly one of the indices:

$$\begin{aligned} a_{+,q+1,i_3} &= p\beta - b_{+,+,i_3}, & a_{+,i_2,s+1} &= p\beta - b_{+,i_2,+}, \\ a_{i_1,+,s+1} &= q\beta - b_{i_1,+,+}, & a_{p+1,+,i_3} &= q\beta - b_{+,+,i_3}, \\ a_{p+1,i_2,+} &= s\beta - b_{+,i_2,+}, & a_{i_1,q+1,+} &= s\beta - b_{i_1,+,+}. \end{aligned}$$

Finally, when exactly two of the indices are outside the original range:

$$a_{+,q+1,s+1} = a_{p+1,+,s+1} = a_{p+1,q+1,+} = \beta.$$

Now for each solution x_{i_1,i_2,i_3} of the 3-way problem with 1-marginals we can recover a **unique** solution y_{i_1,i_2,i_3} of the 3-way problem with 2-marginals that has the same objective function value plus a constant. If x is integral, then y will be integral too when the marginals have integral entries. For this set the value of $y_{i_1,i_2,i_3} := x_{i_1,i_2,i_3}$ when all i_ℓ are in the original range. Using the new 2-marginal equations determine the values of those variables y_{i_1,i_2,i_3} with exactly *one* index outside original range. Thus for fixed i_2, i_3 in the original range:

$$y_{p+1,i_2,i_3} = a_{+,i_2,i_3} - \sum_{i_1=1}^{p} y_{i_1,i_2,i_3} = \beta \sum_{i_1=1}^{p} x_{i_1,i_2,i_3} \geq 0.$$

Next fill the values of those variables y_{i_1,i_2,i_3} with exactly *two* indices outside range. Finally fill the variable $y_{p+1,q+1,s+1}$. It is easy (but tedious) to check that y_{i_1,i_2,i_3} is indeed feasible in the 2-marginal problem.

Now the objective function value is

$$\sum_{i_1=1}^{p+1}\sum_{i_2=1}^{q+1}\sum_{i_3=1}^{s+1} \hat{c}_{i_1,i_2,i_3} y_{i_1,i_2,i_3} = \sum_{i_1=1}^{p}\sum_{i_2=1}^{q}\sum_{i_3=1}^{s} c_{i_1,i_2,i_3} x_{i_1,i_2,i_3} +$$

$$M\left(\sum_{i_1=1}^{p} x_{i_1,q+1,s+1} + \sum_{i_2=1}^{q} x_{p+1,i_2,s+1} + \sum_{i_3=1}^{s} x_{p+1,q+1,i_3}\right)$$

which is equal to

$$\sum_{i_1=1}^{p}\sum_{i_2=1}^{q}\sum_{i_3=1}^{s} c_{i_1,i_2,i_3} x_{i_1,i_2,i_3} + 3M\beta.$$

Conversely, if y is the optimal solution for the 2-marginal problem, the restriction x to those variables with indices $i_1 \leq p, i_2 \leq q, i_3 \leq s$ is an optimal solution of the 1-marginal problem. For this note that because y is optimal the entries of variables with two indices above the original range (e.g. $y_{p+1,q+1,i_3}$) must be zero because their cost is M (a huge constant). Next check x_{i_1,i_2,i_3} is feasible for the 1-marginal problem. Non-negativity is easy: note that

$$\sum_{i_2=1}^{q} y_{p+1,i_2,i_3} = \sum_{i_2=1}^{q}\left(a_{+,i_2,i_3} - \sum_{i_1=1}^{p} y_{i_1,i_2,i_3}\right) = q\beta - b_{+,+,i_3}.$$

Therefore, for the 1-marginal $b_{+,+,i_3}$,

$$\sum_{i_1=1}^{p}\sum_{i_2=1}^{q} y_{i_1,i_2,i_3} = \sum_{i_1=1}^{p+1}\sum_{i_2=1}^{q} y_{i_1,i_2,i_3} - \sum_{i_2=1}^{q} y_{p+1,i_2,i_3} = q\beta - (q\beta - b_{+,+,i_3}) = b_{+,+,i_3}.$$

and the same can be checked for other 1-marginals.

Depending on the application a transportation problem may have a combination of margins that define the polyhedron.

For 3-way transportation problems there are two natural generalizations of 2-way transportation polytopes to 3-way transportation polytopes, whose feasible points are $p \times q \times s$ tables of non-negative reals satisfying certain sum conditions:

- First, consider the 3-way transportation polytope of size $p \times q \times s$ defined by 1-*marginals*: Let $u = (u_1, \dots, u_p) \in \mathbb{R}^p$, $v = (y_1, \dots, y_q) \in \mathbb{R}^q$, and $w = (w_1, \dots, w_s) \in \mathbb{R}^s$ be three vectors. Let P be the polyhedron defined by the following $p + q + s$ equations in the pqs variables $x_{i,j,k} \in \mathbb{R}_{\geq 0}$ $(i \in [p], j \in [q], k \in [s])$:

$$\sum_{j,k} x_{i,j,k} = u_i, \forall i \quad \sum_{i,k} x_{i,j,k} = v_j, \forall j \quad \sum_{i,j} x_{i,j,k} = w_k, \forall k. \tag{3.1}$$

In [**150**], 3-way transportation polytopes defined by all 1-marginals are known as 3-*way axial transportation polytopes.*

- Similarly, a 3-way transportation polytope of size $p \times q \times s$ can be defined by specifying three real-valued matrices U, V, and W of sizes $q \times s$, $p \times s$, and $p \times q$ (respectively). These three matrices specify the line-sums resulting from fixing two of the indices of entries and adding over the remaining index. That is to say, the polyhedron P is defined by the following $pq + ps + qs$ equations, the 2-*marginals*, in the pqs variables $x_{i,j,k} \in \mathbb{R}_{\geq 0}$ satisfying:

$$\sum_i x_{i,j,k} = U_{j,k}, \forall j,k \quad \sum_j x_{i,j,k} = V_{i,k}, \forall i,k \quad \sum_k x_{i,j,k} = W_{i,j}, \forall i,j. \tag{3.2}$$

In [**150**], the 3-way transportation polytopes defined by 2-marginals are called 3-*way planar transportation polytopes.*

3.1. Why d-way transportation polytopes are harder. The 3-way transportation polytopes are very interesting because of the following universality theorem of De Loera and Onn in [**64**] which says that for any rational convex polytope P, there is a 3-way planar transportation polytope T isomorphic to P in a very strong sense.

We say a polytope $P \subset \mathbb{R}^p$ is *representable* as a polytope $T \subset \mathbb{R}^q$ if there is an injection $\sigma : \{1, \ldots, p\} \longrightarrow \{1, \ldots, q\}$ such that the projection $\pi : \mathbb{R}^q \longrightarrow \mathbb{R}^p$

$$x = (x_1, \ldots, x_q) \mapsto \pi(x) = (x_{\sigma(1)}, \ldots, x_{\sigma(p)})$$

is a bijection between T and P and between the sets of integer points $T \cap \mathbb{Z}^q$ and $P \cap \mathbb{Z}^p$.

Note that if P is representable as T then P and T have the same facial structure and all linear or integer programming programs are polynomial-time equivalent. We can state the universality result as follows:

THEOREM 3.3 (Universality [**64**]). *Any polytope $P = \{y \in \mathbb{R}^n_{\geq 0} : Ay = b\}$ with integer $m \times n$ matrix $A = (a_{i,j})$ and integer vector b is polynomial-time representable as a slim $r \times c \times 3$ transportation polytope*

$$T = \left\{ x \in \mathbb{R}^{r \times c \times 3}_{\geq 0} : \sum_i x_{i,j,k} = U_{j,k}\,, \sum_j x_{i,j,k} = V_{i,k}\,, \sum_k x_{i,j,k} = W_{i,j} \right\},$$

with $r = O(m^2(n+L)^2)$ rows and $c = O(m(n+L))$ columns, where $L := \sum_{j=1}^n \max_{i=1}^m \lfloor \log_2 |a_{i,j}| \rfloor$.

The constructive proof of Theorem 3.3 follows three steps.

(1) Decrease the size of the coefficients used in the constraints.
(2) Encode the polytope P as a transportation polytope with 1-margins and with some entries bounded
(3) Encode any transportation polytope with 1-margins and bounded entries into a new transportation polytope with 2-margins

We only explain steps 1 and 2 which already give an interesting corollary.

Step 1: Given $P = \{y \geq 0 : Ay = b\}$ where $A = (a_{i,j})$ is an integer matrix and b is an integer vector. We represent it as a polytope $Q = \{x \geq 0 : Cx = d\}$, in polynomial-time, with a $\{-1, 0, 1, 2\}$-valued matrix $C = (c_{i,j})$ of coefficients. For this use the binary expansion $|a_{i,j}| = \sum_{s=0}^{k_j} t_s 2^s$ with all $t_s \in \{0,1\}$, we rewrite this term as $\pm \sum_{s=0}^{k_j} t_s x_{j,s}$.

For example, the equation $3y_1 - 5y_2 + 2y_3 = 7$ becomes

$$\begin{array}{cccccccccl} 2x_{1,0} & -x_{1,1} & & & & & & = & 0, \\ & & 2x_{2,0} & -x_{2,1} & & & & = & 0, \\ & & & 2x_{2,1} & -x_{2,2} & & & = & 0, \\ & & & & & 2x_{3,0} & -x_{3,1} & = & 0, \\ x_{1,0} & +x_{1,1} & -x_{2,0} & & -x_{2,2} & & +x_{3,1} & = & 7. \end{array}$$

Step 2: Here is a sketch. Each equation $k = 1, \dots, m$ will be encoded in a "horizontal table" plus an extra layer of "slacks". Each variable y_j, $j = 1, \dots, n$ will be encoded in a "vertical box". Other entries are zero. See Figure 11. Given $P = \{y \geq 0 : Ay = b\}$ where $A = (a_{i,j})$ is an $m \times n$ integer matrix and b is an integer vector: we assume that P is bounded and hence a polytope, with an integer upper bound U on the value of any coordinate y_j of any $y \in P$. The sizes of the layers will be given by the numbers $r_j := \max\left(\sum_k \{a_{k,j} : a_{k,j} > 0\}, \sum_k \{|a_{k,j}| : a_{k,j} < 0\}\right)$ and $r := \sum_{j=1}^n r_j$, $R := \{1, \dots, r\}$, $m+1$ and $H := \{1, \dots, m+1\}$. Each equation $k = 1, \dots, m$ is encoded in a "horizontal table" $R \times R \times \{k\}$. The last horizontal table $R \times R \times \{m+1\}$ is included for consistency and its entries can be regarded as "slacks". Each variable y_j, $j = 1, \dots, n$ will be encoded in a "vertical box" $R_j \times R_j \times H$, where $R = \biguplus_{j=1}^n R_j$ is the natural partition of R with $|R_j| = r_j$, namely with $R_j := \{1 + \sum_{l<j} r_l, \dots, \sum_{l \leq j} r_l\}$.

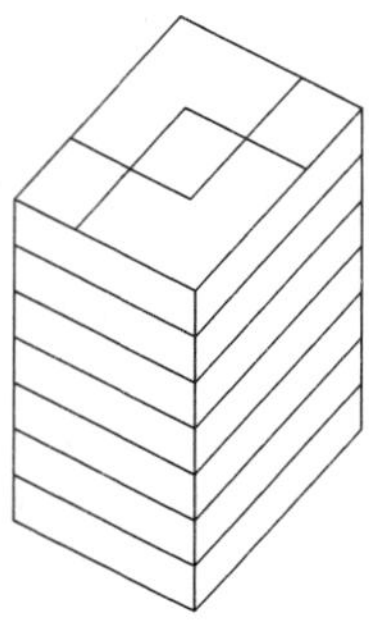

FIGURE 11. Each equation is encoded in one of the 6 horizontal tables, with the seventh table used for slacks

For instance, if we have three variables, with $r_1 = 3, r_2 = 1, r_3 = 2$ then $R_1 = \{1,2,3\}, R_2 = \{4\}, R_3 = \{5,6\}$, and the top view of the matrix $x = (x_{i,j,+})$ is

$$\begin{pmatrix} x_{1,1,+} & x_{1,2,+} & 0 & 0 & 0 & 0 \\ 0 & x_{2,2,+} & x_{2,3+} & 0 & 0 & 0 \\ x_{3,1,+} & 0 & x_{3,3,+} & 0 & 0 & 0 \\ 0 & 0 & 0 & x_{4,4,+} & 0 & 0 \\ 0 & 0 & 0 & 0 & x_{5,5,+} & x_{5,6,+} \\ 0 & 0 & 0 & 0 & x_{6,5,+} & x_{6,6,+} \end{pmatrix} = \begin{pmatrix} y_1 & \bar{y}_1 & 0 & 0 & 0 & 0 \\ 0 & y_1 & \bar{y}_1 & 0 & 0 & 0 \\ \bar{y}_1 & 0 & y_1 & 0 & 0 & 0 \\ 0 & 0 & 0 & U & 0 & 0 \\ 0 & 0 & 0 & 0 & y_3 & \bar{y}_3 \\ 0 & 0 & 0 & 0 & \bar{y}_3 & y_3 \end{pmatrix}.$$

The actual vertical position is decided with respect to the equations that contain a variable. Now, to specify the actual 1-margins: All "vertical" plane-sums are set to the same value U, that is, $u_j := v_j := U$ for $j = 1, \ldots, r$. All entries not in the union $\biguplus_{j=1}^{n} R_j \times R_j \times H$ of the variable boxes will be forbidden. The horizontal plane-sums w are determined as follows: For $k = 1, \ldots, m$, consider the kth equation $\sum_j a_{k,j} y_j = b_k$. Define the index sets $J^+ := \{j : a_{k,j} > 0\}$ and $J^- := \{j : a_{k,j} < 0\}$, and set $w_k := b_k + U \cdot \sum_{j \in J^-} |a_{k,j}|$.

EXAMPLE 3.4. What do the three steps of this construction do, if one starts with the zero-dimensional polytope $P = \{y \mid 2y = 1,\ y \geq 0\}$? In this case, we obtain the 2-margins of a 3-way transportation polytope shown in Figure 12.

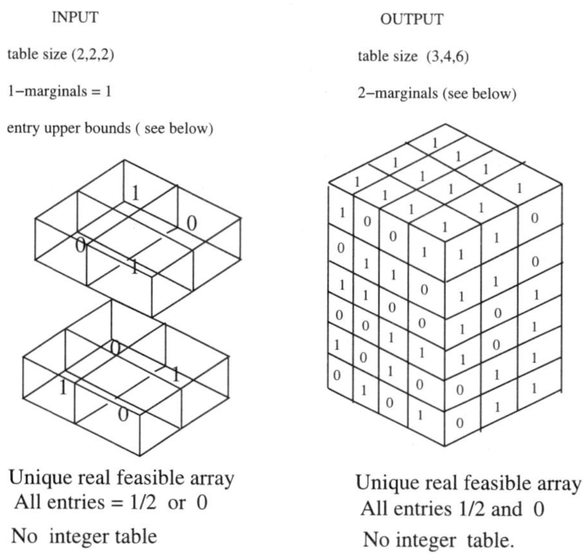

FIGURE 12. Illustration of the universality theorem: A polyhedron consisting of the single point $y = \frac{1}{2}$ is represented by a $3 \times 4 \times 6$ polytope with 2-margins shown at the right of the figure. The left side of the figure shows the encoding using only the first two steps of the algorithm.

3.2. Comparing 2-way and 3-way transportation polytopes. We want to stress some consequences of the construction. First of all, simply from the first two steps above the following interesting theorem follows. Any rational polyhedron is a face of some axial 3-way transportation polytope.

COROLLARY 3.5. *Any rational polytope* $P = \{y \in \mathbb{R}^n \mid Ay = b,\ y \geq 0\}$ *is polynomial-time representable as* **a face** *of a* 3*-way* $r \times c \times 3$ *transportation polytope with* 1*-margins*

$$T = \left\{ x \in \mathbb{R}^{r \times c \times 3}_{\geq 0} \ : \ \sum_{j,k} x_{i,j,k} = u_i,\ \sum_{i,k} x_{i,j,k} = v_j,\ \sum_{i,j} x_{i,j,k} = w_k \right\}.$$

The properties we have seen in Section 2 for classical 2-way transportation polytopes raise the issue whether the analogous questions or properties hold for multi-way transportation polytopes. We will address the following:

(1) **Real feasibility (Vlach Problems [145]):** Is there a simple characterization in terms of the 2-margins of those 3-way transportation polytopes which are empty?

In particular, do any of the conditions on the margins proposed by Schell, Haley, Moravek and Vlach (see pages 374–376 of [**150**]) suffice to guarantee that the polytope is non-empty?

(2) **Dimension:** What are the possible dimensions of a $p \times q \times s$ transportation polytope? Is it always equal to $(p-1)(q-1)(s-1)$?

(3) **Graphs of 3-way transportation polytopes:** Do we have a good bound for the diameter? Is the Linear Hirsch Conjecture true in this case?

(4) **Number of vertices of 3-way transportation polytopes:** Can one estimate minimum and maximum number of vertices possible? Do they have a nice characterization?

(5) **Integer Feasibility Problem:** Given a prescribed collection of integral margins that seem to describe a d-way transportation polytope of size $p_1 \times \cdots \times p_d$, does there exist an integer table with these margins? Can such an integral d-way table be efficiently determined?

(6) **Integer Range Property:** Given a collection of margins coming from d-way table, and an index tuple $(i_1, \ldots, i_d)$, do all integer values inside the range of an interval appear for the coordinate $x_{i_1,\ldots,i_d}$ in the corresponding transportation polytope?

(7) **Graver/Markov basis for 3-way transportation polytopes:** Are the Graver bases for 3-way transportation polytopes as nice as they are for 2-way transportation polytopes? Do 3-way transportation polytopes have the "interval property" for entry values?

Most of these questions had easy solutions for classical transportation polytopes. In the next sections we answer all these questions for multi-way transportation polytopes.

3.2.1. *Feasibility and dimension revisited.* Recall Lemma 2.2 for 2-way transportation polytopes, which gave a simple characterization for a non-empty polytope in terms of its margins. From the equations in (3.1), a similar necessary and sufficient condition for the 3-way axial transportation polytope to be non-empty can be proved:

LEMMA 3.6. *Let P be the $p \times q \times s$ axial 3-way transportation polytope defined by the marginals u, v, and w. The polytope P is non-empty if and only if*

$$\sum_{i=1}^{p} u_i = \sum_{j=1}^{q} v_j = \sum_{k=1}^{s} w_k. \tag{3.3}$$

The proof of this lemma is like Lemma 2.2 for 2-way transportation polytopes, but using a 3-way analogue of the northwest corner rule algorithm.

While similar statements are true for d-way transportation polytopes defined by 1-marginals, the real feasibility problem does not have a known characterization for m-marginals with $m \geq 2$, even for 3-way transportation polytopes. This study is called *real feasibility* or the Vlach Problems (see [**145**]). The conditions on the margins proposed by Schell, Haley, Moravek and Vlach (see [**145**]) are necessary, but not sufficient to guarantee that the polytopes are non-empty. By the universality theorem one cannot expect a simple characterization (in terms of the

2-marginals of 3-way transportation polytope) to decide when they are empty. In fact, due to Theorem 3.3, given a prescribed collection of marginals that seem to describe a d-way transportation polytope of size $p_1 \times \cdots \times p_d$, deciding whether there is an integer table with these margins is an NP-complete problem.

Recall that we had also a nice simple dimension formula for 2-way transportation polytopes. As a consequence of Lemma 3.6, the 3-way axial transportation polytope P defined by (3.1) is completely described by only $p+q+s-2$ independent equations. The maximum possible dimension for $p \times q \times s$ transportation polytopes defined by 1-marginals is $pqs - p - q - s + 2$. For planar 3-way transportation polytopes, one can see that in fact only $pq + ps + qs - p - q - s + 1$ of the defining equations are linearly independent for feasible systems. The maximum possible dimension for $p \times q \times s$ transportation polytopes defined by 2-marginals is $(p-1)(q-1)(s-1)$. Unfortunately, by the universality theorem, the dimension of these polytopes can be any number up to $(p-1)(q-1)(s-1)$.

3.2.2. *Combinatorics of faces revisited.* From the universality theorem one can expect that the f-vector and indeed the entire combinatorial properties of any rational polytope will appear when listing all the f-vectors of 3-way transportation polytopes. Indeed, Shmuel Onn has suggested this as a way to systematically enumerate combinatorial types of polytopes. In [**60**], there was an experimental investigation of the possible polyhedra that arise for small 3-way transportation polytopes. The number of vertices of certain low-dimensional 3-way transportation polytopes have been completely classified:

THEOREM 3.7. *The possible numbers of vertices of non-degenerate* $2 \times 2 \times 2$ *and* $2 \times 2 \times 3$ *axial transportation polytopes are those given in Table* 3.

Moreover, every non-degenerate $2 \times 2 \times 4$ *axial transportation polytope has between* 32 *and* 504 *vertices. Every non-degenerate* $2 \times 3 \times 3$ *axial transportation polytopes has between* 81 *and* 1056 *vertices. The number of vertices of non-degenerate* $3 \times 3 \times 3$ *axial transportation polytopes is at least* 729.

THEOREM 3.8. *The possible numbers of vertices of non-degenerate* $2 \times 2 \times 2$, $2 \times 2 \times 3$, $2 \times 2 \times 4$, $2 \times 2 \times 5$, *and* $2 \times 3 \times 3$ *planar transportation polytopes are those given in Table* 4. *Moreover, every non-degenerate* $2 \times 3 \times 4$ *planar transportation polytope has between* 7 *and* 480 *vertices.*

Size	Dimension	Possible numbers of vertices
$2 \times 2 \times 2$	4	8 11 14
$2 \times 2 \times 3$	7	18 24 30 32 36 38 40 42 44 46 48 50 52 54 56 58 60 62 64 66 68 70 72 74 76 78 80 84 86 96 108

TABLE 3. Numbers of vertices possible in non-degenerate axial transportation polytopes

Again from Theorem 3.3, the graph of every rational convex polyhedron will appear as the graph of some 3-way transportation polytope. In particular if the Hirsch Conjecture is true for 3-way transportation polytopes given by 2-marginals, then it is true for all rational convex polytopes. By Corollary 3.5 the graph of every rational convex polytope is the graph of a face of some 3-way transportation polytope given by 1-margins. Interestingly, in joint work with Onn and Santos

Size	Dimension	Possible numbers of vertices
$2 \times 2 \times 2$	1	2
$2 \times 2 \times 3$	2	3 4 5 6
$2 \times 2 \times 4$	3	4 6 8 10 12
$2 \times 2 \times 5$	4	5 8 11 12 14 15 16 17 18 19 20 21 22 23 24 25 26 27 28 29 30
$2 \times 3 \times 3$	4	5 8 9 11 12 13 14 15 16 17 18 19 20 21 22 23 24 25 26 27 28 29 30 31 32 33 34 35 36 37 38 39 40 41 42 43 44 45 46 47 48 49 50 51 52 53 54 55 56 57 58 59

TABLE 4. Numbers of vertices possible in non-degenerate planar transportation polytopes

(see [**60**]), we proved a quadratic bound on the diameter of 3-way transportation polytopes given by 2-margins:

THEOREM 3.9. *The diameter of every $p \times q \times s$ axial 3-way transportation polytope is at most* $2(p+q+s-2)^2$.

OPEN PROBLEM 3.10. Prove or disprove the Linear Hirsch Conjecture for 3-way axial transportation polytopes: is it true that there is a universal constant k such that the diameter of every $p \times q \times s$ axial 3-way transportation polytope is at most $k(p+q+s)$?

3.2.3. *Integer points revisited.* Recall that finding integral tables with given margins is equivalent to the problem of finding lattice points inside the corresponding transportation polytope. Such problem has many applications [**56**] and some well-known algorithms for counting lattice points can be applied, for example the well-known Barvinok's algorithm [**12, 66**] or other techniques that use generating functions and residue computations [**6, 16**]. But one must remember that the family of transportation polytopes is not easy, even the (simpler) integer feasibility problem is truly difficult because, due to Theorem 3.3, *all* linear and integer programming problems are slim 3-way transportation problems. In other words, *any* linear or integer programming problem is equivalent to one that has a $\{0,1\}$-valued constraint matrix, with exactly three 1's per column in the constraint matrix, and depends only on the right-hand side data.

Again, there is bad news for the integer range property. Unlike Lemma 2.32 for 2-way transportation polytopes, now the values of a variable $x_{i,j,k}$ in a 3-way transportation polytope can have integer gaps. Similarly we saw that 2-way transportation polytopes have a nice *Graver basis*, which we recall is a minimal set of vectors needed to travel between any pair of integer points in the polytope. Unlike the case of 2-way transportation problems and as a consequence of Theorem 3.3, the coefficients in the entries of a Markov basis for d-way transportation polytopes can be arbitrarily large (see [**65**]), not just $0, -1, 1$ as we saw in Lemma 2.33.

Since any integer linear programming problem can be encoded as a slim 3-way transportation problem, the family of 3-way transportation polytopes really varied. The very same family of 3-way transportation problems of $p \times q \times 3$ and specified by 2-margins contains subproblems that admit fully polynomial approximation schemes as well as subproblems that do not have arbitrarily close approximation (unless $NP = P$). For this reason, no purely combinatorial approximation algorithm, i.e., one that does not take into account the 2-margin values, can be devised.

As we had for 2-way transportation polytopes, we have 3-way Birkhoff polytopes, which are much more complicated:

DEFINITION 3.11. The Birkhoff polytope has the following generalizations in the 3-way setting, these are the *multi-way assignment polytopes*:

(1) The *generalized Birkhoff 3-way axial polytope* is the axial 3-way transportation $p \times q \times s$ polytope whose 1-marginals are given by the vectors $u = (qs, \dots, qs) \in \mathbb{R}^p$, $v = (ps, \dots, ps) \in \mathbb{R}^q$, and $w = (pq, \dots, pq) \in \mathbb{R}^s$.
(2) The *generalized Birkhoff 3-way planar polytope* is the planar 3-way transportation $p \times q \times s$ polytope whose 2-marginals are given by the $q \times s$ matrix $U_{j,k} = p$, the $p \times s$ matrix $V_{i,k} = q$, and the $p \times q$ matrix $W_{i,j} = s$.

There is a vibrant study of d-way assignment polytopes. We refer the readers to the recent book [**34**] about assignment problems. We would simply like to mention some results that show how much harder it is to work with them versus the 2-way Birkhoff polytope. First, a now-classical result of Karp about the axial assignment polytope shows it is much more difficult to optimize over d-way assignment polytopes.

THEOREM 3.12 (Karp [**104**]). *The optimization problem*

$$\text{maximize/minimize} \quad \sum c_{i,j,k} x_{i,j,k}$$
$$\text{subject to} \quad \begin{cases} \sum_{j,k} x_{i,j,k} = 1, \\ \sum_{i,k} x_{i,j,k} = 1, \\ \sum_{i,j} x_{i,j,k} = 1, \\ x \in \mathbb{Z}_{\geq 0}^{p \times p \times p} \end{cases}$$

is NP-hard.

THEOREM 3.13 (Crama, Spieksma [**50**]). *For the minimization problem above, no polynomial time algorithm can even achieve a constant performance ratio unless NP=P.*

There are very interesting "universality" results about the coordinates of vertices of the generalized assignment problem with 1-margins.

DEFINITION 3.14. For a vertex x of the d-way 1-margin assignment problem define its *spectrum* to be the vector spectrum(x) with positive and decreasing entries which contains the values of all entries with repetitions deleted.

For example, the spectrum of a permutation matrix is always 1.

Gromova (see [**87**]) gave a characterization of which vectors are in the spectrum. Here are some of her results:

THEOREM 3.15 (Gromova [**87**]). *Given a positive decreasing vector σ of rational numbers its* relation matrix $R(\sigma)$ *is the matrix of all distinct non-negative integer row vectors τ such that $\sigma \cdot \tau = 1$*

(1) *For any positive decreasing vector v with components less than one whose relation matrix is not empty and whose columns are linearly independent and for $k \geq \max(1/v_i)$, there is a vertex of a d-way assignment polytope with 1-margins with spectrum v.*

(2) *Take any positive decreasing vector v with components less than one. It appears as part of the spectrum of some vertex of a 3-way assignment polytope.*

We now discuss d-way assignment polytopes defined by 2-margins. Recall the Birkhoff-von Neumann Theorem (Theorem 2.12) which stated that the $p \times p$ Birkhoff polytope has $p!$ vertices. There is a 3-way analogue of this result: First, recall the $p \times p \times p$ generalized Birkhoff planar polytope (also called the *2-marginals assignment polytope*) is the 3-way transportation polytope of line sums whose 2-margins are given by $U_{j,k} = V_{i,k} = W_{i,j} = 1$. In the case of 2-margins one can see that a solution, which is a 3-way array, has in each planar slice a permutation matrix. This indicates that there is a bijection between the 0/1 vertices of the 3-way 2-marginals $p \times p \times p$ assignment polytope and the possible $p \times p$ *latin squares*. Although their number is not known exactly this is enough to say that the number of 0/1 vertices of the polytope is bounded below by $(p!)^{2p}/p^{p^2}$. Recently, Linial and Luria (see [**112**]) proved that the total number of vertices of the $p \times p \times p$ generalized Birkhoff polytope is at least exponential in the number of Latin squares of order p.

Again we have a hardness result on the planar 3-way assignment polytope:

THEOREM 3.16. *(Dyer, Frieze [**73**]) The linear optimization problem*

$$\text{maximize} \quad \sum c_{i,j,k} x_{i,j,k}$$
$$\text{subject to} \quad \begin{cases} \sum_i x_{i,j,k} = 1, \\ \sum_j x_{i,j,k} = 1, \\ \sum_k x_{i,j,k} = 1, \\ x \in \mathbb{Z}_{\geq 0}^{p \times p \times p} \end{cases}$$

is in general NP-hard, even when $c_{i,j,k} \in \{0, 1\}$. However, when $c_{i,j,k} = c_{i,j,l}$ for all l, k then the problem is polynomially solvable.

Though this maximization problem is NP-hard, we note that Nishizeki and Chiba (see [**122**]) showed that a PTAS exists.

4. Further research directions and more open problems

There are several fascinating areas of research where tables with prescribed sums of their entries play a role. In this last section we would like to take a quick look at some of these areas and highlight some very nice open questions.

4.1. 0-1 tables. We have seen some results like Birkhoff's theorem on permutation matrices that deal specifically with 0-1 tables. Interesting problems about 0-1 tables appear naturally in combinatorial representation theory (see [**5**, **127**]) and number theory (see [**3**]). Analyzing the properties of 0-1 tables is a classic area of research in combinatorial matrix theory. This field combines techniques from combinatorics and group theory and it is so large we do not even attempt to summarize the results available. The reader should consult the books [**25**, **31**]. As a taste of the richness of the field of combinatorial matrix theory let us just talk about results known for subpolytopes of the Birkhoff polytope. One can consider permutation polytopes, obtained as the convex hull of some vertices of a Birkhoff polytope. (Note that this notion of permutation polytope is distinct from the permutation polytopes of Billera and Sarangarajan in [**17**].) In [**123**], Onn analyzed the geometry, complexity and combinatorics of permutation polytopes. In [**14**], Baumeister et al. studied the faces and combinatorial types that appear in small permutation polytopes. Brualdi (see [**24**]) investigated the faces of the convex polytope of doubly stochastic matrices which are invariant under a fixed row and column permutation. The *pth tridiagonal Birkhoff polytope* is the convex hull of the vertices of the Birkhoff polytope whose support entries are in $\{(i,j) \in [p] \times [p] \mid |i-j| \leq 1\}$. In [**53**], da Fonseca et al. counted the number of vertices of tridiagonal Birkhoff polytopes. In [**47**], Costa et al. presented a formula counting the number of faces tridiagonal Birkhoff polytopes. Volumes of permutation polytopes were studied in [**33**].

Costa et al. (see [**45**]) defined a *pth acyclic Birkhoff polytope* to be any polytope that is the convex hull of the set of matrices whose support corresponds to (some subset of) a fixed tree graph's edges (including loops). In [**46**], Costa et al. counted the faces of acyclic Birkhoff polytopes. In [**45**], Costa et al. proved an upper bound on the diameter of acyclic Birkhoff polytopes, which generalized the diameter result of Dahl in [**54**].

Let the *pth even Birkhoff polytope* be the convex hull of the $\frac{1}{2}p!$ permutation matrices corresponding to even permutations. In [**52**], Cunningham and Wang confirmed a conjecture of Brualdi and Liu (see [**30**]) that the pth even Birkhoff polytope cannot be described as the solution set of polynomially many linear inequalities. In [**99**], Hood and Perkinson described some of the facets of the even Birkhoff polytope and proved a conjecture of Brualdi and Liu (see [**30**]) that the number of facets of the pth even Birkhoff polytope is not polynomial in p. In [**146**], von Below showed that the condition of Mirsky given in [**117**] is not sufficient for determining membership of a point in an even Birkhoff polytope. Cunningham and Wang (see [**52**]) also investigated the membership problem for the even Birkhoff polytope. In [**147**], von Below and Rénier described even and odd diagonals in even Birkhoff polytopes. In [**42**], Cho and Nam introduced a signed analogue of the Birkhoff polytope.

The 0-1 points of transportation polytopes also have a strong connection to *discrete tomography*, which considers the problem of reconstructing binary images (or finite subsets of objects placed in a lattice) from a small number of their projections. The connection to tables is clear as one can think of the position of the objects in points in a grid as the placement of 0's and 1's in entries of a table. This is a very active field of research. See [**4, 32, 78–85, 93, 94**] and the references therein. These reconstruction problems are important in CAT scanner development, electron microscope image reconstruction, and quality control in semiconductor production (see, e.g., [**4, 78, 79**] and the references therein).

In light of this discussion, the following open problem is interesting:

OPEN PROBLEM 4.1. What is the complexity of counting all 2-way 0-1 tables for given margins?

In the next section, we discuss what is known about enumerating contingency tables in general.

4.2. Enumeration, sampling and optimization. We have seen that counting contingency tables is quite important in combinatorics and statistics. In [**63**],

De Loera and Onn gave a complete description of the computational complexity of existence, counting, and entry-security in multi-way table problems. The following theorem summarizes what is known about counting (specified in terms of binary encoding or unary encoding of data):

THEOREM 4.2. *The computational complexity of the counting problem for integral 3-way tables of size $p \times q \times s$ with $2 \leq p \leq q \leq s$ and all 2-marginals specified is provided by the following table:*

	p, q, s fixed	*p, q fixed, s variable*	*p fixed, q, s variable*	*p, q, s variable*
unary 2-marginals	*P*	*P*	*#PC*	*#PC*
binary 2-marginals	*P*	*#PC*	*#PC*	*#PC*

Using the highly-structured Graver bases of transportation polytopes with special restrictions one can do some polynomial-time optimization on highly difficult problems: E.g., De Loera, Hemmecke, Onn, and Weismantel (see [**59**]) proved there is a polynomial time algorithm that, given s and fixing p and q, solves integer programming problems of 3-way transportation polytopes of size $p \times q \times s$ defined by 2-marginals, over any integer objective. Later on, in [**58**] De Loera, Hemmecke, Onn, Rothblum, and Weismantel presented a polynomial oracle-time algorithm to solve convex integer maximization over 3-way planar transportation polytopes, if two of the margin sizes remain fixed. More recently (in [**96**]) Hemmecke, Onn, and Weismantel proved a similar result for convex integer minimization.

4.3. More open problems on transportation polytopes. We will also mention some more conjectures and open problems on transportation polytopes, and where applicable, give an update on problems where there are solutions and partial answers. We hope this will help to increase the interest in this subject.

CONJECTURE 4.3. *It is impossible to have $p \times q \times s$ non-degenerate 3-way transportation polytopes, specified by 2-margin matrices U, V, W, whose number f_0 of vertices satisfies the inequalities $(p-1)(q-1)(s-1)+1 < f_0(M(U,V,W)) < 2(p-1)(q-1)(s-1)$?*

This conjecture is true when $p, q, s \leq 3$.

OPEN PROBLEM 4.4. Is it true that the graph of any 2-way $p \times q$ transportation polytope is Hamiltonian?

Hamiltonicity of the graph is known to hold for small values of p and q.

OPEN PROBLEM 4.5. Suppose $\phi_1(p,q), \phi_2(p,q), \ldots, \phi_{t_{p,q}}(p,q)$ are all possible values of the number of vertices of $p \times q$ transportation polytopes. Give a formula for $t_{p,q}$.

This is related to the problem of enumerating all triangulations or chambers of a vector configuration.

CONJECTURE 4.6. *All integer numbers between 1 and $p+q-1$, and only these, are realized as the diameters of $p \times q$ transportation polytopes.*

OPEN PROBLEM 4.7. What are the possible number of facets for 3-way $p \times q \times s$ non-degenerate transportation polytopes given by 2-margins?

OPEN PROBLEM 4.8. What is the largest possible number of vertices in a 3-way $p \times q \times s$ transportation polytope?

Recall the $p \times q \times s$ *generalized central transportation polytope* is the 3-way transportation polytope of line sums whose 2-margins are given by the $q \times s$ matrix $U_{j,k} = p$, the $p \times s$ matrix $V_{i,k} = q$, and the $p \times q$ matrix $W_{i,j} = s$. Yemelichev, Kovalev, Kratsov stated in [**150**] the conjecture that the generalized central transportation polytope had the largest number of vertices among 3-way transportation polytopes. This conjecture was proved to be false in [**60**]. Here are explicit 2-marginals for a $3 \times 3 \times 3$ transportation polytope which has more vertices (270 vertices) than the generalized central transportation polytope, with only 66 vertices:

164424	324745	127239	163445	49395	403568	184032	123585	269245
262784	601074	9369116	1151824	767866	8313284	886393	6722333	935582
149654	7618489	1736281	1609500	6331023	1563901	1854344	302366	9075926

Acknowlegements

We thank Raymond Hemmecke, Fu Liu, Shmuel Onn, Francisco Santos, and Ruriko Yoshida for their suggestions and joint work on transportation problems. We are also grateful to them and Andreas Alpers, Matthias Beck, Steffen Borgwardt, Persi Diaconis, Peter Gritzmann, Igor Pak, Seth Sullivant, and Ernesto Vallejo for suggestions. The first author is grateful for the support received from NSF grant DMS-0914107. The first author is also grateful to the Technische Universität München for the hospitality received during the days while writing this survey.

References

[1] M. M. Ahmed. *Algebraic Combinatorics of Magic Squares.* Ph.D. thesis, University of California, Davis, 2004.

[2] M. M. Ahmed, *Polytopes of magic labelings of graphs and the faces of the Birkhoff polytope*, Ann. Comb. **12** (2008), no. 3, 241–269, DOI 10.1007/s00026-008-0349-y. MR2447256 (2009i:05202)

[3] A. Alpers and R. Tijdeman, *The two-dimensional Prouhet-Tarry-Escott problem*, J. Number Theory **123** (2007), no. 2, 403–412, DOI 10.1016/j.jnt.2006.07.001. MR2301222 (2008c:11028)

[4] A. Alpers and P. Gritzmann, *On stability, error correction, and noise compensation in discrete tomography*, SIAM J. Discrete Math. **20** (2006), no. 1, 227–239 (electronic), DOI 10.1137/040617443. MR2257258 (2007e:11149)

[5] D. Avella-Alaminos and E. Vallejo, *Kronecker products and the RSK correspondence*, Discrete Math. **312** (2012), no. 8, 1476–1486, DOI 10.1016/j.disc.2012.01.006. MR2889506

[6] W. Baldoni-Silva, J. A. De Loera, and M. Vergne, *Counting integer flows in networks*, Found. Comput. Math. **4** (2004), no. 3, 277–314, DOI 10.1007/s10208-003-0088-8. MR2078665 (2005i:05077)

[7] M. Baïou and M. Balinski, *The stable allocation (or ordinal transportation) problem*, Math. Oper. Res. **27** (2002), no. 3, 485–503, DOI 10.1287/moor.27.3.485.310. MR1926655 (2003d:90013)

[8] M. L. Balinski and R. E. Gomory. A primal method for the assignment and transportation problem. *Mgmt Sci.*, 10:578–593, 1964.

[9] M. L. Balinski, *The Hirsch conjecture for dual transportation polyhedra*, Math. Oper. Res. **9** (1984), no. 4, 629–633, DOI 10.1287/moor.9.4.629. MR769400 (86c:90038)

[10] M. L. Balinski and F. J. Rispoli, *Signature classes of transportation polytopes*, Math. Programming **60** (1993), no. 2, Ser. A, 127–144, DOI 10.1007/BF01580606. MR1239594 (94h:90037)

[11] A. Barvinok, *Asymptotic estimates for the number of contingency tables, integer flows, and volumes of transportation polytopes*, Int. Math. Res. Not. IMRN **2** (2009), 348–385, DOI 10.1093/imrn/rnn133. MR2482118 (2010c:52019)

[12] A. I. Barvinok, *A polynomial time algorithm for counting integral points in polyhedra when the dimension is fixed*, Math. Oper. Res. **19** (1994), no. 4, 769–779, DOI 10.1287/moor.19.4.769. MR1304623 (96c:52026)

[13] A. Barvinok and K. Woods, *Short rational generating functions for lattice point problems*, J. Amer. Math. Soc. **16** (2003), no. 4, 957–979 (electronic), DOI 10.1090/S0894-0347-03-00428-4. MR1992831 (2004e:05009)

[14] B. Baumeister, C. Haase, B. Nill, and A. Paffenholz, *On permutation polytopes*, Adv. Math. **222** (2009), no. 2, 431–452, DOI 10.1016/j.aim.2009.05.003. MR2538016 (2010j:52042)

[15] M. Beck, M. Cohen, J. Cuomo, and P. Gribelyuk, *The number of "magic" squares, cubes, and hypercubes*, Amer. Math. Monthly **110** (2003), no. 8, 707–717, DOI 10.2307/3647853. MR2023999 (2004k:05009)

[16] M. Beck and D. Pixton, *The Ehrhart polynomial of the Birkhoff polytope*, Discrete Comput. Geom. **30** (2003), no. 4, 623–637, DOI 10.1007/s00454-003-2850-8. MR2013976 (2004g:52015)

[17] L. J. Billera and A. Sarangarajan, *The combinatorics of permutation polytopes*, Formal power series and algebraic combinatorics (New Brunswick, NJ, 1994), DIMACS Ser. Discrete Math. Theoret. Comput. Sci., vol. 24, Amer. Math. Soc., Providence, RI, 1996, pp. 1–23. MR1363504 (96m:52014)

[18] G. Birkhoff, *Three observations on linear algebra* (Spanish), Univ. Nac. Tucumán. Revista A. **5** (1946), 147–151. MR0020547 (8,561a)

[19] E. D. Bolker, *Transportation polytopes*, J. Combinatorial Theory Ser. B. **13** (1972), 251–262. MR0311297 (46 #10389)

[20] N. Bonifas, M. Di Summa, F. Eisenbrand, N. Hähnle, and M. Niemeier, *On sub-determinants and the diameter of polyhedra*, Computational geometry (SCG'12), ACM, New York, 2012, pp. 357–362, DOI 10.1145/2261250.2261304. MR3050652

[21] S. Borgwardt. On the Diameter of Partition Polytopes and Vertex-Disjoint Cycle Cover, Mathematical Programming, Series A, 2011.

[22] F. Bourgeois and J.-C. Lassalle, *An extension of the Munkres algorithm for the assignment problem to rectangular matrices*, Comm. ACM **14** (1971), 802–804. MR0316080 (47 #4628)

[23] G. Brightwell, J. van den Heuvel, and L. Stougie, *A linear bound on the diameter of the transportation polytope*, Combinatorica **26** (2006), no. 2, 133–139, DOI 10.1007/s00493-006-0010-5. MR2223631 (2008c:90052)

[24] R. A. Brualdi, *Convex polytopes of permutation invariant doubly stochastic matrices*, J. Combinatorial Theory Ser. B **23** (1977), no. 1, 58–67. MR0463214 (57 #3170)

[25] R. A. Brualdi, D. Cvetković. *A Combinatorial Approach to Matrix Theory and Its Applications.* CRC Press, 2008.

[26] R. A. Brualdi and P. M. Gibson, *Convex polyhedra of doubly stochastic matrices. I. Applications of the permanent function*, J. Combinatorial Theory Ser. A **22** (1977), no. 2, 194–230. MR0437562 (55 #10486)

[27] R. A. Brualdi and P. M. Gibson, *Convex polyhedra of double stochastic matrices. II. Graph of U_n*, J. Combinatorial Theory Ser. B **22** (1977), no. 2, 175–198. MR0437563 (55 #10487)

[28] R. A. Brualdi and P. M. Gibson, *Convex polyhedra of doubly stochastic matrices. III. Affine and combinatorial properties of U_n*, J. Combinatorial Theory Ser. A **22** (1977), no. 3, 338–351. MR0437564 (55 #10488)

[29] R. A. Brualdi and P. M. Gibson, *Convex polyhedra of doubly stochastic matrices. IV*, Linear Algebra and Appl. **15** (1976), no. 2, 153–172. MR0447291 (56 #5606)

[30] R. A. Brualdi and B. L. Liu, *The polytope of even doubly stochastic matrices*, J. Combin. Theory Ser. A **57** (1991), no. 2, 243–253, DOI 10.1016/0097-3165(91)90048-L. MR1111560 (92k:15043)

[31] R. A. Brualdi and H. J. Ryser, *Combinatorial matrix theory*, Encyclopedia of Mathematics and its Applications, vol. 39, Cambridge University Press, Cambridge, 1991. MR1130611 (93a:05087)

[32] S. Brunetti, A. Del Lungo, P. Gritzmann, and S. de Vries, *On the reconstruction of binary and permutation matrices under (binary) tomographic constraints*, Theoret. Comput. Sci. **406** (2008), no. 1-2, 63–71, DOI 10.1016/j.tcs.2008.06.014. MR2453877 (2009i:05049)

[33] K. Burgraff, J. A. De Loera, and M. Omar. On volumes of permutation polytopes, In *Discrete Geometry and Optimization*, editors (K. Bezdek, A. Deza, and Y. Ye), vol. 69 of Fields Institute Communications,

[34] R. Burkard, M. Dell'Amico, and S. Martello, *Assignment problems*, Society for Industrial and Applied Mathematics (SIAM), Philadelphia, PA, 2009. MR2488749 (2010c:90002)

[35] L. Buzzigoli and A. Giusti. An algorithm to calculate the lower and upper bounds of the elements of an array given its marginals. In *Statistical Data Protection: Proceedings of the Conference*, pages 131–147. EUROSTAT, 1999.

[36] E. R. Canfield and B. D. McKay. The asymptotic volume of the Birkhoff polytope. Available at `http://arxiv.org/abs/0705.2422`. 2007.

[37] L. Carlitz, *Enumeration of symmetric arrays*, Duke Math. J. **33** (1966), 771–782. MR0201332 (34 #1216)

[38] C. S. Chan and D. P. Robbins, *On the volume of the polytope of doubly stochastic matrices*, Experiment. Math. **8** (1999), no. 3, 291–300. MR1724161 (2000k:15048)

[39] Y. Chen, *Conditional inference on tables with structural zeros*, J. Comput. Graph. Statist. **16** (2007), no. 2, 445–467, DOI 10.1198/106186007X209226. MR2370949

[40] Y. Chen, P. Diaconis, S. P. Holmes, and J. S. Liu, *Sequential Monte Carlo methods for statistical analysis of tables*, J. Amer. Statist. Assoc. **100** (2005), no. 469, 109–120, DOI 10.1198/016214504000001303. MR2156822 (2006f:62062)

[41] Y. Chen, I. H. Dinwoodie, and S. Sullivant, *Sequential importance sampling for multiway tables*, Ann. Statist. **34** (2006), no. 1, 523–545, DOI 10.1214/009053605000000822. MR2275252 (2008i:62075)

[42] S. Cho and Y. Nam, *Convex polytopes of generalized doubly stochastic matrices*, Commun. Korean Math. Soc. **16** (2001), no. 4, 679–690. MR1862369 (2002g:15047)

[43] S. Chowdhury, G. T. Duncan, R. Krishnan, S. F. Roehrig, and S. Mukherjee. Disclosure detection in multivariate categorical databases: Auditing confidentiality protection through two new matrix operators. *Mgmt Sci.*, 45:1710–1723, 1999.

[44] Christandl, M.; Doran, B.; Walter, M., Computing Multiplicities of Lie Group Representations, Foundations of Computer Science (FOCS), 2012 IEEE 53rd Annual Symposium on, 20-23 Oct. 2012 639–648.

[45] L. Costa, C. M. da Fonseca, and E. A. Martins, *The diameter of the acyclic Birkhoff polytope*, Linear Algebra Appl. **428** (2008), no. 7, 1524–1537, DOI 10.1016/j.laa.2007.09.028. MR2388636 (2009d:15054)

[46] L. Costa, C. M. da Fonseca, and E. A. Martins, *Face counting on an acyclic Birkhoff polytope*, Linear Algebra Appl. **430** (2009), no. 4, 1216–1235, DOI 10.1016/j.laa.2008.10.015. MR2489389 (2010e:05011)

[47] L. Costa, C. M. da Fonseca, and E. A. Martins, *The number of faces of the tridiagonal Birkhoff polytope* (Russian, with Russian summary), Sovrem. Mat. Prilozh., Optimalnoe Upravlenie **63** (2009), 87–97, DOI 10.1007/s10958-009-9607-8; English transl., J. Math. Sci. (N. Y.) **161** (2009), no. 6, 867–877. MR2760276 (2012f:52024)

[48] L. H. Cox, *Bounds on entries in 3-dimensional contingency tables subject to given marginal totals*, Inference control in statistical databases, Lecture Notes in Comput. Sci., vol. 2316, Springer, Berlin, 2002, pp. 21–33, DOI 10.1007/3-540-47804-3_3. MR1962210

[49] L. H. Cox, *On properties of multi-dimensional statistical tables*, J. Statist. Plann. Inference **117** (2003), no. 2, 251–273, DOI 10.1016/S0378-3758(02)00392-0. MR2004659 (2004h:62104)

[50] Y. Crama, F. C. R. Spieksma. Approximation algorithms for three-dimensional assignment problems with triangle inequalities. *Eur. J. Oper. Res.* 60:273–279, 1992.

[51] M. Cryan, M. Dyer, H. Müller, and L. Stougie, *Random walks on the vertices of transportation polytopes with constant number of sources*, Proceedings of the Fourteenth Annual ACM-SIAM Symposium on Discrete Algorithms (Baltimore, MD, 2003), ACM, New York, 2003, pp. 330–339. MR1974936

[52] W. H. Cunningham and Y. Wang, *On the even permutation polytope*, Linear Algebra Appl. **389** (2004), 269–281, DOI 10.1016/j.laa.2004.03.016. MR2080410 (2006a:90174)

[53] C. M. da Fonseca and E. Marques de Sá, *Fibonacci numbers, alternating parity sequences and faces of the tridiagonal Birkhoff polytope*, Discrete Math. **308** (2008), no. 7, 1308–1318, DOI 10.1016/j.disc.2007.03.077. MR2382366 (2008k:05011)

[54] G. Dahl, *Tridiagonal doubly stochastic matrices*, Linear Algebra Appl. **390** (2004), 197–208, DOI 10.1016/j.laa.2004.04.017. MR2083656 (2006b:15032)

[55] G. Dahl, *Transportation matrices with staircase patterns and majorization*, Linear Algebra Appl. **429** (2008), no. 7, 1840–1850, DOI 10.1016/j.laa.2008.05.019. MR2444364 (2009j:15075)

[56] J. A. De Loera, *The many aspects of counting lattice points in polytopes*, Math. Semesterber. **52** (2005), no. 2, 175–195, DOI 10.1007/s00591-005-0094-9. MR2159956 (2006c:52015)

[57] J. A. De Loera, R. Hemmecke, and M. Köppe, *Algebraic and geometric ideas in the theory of discrete optimization*, MOS-SIAM Series on Optimization, vol. 14, Society for Industrial and Applied Mathematics (SIAM), Philadelphia, PA; Mathematical Optimization Society, Philadelphia, PA, 2013. MR3024570

[58] J. A. De Loera, R. Hemmecke, S. Onn, U. G. Rothblum, and R. Weismantel, *Convex integer maximization via Graver bases*, J. Pure Appl. Algebra **213** (2009), no. 8, 1569–1577, DOI 10.1016/j.jpaa.2008.11.033. MR2517994 (2010f:90138)

[59] J. A. De Loera, R. Hemmecke, S. Onn, and R. Weismantel, *n-fold integer programming*, Discrete Optim. **5** (2008), no. 2, 231–241, DOI 10.1016/j.disopt.2006.06.006. MR2408419 (2009d:90076)

[60] J. A. De Loera, E. D. Kim, S. Onn, and F. Santos, *Graphs of transportation polytopes*, J. Combin. Theory Ser. A **116** (2009), no. 8, 1306–1325, DOI 10.1016/j.jcta.2009.03.010. MR2568801 (2010m:05094)

[61] J. A. De Loera and S. Klee, *Transportation problems and simplicial polytopes that are not weakly vertex-decomposable*, Math. Oper. Res. **37** (2012), no. 4, 670–674, DOI 10.1287/moor.1120.0554. MR2997897

[62] J. A. De Loera, F. Liu, and R. Yoshida, *A generating function for all semi-magic squares and the volume of the Birkhoff polytope*, J. Algebraic Combin. **30** (2009), no. 1, 113–139, DOI 10.1007/s10801-008-0155-y. MR2519852 (2010h:52017)

[63] J. De Loera and S. Onn, *The complexity of three-way statistical tables*, SIAM J. Comput. **33** (2004), no. 4, 819–836 (electronic), DOI 10.1137/S0097539702403803. MR2065334 (2005b:68140)

[64] J. A. De Loera and S. Onn, *All linear and integer programs are slim 3-way transportation programs*, SIAM J. Optim. **17** (2006), no. 3, 806–821 (electronic), DOI 10.1137/040610623. MR2257210 (2007f:90069)

[65] J. A. De Loera and S. Onn, *Markov bases of three-way tables are arbitrarily complicated*, J. Symbolic Comput. **41** (2006), no. 2, 173–181, DOI 10.1016/j.jsc.2005.04.010. MR2197153

[66] J. A. De Loera, R. Hemmecke, J. Tauzer, and R. Yoshida, *Effective lattice point counting in rational convex polytopes*, J. Symbolic Comput. **38** (2004), no. 4, 1273–1302, DOI 10.1016/j.jsc.2003.04.003. MR2094541 (2005i:52020)

[67] P. Diaconis and A. Gamburd, *Random matrices, magic squares and matching polynomials*, Electron. J. Combin. **11** (2004/06), no. 2, Research Paper 2, 26. MR2120097 (2005k:05015)

[68] P. Diaconis and A. Gangolli, *Rectangular arrays with fixed margins*, Discrete probability and algorithms (Minneapolis, MN, 1993), IMA Vol. Math. Appl., vol. 72, Springer, New York, 1995, pp. 15–41, DOI 10.1007/978-1-4612-0801-3_3. MR1380519 (97e:05013)

[69] A. Dobra and S. E. Fienberg, *Bounds for cell entries in contingency tables given marginal totals and decomposable graphs*, Proc. Natl. Acad. Sci. USA **97** (2000), no. 22, 11885–11892 (electronic), DOI 10.1073/pnas.97.22.11885. MR1789526 (2001g:62038)

[70] G. T. Duncan, S. E. Fienberg, R. Krishnan, R. Padman, and S. F. Roehrig. Disclosure limitation methods and information loss for tabular data. In P. Doyle, J. I. Land, J. M. Theeuwes, and L. V. Zayatz, editors, *Confidentiality, Disclosure and Data Access: Theory and Practical Applications for Statistical Agencies*. North-Holland, 2001.

[71] G. T. Duncan, R. Krishnan, R. Padman, and S. F. Roehrig. Disclosure detection in multiple linked categorical datafiles: A unified network approach. In *Statistical Data Protection: Proceedings of the Conference*, pages 149–162. EUROSTAT, 1999.

[72] M. Dyer and A. Frieze, *Random walks, totally unimodular matrices, and a randomised dual simplex algorithm*, Math. Programming **64** (1994), no. 1, Ser. A, 1–16, DOI 10.1007/BF01582563. MR1274170 (95c:90068)

[73] M. E. Dyer and A. M. Frieze, *Planar 3DM is NP-complete*, J. Algorithms **7** (1986), no. 2, 174–184, DOI 10.1016/0196-6774(86)90002-7. MR840769 (87h:68047)

[74] M. Dyer, R. Kannan, and J. Mount, *Sampling contingency tables*, Random Structures Algorithms **10** (1997), no. 4, 487–506, DOI 10.1002/(SICI)1098-2418(199707)10:4⟨487::AID-RSA4⟩3.0.CO;2-Q. MR1608222 (98i:68124)

[75] J. Edmonds, *Paths, trees, and flowers*, Canad. J. Math. **17** (1965), 449–467. MR0177907 (31 #2165)

[76] J. Edmonds and R. M. Karp, *Theoretical improvements in algorithmic efficiency for network flow problems*, Combinatorial Structures and their Applications (Proc. Calgary Internat. Conf., Calgary, Alta., 1969), Gordon and Breach, New York, 1970, pp. 93–96. MR0266680 (42 #1583)

[77] S. E. Fienberg. Fréchet and Bonferroni bounds for multi-way tables of counts with applications to disclosure limitation. In *Statistical Data Protection: Proceedings of the Conference*, pages 115–129. EUROSTAT, 1999.

[78] R. J. Gardner and P. Gritzmann, *Discrete tomography: determination of finite sets by X-rays*, Trans. Amer. Math. Soc. **349** (1997), no. 6, 2271–2295, DOI 10.1090/S0002-9947-97-01741-8. MR1376547 (97h:52021)

[79] R. J. Gardner and P. Gritzmann, *Successive determination and verification of polytopes by their X-rays*, J. London Math. Soc. (2) **50** (1994), no. 2, 375–391, DOI 10.1112/jlms/50.2.375. MR1291744 (95h:52004)

[80] R. J. Gardner, P. Gritzmann, and D. Prangenberg, *On the computational complexity of reconstructing lattice sets from their X-rays*, Discrete Math. **202** (1999), no. 1-3, 45–71, DOI 10.1016/S0012-365X(98)00347-1. MR1694525 (2000f:92008)

[81] P. Gritzmann and B. Langfeld, *On the index of Siegel grids and its application to the tomography of quasicrystals*, European J. Combin. **29** (2008), no. 8, 1894–1909, DOI 10.1016/j.ejc.2008.01.017. MR2463165 (2009j:52022)

[82] P. Gritzmann, B. Langfeld, and M. Wiegelmann, *Uniqueness in discrete tomography: three remarks and a corollary*, SIAM J. Discrete Math. **25** (2011), no. 4, 1589–1599, DOI 10.1137/100803262. MR2873205

[83] P. Gritzmann, D. Prangenberg, S. de Vries, M. Wiegelmann. Success and Failure of Certain Reconstruction and Uniqueness Algorithms in Discrete Tomography. *Intl. J. Imaging Syst. Tech.*, 9:101–109, 1998.

[84] P. Gritzmann, S. de Vries, and M. Wiegelmann, *Approximating binary images from discrete X-rays*, SIAM J. Optim. **11** (2000), no. 2, 522–546 (electronic), DOI 10.1137/S105262349935726X. MR1789551 (2001j:90131)

[85] P. Gritzmann and S. de Vries, *On the algorithmic inversion of the discrete Radon transform*, Theoret. Comput. Sci. **281** (2002), no. 1-2, 455–469, DOI 10.1016/S0304-3975(02)00023-3. Selected papers in honour of Maurice Nivat. MR1909584 (2003f:68056)

[86] J. E. Graver, *On the foundations of linear and integer linear programming. I*, Math. Programming **9** (1975), no. 2, 207–226. MR0386673 (52 #7525)

[87] M. B. Gromova, *The Birkhoff-von Neumann theorem for polystochastic matrices* (Russian), Operations research and statistical simulation, No. 2 (Russian), Izdat. Leningrad. Univ., Leningrad, 1974, pp. 3–15, 149. MR0379543 (52 #448)

[88] B. Grünbaum, *Convex polytopes*, 2nd ed., Graduate Texts in Mathematics, vol. 221, Springer-Verlag, New York, 2003. Prepared and with a preface by Volker Kaibel, Victor Klee and Günter M. Ziegler. MR1976856 (2004b:52001)

[89] K. Haley. The multi-index problem. *Oper. Res.*, 11(3):368–379, 1963.

[90] K. B. Haley. Note on the letter by Morávek and Vlach. *Opns. Res.*, 15:545–546, 1967.

[91] E. Q. Halleck. *Magic squares subclasses as linear Diophantine systems.* Ph.D. thesis, University of California, San Diego, 2000.

[92] D. J. Hartfiel, *Full patterns in truncated transportation polytopes*, Linear and Multilinear Algebra **30** (1991), no. 3, 213–221, DOI 10.1080/03081089108818105. MR1126443 (92h:52011)

[93] *Discrete tomography*, Applied and Numerical Harmonic Analysis, Birkhäuser Boston, Inc., Boston, MA, 1999. Foundations, algorithms, and applications; Edited by Gabor T. Herman and Attila Kuba. MR1722457 (2000h:92015)

[94] *Advances in discrete tomography and its applications*, Applied and Numerical Harmonic Analysis, Birkhäuser Boston, Inc., Boston, MA, 2007. Edited by Gabor T. Herman and Attila Kuba. MR2452005 (2009e:94026)

[95] R. Hemmecke, *On the computation of Hilbert bases of cones*, Mathematical software (Beijing, 2002), World Sci. Publ., River Edge, NJ, 2002, pp. 307–317. MR1932617

[96] R. Hemmecke, S. Onn, and R. Weismantel. A polynomial oracle-time algorithm for convex integer minimization. *Math. Program. Ser. A*, 2009.
[97] F. L. Hitchcock, *The distribution of a product from several sources to numerous localities*, J. Math. Phys. Mass. Inst. Tech. **20** (1941), 224–230. MR0004469 (3,11i)
[98] A. Hoffman, *What the transportation problem did for me*, Ann. Oper. Res. **149** (2007), 117–120, DOI 10.1007/s10479-006-0108-6. MR2313369
[99] J. Hood and D. Perkinson, *Some facets of the polytope of even permutation matrices*, Linear Algebra Appl. **381** (2004), 237–244, DOI 10.1016/j.laa.2003.11.015. MR2039809 (2004m:52024)
[100] C. Hurkens. Personal communication. 2007.
[101] R. W. Irving and M. R. Jerrum, *Three-dimensional statistical data security problems*, SIAM J. Comput. **23** (1994), no. 1, 170–184, DOI 10.1137/S0097539790191010. MR1259001 (95k:68052)
[102] W. Junginger. *Mehrdimensionale Transportprobleme*, Habilitationsschrift, Universität Stuttgart, Stuttgart, 1976.
[103] L. Kantorovitch, *On the translocation of masses*, C. R. (Doklady) Acad. Sci. URSS (N.S.) **37** (1942), 199–201. MR0009619 (5,174d)
[104] R. M. Karp. Reducibility Among Combinatorial Problems. *Complexity of Computer Computations* 85-103, 1972.
[105] E. Klafszky and T. Terlaky, *Variants of the Hungarian method for solving linear programming problems*, Optimization **20** (1989), no. 1, 79–91, DOI 10.1080/02331938908843416. MR977221 (90a:90116)
[106] V. Klee and C. Witzgall, *Facets and vertices of transportation polytopes*, Mathematics of the Decision Sciences, Part I (Seminar, Stanford, Calif., 1967), Amer. Math. Soc., Providence, R.I., 1968, pp. 257–282. MR0235832 (38 #4134)
[107] D. Kőnig. Gráfok és alkalmazásuk a determinánsok és a halmazok elméletére. *Matematikai és Természettudományi Értesítő*, 34:104–119, 1916.
[108] D. König, *Über Graphen und ihre Anwendung auf Determinantentheorie und Mengenlehre* (German), Math. Ann. **77** (1916), no. 4, 453–465, DOI 10.1007/BF01456961. MR1511872
[109] T. C. Koopmans. Optimum Utilization of the Transportation System. In *Proc. International Statistical Conferences* 5:136–146, 1947.
[110] H. W. Kuhn, *The Hungarian method for the assignment problem*, Naval Res. Logist. Quart. **2** (1955), 83–97. MR0075510 (17,759d)
[111] H. W. Kuhn, *Variants of the Hungarian method for assignment problems*, Naval Res. Logist. Quart. **3** (1956), 253–258 (1957). MR0091857 (19,1024c)
[112] N. Linial and Z. Luria, *On the vertices of the d-dimensional Birkhoff polytope*, Discrete Comput. Geom. **51** (2014), no. 1, 161–170, DOI 10.1007/s00454-013-9554-5. MR3148654
[113] F. Liu, *Perturbation of transportation polytopes*, J. Combin. Theory Ser. A **120** (2013), no. 7, 1539–1561, DOI 10.1016/j.jcta.2013.05.003. MR3092682
[114] L. Lovász and M. D. Plummer. *Matching Theory.* North-Holland, Amsterdam, 1986.
[115] J. Matoušek, *Using the Borsuk-Ulam theorem*, Universitext, Springer-Verlag, Berlin, 2003. Lectures on topological methods in combinatorics and geometry; Written in cooperation with Anders Björner and Günter M. Ziegler. MR1988723 (2004i:55001)
[116] C. R. Mehta and N. R. Patel, *A network algorithm for performing Fisher's exact test in* $r \times c$ *contingency tables*, J. Amer. Statist. Assoc. **78** (1983), no. 382, 427–434. MR711119 (85g:62093)
[117] L. Mirsky, *Even doubly-stochastic matrices*, Math. Ann. **144** (1961), 418–421. MR0154873 (27 #4817)
[118] J. Morávek and M. Vlach. On the necessary conditions for the existence of the solution of the multi-index transportation problem. *Oper. Res.*, 15(3):471–474, 1967.
[119] J. Morávek and M. Vlach, *On necessary conditions for a class of systems of linear inequalities.* (English, with Czech and Russian summaries), Apl. Mat. **13** (1968), 299–303. MR0240119 (39 #1473)
[120] T. S. Motzkin. The multi-index transportation problem. *Bull. Amer. Math. Soc.*, 58:494, 1952.
[121] J. Munkres, *Algorithms for the assignment and transportation problems*, J. Soc. Indust. Appl. Math. **5** (1957), 32–38. MR0093429 (19,1244f)

[122] T. Nishizeki and N. Chiba, *Planar graphs: theory and algorithms*, North-Holland Mathematics Studies, vol. 140, North-Holland Publishing Co., Amsterdam, 1988. Annals of Discrete Mathematics, 32. MR941967 (89f:05068)

[123] S. Onn, *Geometry, complexity, and combinatorics of permutation polytopes*, J. Combin. Theory Ser. A **64** (1993), no. 1, 31–49, DOI 10.1016/0097-3165(93)90086-N. MR1239510 (94j:52020)

[124] S. Onn, *Nonlinear discrete optimization*, Zurich Lectures in Advanced Mathematics, European Mathematical Society (EMS), Zürich, 2010. An algorithmic theory. MR2724387 (2011m:90002)

[125] I. Pak, *Four questions on Birkhoff polytope*, Ann. Comb. **4** (2000), no. 1, 83–90, DOI 10.1007/PL00001277. MR1763951 (2001g:52017)

[126] I. Pak, *On the number of faces of certain transportation polytopes*, European J. Combin. **21** (2000), no. 5, 689–694, DOI 10.1006/eujc.1999.0392. MR1771982 (2001f:90077)

[127] I. Pak, *Hook length formula and geometric combinatorics*, Sém. Lothar. Combin. **46** (2001/02), Art. B46f, 13 pp. (electronic). MR1877632 (2003c:05222)

[128] D. Pixton. The Volumes of Birkhoff polytopes for $n \leq 10$. Available at `http://www.math.binghamton.edu/dennis/Birkhoff/volumes.html`.

[129] J. S. Provan and L. J. Billera, *Decompositions of simplicial complexes related to diameters of convex polyhedra*, Math. Oper. Res. **5** (1980), no. 4, 576–594, DOI 10.1287/moor.5.4.576. MR593648 (82c:52010)

[130] M. Queyranne and F. C. R. Spieksma. Multi-index transportation problems. In C. A. Floudas and P. Pardalos, editors, *Encyclopedia of Optimization*, volume III, pages 450–456. Kluwer, 2001.

[131] F. Santos, *A counterexample to the Hirsch conjecture*, Ann. of Math. (2) **176** (2012), no. 1, 383–412, DOI 10.4007/annals.2012.176.1.7. MR2925387

[132] E. Schell. Distribution of a product by several properties. In *Proceedings of the 2nd symposium on Linear Programming*, volume 2, Washington, 1955.

[133] A. Schrijver, *Theory of linear and integer programming*, Wiley-Interscience Series in Discrete Mathematics, John Wiley & Sons, Ltd., Chichester, 1986. A Wiley-Interscience Publication. MR874114 (88m:90090)

[134] G. Smith, *Further necessary conditions for the existence of a solution to the multi-index problem*, Operations Res. **21** (1973), 380–386. Mathematical programming and its applications. MR0373592 (51 #9792)

[135] G. Smith, *A procedure for determining necessary and sufficient conditions for the existence of a solution to the multi-index problem* (English, with Czech summary), Apl. Mat. **19** (1974), 177–183. MR0349219 (50 #1713)

[136] G. Smith, *On the Morávek and Vlach conditions for the existence of a solution to the multi-index problem*, Apl. Mat. **20** (1975), no. 6, 432–435. MR0446492 (56 #4817)

[137] R. P. Stanley, *Linear homogeneous Diophantine equations and magic labelings of graphs*, Duke Math. J. **40** (1973), 607–632. MR0317970 (47 #6519)

[138] R. P. Stanley, *Magic labelings of graphs, symmetric magic squares, systems of parameters, and Cohen-Macaulay rings*, Duke Math. J. **43** (1976), no. 3, 511–531. MR0444514 (56 #2865)

[139] E. Steinitz. *Über die Konstruction der Configurationen n (sub 3)*. Ph.D. thesis, Universität Breslau, 1894.

[140] L. Stougie. A polynomial bound on the diameter of the transportation polytope, 2002.

[141] N. Tomizawa, *On some techniques useful for solution of transportation network problems*, Networks **1** (1971/72), 173–194. MR0297347 (45 #6404)

[142] Transportation Polytope Database. Available at `http://www.math.ucdavis.edu/~deloera/TRANSPORT_DATABASE/`

[143] E. Vallejo, *A stability property for coefficients in Kronecker products of complex S_n characters*, Electron. J. Combin. **16** (2009), no. 1, Note 22, 8. MR2529813 (2011b:20040)

[144] J. van den Heuvel and L. Stougie. A quadratic bound on the diameter of the transportation polytope.

[145] M. Vlach, *Conditions for the existence of solutions of the three-dimensional planar transportation problem*, Discrete Appl. Math. **13** (1986), no. 1, 61–78, DOI 10.1016/0166-218X(86)90069-7. MR829339 (87e:90065)

[146] J. von Below, *On a theorem of L. Mirsky on even doubly-stochastic matrices*, Discrete Math. **55** (1985), no. 3, 311–312, DOI 10.1016/S0012-365X(85)80007-8. MR802669 (86j:15017)

[147] J. von Below and S. Rénier, *Even and odd diagonals in doubly stochastic matrices*, Discrete Math. **308** (2008), no. 17, 3917–3929, DOI 10.1016/j.disc.2007.07.097. MR2418095 (2009d:15053)

[148] J. von Neumann. A certain zero-sum two-person game equivalent to an optimal assignment problem. *Ann. Math. Studies*, 28:5–12, 1953.

[149] J. Xi, R. Yoshida, D. Haws. Estimating the number of zero-one multi-way tables via sequential importance sampling

[150] V. A. Yemelichev, M. M. Kovalëv, and M. K. Kravtsov, *Polytopes, graphs and optimisation*, Cambridge University Press, Cambridge, 1984. Translated from the Russian by G. H. Lawden. MR744197 (85b:52008)

[151] G. M. Ziegler. *Lectures on Polytopes*. Number 152 in Graduate Texts in Mathematics. Springer-Verlag, New York, NY, 1994.

Department of Mathematics, University of California, Davis, One Shields Avenue, Davis, California 95616

E-mail address: `deloera@math.ucdavis.edu`

Department of Mathematics, University of Wisconsin-La Crosse, 1725 State Street, La Crosse, Wisconsin 54601

E-mail address: `ekim@uwlax.edu`

Contemporary Mathematics
Volume **625**, 2014
http://dx.doi.org/10.1090/conm/625/12492

A Tree Sperner Lemma

Andrew Niedermaier, Douglas Rizzolo, and Francis Edward Su

ABSTRACT. In this paper we prove a combinatorial theorem for finite labellings of trees, and show that it is equivalent to a theorem for finite covers of metric trees and a fixed point theorem on metric trees. We trace how these connections mimic the equivalence of the Brouwer fixed point theorem with the classical KKM lemma and Sperner's lemma. We also draw connections to a KKM-type theorem about infinite covers of metric trees and fixed point theorems for non-compact metric trees. Finally, we develop a new KKM-type theorem for cycles, and discuss interesting social consequences, including an application in voting theory.

1. Introduction

The Brouwer fixed point theorem is a celebrated topological result that says every continuous map of an n-ball to itself has a fixed point. It is known [**6**, **11**] to be equivalent to a set-covering result known as the KKM lemma, and a combinatorial result known as Sperner's lemma. Aside from their intrinsic interest, these equivalent formulations have led to simpler methods for proving the Brouwer result as well as practical algorithms for finding fixed points of highly non-linear functions (see e.g., [**9**]). Other topological theorems such as the Borsuk-Ulam theorem also admit similar set-covering and combinatorial formulations [**7**].

Like the n-ball, a finite tree as a topological space also has the *fixed point property*: every continuous map of a tree to itself has a fixed point. A primary goal of this paper is to explore a combinatorial analogue of the fixed point property for trees and draw a connection to a set-covering analogue, akin to the Sperner and KKM lemma analogues of the Brouwer theorem.

Our new combinatorial analogue is Theorem 1 which we call the Tree Sperner Lemma because of its similarities with Sperner's Lemma. We show that this is equivalent to a new Tree Fixed Vertex-Edge Theorem (Theorem 2) involving functions defined only on the vertex set of a combinatorial tree. These are proved in Section 2 and are easy to establish, but lead to simpler proofs of some known results about metric trees: a Tree KKM Theorem (Theorem 5) in Section 4 and a Tree Fixed Point Theorem (Theorem 7) in Section 5. Theorem 10 shows that these four results are equivalent.

2010 *Mathematics Subject Classification.* Primary 05C10; Secondary 55M20, 05C05.

The second author acknowledges partial support by NSF grant DMS-1204840 and NSF Graduate Research Fellowship grant DGE-1106400. The third author acknowleges partial support by NSF Grants DMS-0301129, DMS-0701308, DMS-1002938.

Because finite trees are compact and acyclic, the fixed point property for trees follows from the Lefschetz fixed point theorem just like the Brouwer theorem does. However, this approach gives little insight into the location of a fixed point, and the Lefschetz theorem is not easy to prove. By contrast, our Tree Sperner Lemma (i) gives an accessible proof of the fixed point result for trees, (ii) suggests where the corresponding fixed point is and a constructive procedure for finding it, and (iii) is of intrinsic interest due to its similarity with Sperner's lemma.

Moreover, the Tree Sperner Lemma applies more generally to infinite trees, as long as the label set is finite. So we also show in Section 6 how it can be used to prove a known KKM-type result about infinite covers of trees (Theorem 12), as well as a fixed point result for compact maps of infinite trees (Theorem 13).

Finally, we use the Tree KKM Theorem in Section 7 to prove a new KKM-type result for covers of cycles (Theorem 14). Along the way we also consider interesting social interpretations of our results, including applications to voting theory.

2. A Tree Sperner Lemma

The usual Sperner's lemma starts with a triangulated n-simplex Δ whose vertices have a *Sperner labeling*:

- each main vertex of Δ has a distinct label (chosen from $n+1$ labels), and
- each vertex v of the triangulation is assigned a label of one of the main vertices spanning the minimal face of Δ that v is on.

For instance, if v is on the edge of Δ spanned by a and b, then v must be labelled either a or b.

For such a labelling, Sperner's lemma asserts that there must be a *fully-labelled* simplex, i.e., one with all $n+1$ labels. In Theorem 1, we develop an analogous combinatorial theorem for *proper* labellings of n-vertex trees by n labels that will assert the existence of an edge with all n labels.

For this result, we view trees as combinatorial (i.e., connected acyclic graphs specified by vertices and edges), although in subsequent sections we shall consider the implications of our result for metric trees (tree-like metric spaces).

Let $T = (V, E)$ be a tree with vertex set V and edge set E. To avoid trivialities, we assume V has at least two vertices. If V is finite, we say T is a *finite* tree; otherwise T is *infinite*. Note that even for an infinite tree, between any two vertices u and w, there is a finite chain of edges that connect u and w and this path of edges is unique.

If v is a vertex, then let $T \setminus v$ denote the graph that results from removing v from V and all the edges incident to v from E. This new graph may have several connected components. Similarly, for an element $e = \{v_1, v_2\} \in E$, we let $T \setminus e$ denote the graph that results from removing e from E. We note that, since T is a tree, $T \setminus e$ has exactly two components.

Let A be a subset of V which we call the *labels*; every vertex of V will be assigned a collection of labels by a labelling function ℓ. Let 2^A denote the power set of A, i.e., the set of all subsets of A.

We call a labelling $\ell : V \to 2^A$ *proper* if:

- for each $a \in A$, $\ell(a)$ contains a, and
- if $v \in V$ lies on the unique path between $a, b \in A$, then $\ell(v)$ contains either a or b.

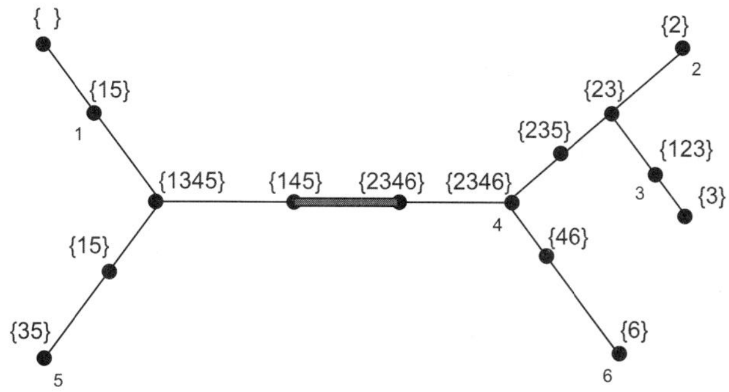

FIGURE 1. A proper labelling. The non-bracketed numbers mark the vertices that are labels and the bracketed numbers denote the labels of each vertex. The thickened edge is a fully-labelled edge.

Note how these conditions are analogous to those of the usual Sperner's lemma. It is easy to verify that they are equivalent to this one condition:

- for each $v \in V$, if the set $A \setminus \ell(v)$ is nonempty, then it consists of vertices that all lie in one component of $T \setminus v$.

We may think of this condition as saying that $\ell(v)$ is missing labels from at most one component of $T \setminus v$. Similarly, if e is an edge incident to v, then $\ell(v)$ is missing labels from at most one component of $T \setminus e$. Figure 1 shows an example of a tree with a proper labelling.

Let e be an edge with endpoints x, y. We call that edge *fully-labelled* if $\ell(x) \cup \ell(y) = A$, i.e., if the edge contains all labels. The proper labelling in Figure 1 has a fully-labelled edge. This exemplifies our theorem, which may be viewed as an analogue of Sperner's lemma for trees:

THEOREM 1 (Tree Sperner Lemma). *Let $T = (V, E)$ be a tree, let A be a finite subset of V, and let $\ell : V \to 2^A$ be a proper labelling. Then T contains a fully-labelled edge.*

The finiteness of A is essential, as one may see by considering the *integer tree*: whose vertices are the integers and whose edges connect successive integers (see Figure 3). Now let $A = V$ and label each vertex n by all the integers less than or equal to n. This labelling is proper, but it does not have a fully-labelled edge.

PROOF. It suffices to consider the case where V is also finite, for if not, we may restrict our attention to the finite subtree of T spanned by the vertices of A, noting that any fully-labelled edge in this subtree is fully-labelled in the whole tree.

If there exists v in V with $\ell(v) = A$, then desired conclusion holds: any edge containing v is fully-labelled. Otherwise, we can construct a "successor" function $S : V \to V$. For each v in V, the labels $A \setminus \ell(v)$ come from exactly one component of $T \setminus v$. In that component, let $S(v)$ be the unique vertex that shares an edge with v in T.

Consider the sequence $\{v_n\}$ defined by choosing some $v_1 \in A$ and letting $v_k = S(v_{k-1})$ for $k > 1$. Since V is finite, this sequence must be eventually periodic. Since T is a tree, this period cannot be of length greater than two. Thus $S(x) = y$ and $S(y) = x$ for some x, y that are endpoints of an edge e.

We claim e is fully-labelled. Since $S(y) = x$, the labels $A \setminus \ell(y)$ are in the component of $T \setminus y$ containing x. Since $S(x) = y$, the labels $A \setminus \ell(x)$ are in the component of $T \setminus x$ containing y. Then $A \setminus \ell(y)$ and $A \setminus \ell(x)$ are disjoint because they are in different components of $T \setminus e$. Hence $\ell(x) \cup \ell(y) = A$ so that e is a fully-labelled edge. □

Note that the above successor function yields a systematic method for locating a fully-labelled edge, by iterating S until one reaches a vertex for which S is not defined (and therefore has all labels), or until the sequence repeats. This avoids exhaustively checking vertices, which can be problematic if V is infinite.

Theorem 1 yields an interesting corollary for functions whose domain and range are vertices of T.

THEOREM 2 (Tree Fixed Vertex-Edge Theorem). *Suppose $T = (V, E)$ is a tree and $f : V \to V$ is a function on vertices with finite range. Then either f fixes some vertex, or there is an edge $e \in E$ with endpoints x and y such that e is on the path from $f(x)$ to $f(y)$.*

Theorem 2 may be viewed as a kind of "discrete" fixed point theorem for trees, because it says that either there is a fixed vertex, or some edge must be covered by the path between the images of its endpoints. Compare it to the continuous version we prove later in Theorem 7.

PROOF. Let A be the range of f. Suppose f does not fix any vertex. Consider a labelling $\ell : V \to 2^A$ such that for each vertex v, $\ell(v)$ is the set of all vertices of A that are not in the component of $T \setminus v$ containing $f(v)$ (including v if $v \in A$). The labelling is by definition proper. Theorem 1 implies that there is a fully-labelled edge e with endpoints x and y.

Since by definition $f(y) \notin \ell(y)$, we must have $f(y) \in \ell(x)$. Thus $f(y)$ is not in the component of $T \setminus x$ containing $f(x)$, so x must be on the path between $f(x)$ and $f(y)$. Similarly, $f(x) \notin \ell(x)$ implies $f(x) \in \ell(y)$. Thus $f(x)$ is not in the component of $T \setminus y$ containing $f(y)$, so y must be on the path between $f(x)$ and $f(y)$. Thus e must be on the path between $f(x)$ and $f(y)$, as desired. □

Moreover,

THEOREM 3. *Theorem 2 is equivalent to Theorem 1.*

PROOF. Having already shown Theorem 1 implies Theorem 2, we now show the converse. Suppose that $\ell : V \to 2^A$ is a proper labelling of T, with A finite. If there is a vertex v such that $\ell(v) = A$, then any edge containing v is fully-labelled. Otherwise, for each vertex v define $f(v)$ to be a label from A which is not in $\ell(v)$ (if there are several options, choose one). Clearly $f : V \to A$ has no fixed points, so Theorem 2 implies that there is an edge $e = \{x, y\}$ such that e is on the path from $f(x)$ to $f(y)$.

We claim that e is fully-labelled. Since $f(x)$ and $f(y)$ are in different components of $T \setminus e$ and ℓ is proper, then the definition of f shows that $\ell(x)$ and $\ell(y)$ are missing labels from different components of $T \setminus e$. So there are no vertices of A that are missing from both $\ell(x)$ and $\ell(y)$, i.e., $\ell(x) \cup \ell(y) = A$ and e is a fully-labelled edge. □

3. Metric Trees and Segmentations

Theorems 1 and 2 have several applications to metric trees, which are essentially combinatorial trees realized as metric spaces by replacing edges with line segments isometric to a compact interval of $\mathbb{R}$. We make precise in this section what we mean by *metric tree* as well as the concept of a subdivision called a *segmentation*, but there are no surprises here, so this brief section may be skimmed if desired.

A *metric tree* is a triple $T = (V, E, X)$, with a vertex set V (that may be finite or infinite), an edge set E, and *underlying metric space* X. Here, (V, E) specifies a combinatorial tree and the metric space X is obtained from (V, E) by realizing every edge e as an isometric copy of some closed interval $[0, L_e]$ and gluing the realized edges according to the instructions in (V, E). The number L_e is called the *length* of edge e.

Since (V, E) has no cycles, between any two points x, y in X there is a unique non-self-intersecting path between x and y. There is a natural metric on X: let $d(x, y)$ be the length of this path between x and y, i.e., the sum of the lengths of the edges (or partial edges) along this unique path. It will be useful to note if z is on the path between x and y, then the triangle inequality becomes an equality: $d(x, y) = d(x, z) + d(z, y)$.

We remark that our definition of metric tree differs from others in the literature because it retains the combinatorial structure specified by V and E.

Given a tree $T = (V, E, X)$ it will be convenient at times to consider a *segmentation* of T, which is another metric tree obtained from T by finite subdivision of its edges. In particular, $T' = (V', E', X)$ is a *segmentation* of $T = (V, E, X)$ if: (1) $V' = V \cup V^*$ where V^* is a collection of points $\{v_\alpha\}$ from X so that at most a finite number of the v_α come from a realized edge e_X, $e \in E$, and (2) E' is the collection of edges obtained from E in the natural way (by deleting edges in which elements of V^* appear and including edges of the implied subdivision along that edge). Note that the metric spaces for T' and T are the same, so the set of continuous functions on T and T' are the same.

The *size* of a segmentation $T' = (V', E', X)$ is defined by $size(T') = \sup_{e \in E'} L_e$, and bounds the size of the longest edge. Clearly every tree has an arbitrarily small segmentation.

Note also that every point in a metric tree T that is not a leaf (a vertex of degree 1) is a *cut point*: its removal "cuts" T into more than one path-connected component.

In what follows, all trees T are metric trees.

4. KKM Covers of Trees

The usual KKM lemma [**6**] starts by considering an n-simplex Δ with a *KKM cover* by $n + 1$ closed sets $\{C_i\}$: these have the properties that

- each main vertex v_i has an associated closed set C_i that contains it
- the face of Δ spanned by $\{v_{i_1}, \dots, v_{i_k}\}$ is covered by $\{C_{i_1}, \dots, C_{i_k}\}$.

Under these conditions, the KKM lemma says there exists a point in the intersection of all $n + 1$ sets.

Just as the traditional Sperner's Lemma implies the classical KKM theorem for covers of simplices by closed sets, Theorem 1 implies a KKM theorem for metric trees.

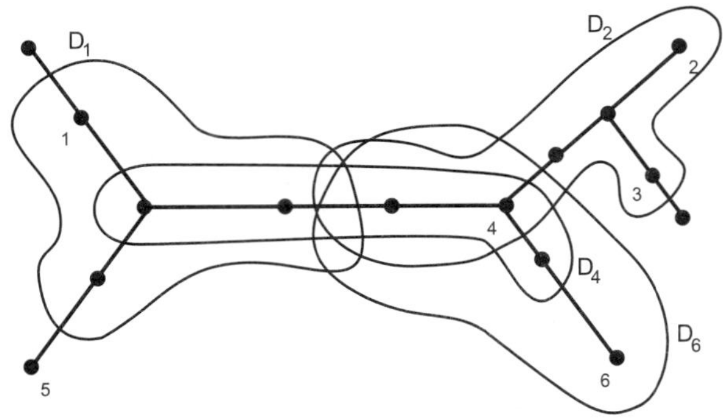

FIGURE 2. A KKM cover of a tree relative to the vertices $\{1, 2, 4, 6\}$. The sets of a KKM cover do not have to be connected sets (although they are in this diagram).

Let $T = (V, E, X)$ be a tree, and A a subset of points in X. Call a family of closed sets $\mathscr{D} = \{D_a : a \in A\}$ a *KKM cover of* T *relative to* A if:

- each $a \in D_a$, and
- for any two points a and b in A, the path between them is contained in $D_a \cup D_b$.

We call the last condition the *path-covering property* of $\mathscr{D}$; it is analogous to the face-covering property of KKM covers of simplices. See Figure 2.

If $A = V$, we may simply say $\mathscr{D}$ is a *KKM cover* of T, and the the sets of $\mathscr{D}$ will cover the underlying metric space X. (Note that if $A \neq V$, then a KKM cover of T relative to A may not cover X, but it will cover the subtree spanned by A.)

Each KKM cover $\mathscr{D}$ relative to a (finite or infinite) subset of vertices defines a *membership labeling* by assigning to a vertex v the indices of all the sets of $\mathscr{D}$ that contain it. (The label set may be empty if v is not covered by the sets of $\mathscr{D}$.) This labeling is proper:

LEMMA 4. *Let* $T = (V, E, X)$ *be a tree, let* $A \subset V$ *be a subset of vertices, and let* $\mathscr{D} = \{D_a : a \in A\}$ *be a KKM cover of* T *relative to* A*. Then the labelling* $\ell : V \to 2^A$ *defined by* $\ell(v) = \{a : v \in D_a\}$ *is proper.*

As an example, in Figure 2, all vertices in the set D_1 will have 1 in their label set. Similarly, because vertex 4 is in sets D_2 and D_4 and D_6, $\ell(4)$ will contain 1, 2, and 4. The leaf at top left will have an empty label set. The reader may notice that Figure 1 gives the membership labelling for the KKM cover in Figure 2 if labels 3 and 5 were removed from every label set.

PROOF. Fix a vertex v in V, and suppose there were two vertices a and b in A that are not in $\ell(v)$. Then by definition v is not in D_a nor in D_b. Since $a \in D_a$ and $b \in D_b$, then neither a nor b can be the vertex v.

If a and b were in different components of $T \setminus \{v\}$, then v must lie on the unique path between a and b in T and the path-covering property of $\mathscr{D}$ would imply $v \in D_a \cup D_b$, a contradiction. Therefore any vertices not in $\ell(v)$ must lie in one component of $T \setminus \{v\}$, as desired. $\square$

We now use our Tree Sperner Lemma to prove a known theorem about finite KKM covers of trees. Although it follows from the results in [**2**] and [**5**], our proof is more elementary.

THEOREM 5 (Tree KKM Theorem). *Let $T = (V, E, X)$ be a metric tree, A a finite subset of points of X, and suppose $\mathscr{D} = \{D_a : a \in A\}$ is a KKM cover of T relative to A. Then*

$$\bigcap_{a \in A} D_a \neq \emptyset.$$

PROOF. We may as well assume that V is finite, for otherwise we may restrict our attention to the subtree K spanned by a finite set of edges that contain A, which contains a finite number of vertices. Any KKM cover of T relative to A will also restrict to a KKM cover of K relative to A, and a nonempty intersection of the KKM cover of K would imply a nonempty intersection of the KKM cover of T.

Suppose, by way of contradiction, that the intersection $\cap_{a \in A} D_a$ were empty. Then the set of complements $\mathscr{C} = \{T \setminus D_a : a \in A\}$ is an open cover of T. Since V is finite, X is compact and this cover has a Lebesgue number δ. Let $T' = (V', E', X)$ be a segmentation of T with $size(T') < \delta$ chosen so that A is a subset of the vertices of V'.

Consider the membership labelling $\ell : V' \to 2^A$ defined by $\ell(v') = \{a : v' \in D_a\}$. Lemma 4 shows that ℓ is a proper labelling. By the Tree Sperner Lemma, there exists a fully-labelled edge $e \in T'$ with endpoints x and y such that $\ell(x) \cup \ell(y) = A$. Thus, for all $a \in A$, either $x \in D_a$ or $y \in D_a$ (or both). However, since $size(T') < \delta$, the Lebesgue number property guarantees that $e \subseteq T \setminus D_a$ for some a, implying that $e \cap D_a = \emptyset$, a contradicting that e was fully-labelled. Therefore, we conclude that $\cap_{a \in A} D_a \neq \emptyset$. □

Note also that the sets of a KKM cover do not have to be connected (though they are in Figure 2. However, if a tree is covered by sets that are connected as well as pairwise intersecting, then it is a KKM cover!

THEOREM 6 (Tree KKM for Connected Sets). *Let $\mathscr{D} = \{D_1, D_2, ..., D_k\}$ be a finite collection of closed, connected sets that cover a metric tree $T = (V, E, X)$ such that each pair $D_i \cap D_j$ is nonempty. Then there is a point x in all the sets of $\mathscr{D}$.*

PROOF. Choose points $a_i \in D_i$ for each i, and put them in a set A. To show $\mathscr{D}$ is a KKM cover of T relative to A, it remains to show the path-covering property.

If for some pair a_i and a_j in A, the path-covering property did not hold, then the path between a_i and a_j would contain a point y that is not covered by $D_i \cup D_j$. Then $X \setminus \{y\}$ would have two connected components that would separate a_i from a_j. Then D_i and D_j must lie in different components because each is connected. But then they could not be pairwise intersecting, a contradiction. □

We indicate some implications of Theorem 5 below, and note that the proof of Theorem 1 would suggest associated constructive algorithms.

Pizza Delivery. Suppose you are starting a pizza delivery business and you desire a good location for your store. Your city has several neighborhoods connected by a tree of roads. For each neighborhood i, there is a "deliverability" set

D_i: the set of all locations on the tree with an acceptable commute to neighborhood i. These sets are naturally closed and connected, as in Figure 2. Then the Tree KKM Theorem for Connected Sets (Theorem 6) says that if for every pair of neighborhoods i and j there is a common acceptable location to place your store, then there will be a location with an acceptable commute to all neighborhoods.

Grand Central Station. Suppose several cities are connected by a tree of train tracks. Each city has its own train authority, and suppose it is possible to get from city i to city j using only those cities' trains (possibly switching several times). Then Theorem 5 shows that there must be a location through which trains from all cities must pass, i.e., there is a location where one could place a Grand Central Station.

Note that something further is true if we make some intuitive assumptions about the structure of the tree of train tracks. It seems reasonable to suppose that each vertex of this tree is a station and that trains only change directions at stations. With these assumptions the following result becomes apparent: there must already be a station at which trains from all of the cities stop, i.e., a Grand Central Station already exists. If the point guaranteed by Theorem 5 is not a station, then trains from each city also must pass through the two nearest stations on either side of this point (because trains only change directions at stations).

5. A Fixed Point Theorem for Finite Trees

Just as the KKM Theorem and Sperner's Lemma imply Brouwer's Theorem on simplices, we can use our previous theorems to prove a classical fixed point theorem for trees.

THEOREM 7. *(The Tree Fixed Point Theorem) Let $T = (V, E, X)$ be a metric tree, V be finite, and $f : T \to T$ be a continuous function. Then f has a fixed point.*

As already noted, this result follows from the Lefschetz fixed point theorem, but the proofs we give below have an advantage in being elementary. The first is a non-constructive proof using the Tree KKM Theorem, the second is a constructive approach using the Tree Sperner Lemma.

We shall often appeal to a special family of closed sets associated to any continuous self-map on a tree. Given $T = (V, E, X)$ be a metric tree, and $A \subset V$ a subset of vertices, and $f : T \to T$ be a continuous function, let $\mathscr{D}_{f,A} = \{D_a : a \in A\}$ be the family of sets defined by

$$D_a = \{x \in T : d(x, a) \leq d(f(x), a)\}.$$

To interpret, D_a contains the set of all points that stay the same distance or move away from a. We now show this family $\mathscr{D}_{f,A}$ is a KKM cover, so we shall refer to it as the *move-away KKM cover associated to f and A.*

THEOREM 8. *The family $\mathscr{D}_{f,A}$ defined above is a KKM cover of T relative to A.*

PROOF. It is apparent from the definition that $a \in D_a$ for each $a \in A$.

We now show that each D_a is closed. Let $\{x^k\}$ be a sequence in D_a that converges to x in T. Since $x^k \in D_a$, we have $d(x^k, a) \leq d(f(x^k), a)$. Since f is continuous, we see that $f(x^k) \to f(x)$.

Let $\epsilon > 0$ be given. Choose N such that $k \geq N$ implies that $d(x^k, x) < \epsilon$ and $d(f(x^k), f(x)) < \epsilon$. Then

$$\begin{aligned} d(x,a) &\leq d(x,x^k) + d(x^k,a) \\ &\leq d(x,x^k) + d(f(x^k),a) \\ &\leq d(x,x^k) + d(f(x^k),f(x)) + d(f(x),a) \\ &< 2\epsilon + d(f(x),a). \end{aligned}$$

Since this is true for all $\epsilon > 0$ we conclude that $d(x,a) \leq d(f(x),a)$, so $x \in D_a$. Therefore D_a is closed.

Finally, we show the path-covering property of $\mathscr{D}_{f,A}$. If a and b are in A, consider x on the path between them. If $f(x) = x$, then x doesn't move, so $x \in D_a \cup D_b$ as desired. Otherwise, suppose that $f(x)$ is not in the same component of $T \setminus \{x\}$ as a. Then the path from a to $f(x)$ and must contain x. It follows that the path from a to x is contained in the path from a to $f(x)$, so we conclude that $d(a,x) \leq d(a, f(x))$, hence $x \in D_a$. By similar argument, if $f(x)$ is not in the same component of $T \setminus \{x\}$ as b, then $x \in D_b$. Either way, $x \in D_a \cup D_b$. □

We can now give a quick proof of the Tree Fixed Point Theorem:

FIRST PROOF OF THEOREM 7. Consider the move-away KKM cover $\mathscr{D}_{f,V}$ associated to f and V. Since V is finite, the Tree KKM Theorem shows there exists a point x in all sets of $\mathscr{D}_{f,V}$. By definition this means $d(x,v) \leq d(f(x),v)$ for all $v \in V$.

Suppose that $x \neq f(x)$. If $f(x)$ were a vertex v, then this would contradict $x \in D_v$. So $f(x)$ is not a vertex and is thus a cut point. Choose $w \in V$ such that w and x are in different components of $T \setminus \{f(x)\}$. Then the path from w to $f(x)$ is contained in the path from w to x. Since $f(x) \neq x$, this implies that $d(x,w) > d(f(x),w)$, so that $x \notin D_w$, a contradiction. Hence $f(x) = x$. □

A second proof of the Tree Fixed Point Theorem relies on the following rather standard lemma that we include for completeness. An ϵ-fixed point for f is an approximate fixed point x such that $d(x, f(x)) < \epsilon$.

LEMMA 9 (Epsilon Fixed Point Theorem). *Suppose that K is a compact subset of the metric space (X,d) and that $f : K \to K$ is continuous. If f has an ϵ-fixed point for every $\epsilon > 0$ then f has a fixed point.*

PROOF. Let $\{a_n\}$ be a sequence of $1/n$-fixed points, that is, $d(a_n, f(a_n)) < 1/n$ for all n. Since K is compact, $\{a_n\}$ has a convergent subsequence $\{a'_n\}$ converging to $x \in K$. Let $\epsilon > 0$. Since $a'_n \to x$ there exists N_1 such that $n \geq N_1$ implies that $d(a'_n, x) < \epsilon/2$. Let $N = \max(N_1, 2/\epsilon)$. Then $n \geq N$ implies that

$$d(x, f(a'_n)) \leq d(x, a'_n) + d(a'_n, f(a'_n)) < \epsilon,$$

so that $f(a'_n) \to x$. However, since f is continuous, we have also that $f(a'_n) \to f(x)$. Hence $f(x) = x$, and x is a desired fixed point. □

Now if we construct the membership labeling associated with the move-away KKM cover in the prior proof, the Tree Sperner Lemma gives a fully-labelled edge. The next proof of the Tree Fixed Point Theorem shows that a point in this edge is an ϵ-fixed point, which means that locating such an edge, as the Tree Sperner Lemma allows us to do, will allow us to find approximate fixed points.

SECOND PROOF OF THEOREM 7. By Lemma 9, it is sufficient to show that f has an ϵ-fixed point for all $\epsilon > 0$.

Fix $\epsilon > 0$. Since V is finite, X is compact, hence f is uniformly continuous. So there exists $\gamma > 0$ such that if $d(x, y) < \gamma$ then $d(f(x), f(y)) < \epsilon/2$. Let $\delta = \min(\gamma, \epsilon/2)$ and let $T' = (V', E', X)$ be a segmentation of $T = (V, E, X)$ with $size(T') < \delta$.

Let $\ell : V' \to 2^V$ be defined by

$$\ell(v') = \{v \in V : d(v', v) \leq d(f(v'), v)\}.$$

Note that this is just the membership labelling defined in Lemma 4 for the move-away KKM cover $\mathscr{D}_{f,V}$ in the prior proof. By the Tree Sperner Theorem, there is a fully-labelled edge $e' \in E'$ with endpoints y and z. We claim that y is the desired ϵ-fixed point.

Suppose e' intersects its image $f(e')$ in some point w. Then

$$d(y, f(y)) \leq d(y, w) + d(w, f(y)) < \delta + \frac{\epsilon}{2} \leq \frac{\epsilon}{2} + \frac{\epsilon}{2} \leq \epsilon,$$

where the second inequality uses the fact that w is a point in e' as well as in $f(e')$. So in this case, y is an ϵ-fixed point.

Otherwise, if $e' \cap f(e')$ were empty, then since f is continuous, the image $f(e')$ is connected and contained in one component of the set $X \setminus e$. The point $f(z)$ is contained in some edge e of E. If $f(z) \in V$ then let $v = f(z)$, else let v be the unique endpoint of e that is in a different component of $X \setminus f(z)$ than e'. Either way, $f(z)$ is on the path from v to z, and $f(z) \neq z$. So $v \notin \ell(z)$ because $f(z)$ is strictly closer to v than z is. But then $v \in \ell(y)$ because e' is fully labelled, so $d(v, y) \leq d(v, f(y))$. Since either $f(z) = v$ or $f(z)$ separates v from y, we see that $f(z)$ is on the path from v to y. Hence

$$d(v, f(z)) + d(f(z), y) = d(v, y) \leq d(v, f(y)) \leq d(v, f(z)) + d(f(z), f(y)),$$

which implies that $d(y, f(z)) \leq d(f(y), f(z))$. But $d(f(y), f(z)) < \epsilon/2$. So

$$d(y, f(y)) \leq d(y, f(z)) + d(f(z), f(y)) < \frac{\epsilon}{2} + \frac{\epsilon}{2} < \epsilon.$$

Again, y is an ϵ-fixed point. □

The constructive nature of the Tree Sperner Lemma suggests a method for locating an approximate fixed point in a systematic way. Namely, we can choose a segmentation of sufficiently small size, start at any vertex, and "move in the direction of the missing labels" as suggested by the proof of Theorem 1. This will eventually lead to a fully-labelled edge whose endpoints are approximate fixed points.

One can also use this method to converge to an actual fixed point. In much the same way that homotopy algorithms for Sperner's Lemma can be used to "home in" on a fixed point by changing the size of a triangulation as one proceeds. In the case of Sperner's lemma, one can, for example, define a triangulation on $\Delta \times \mathbb{R}$ that interpolates between triangulations on Δ with different mesh sizes; these are called *homotopy algorithms* (e.g., see [**9**], [**10**]). Similarly, one can construct homotopy algorithms for trees in much the same way, by defining a triangulation on $T \times \mathbb{R}$.

We note that Theorem 7 implies the Tree Fixed Vertex-Edge Theorem (Theorem 2), and thus:

THEOREM 10. *The following are all equivalent:*

(1) *Tree Sperner Lemma (Theorem* 1*),*
(2) *Tree Fixed Vertex-Edge Theorem (Theorem* 2*),*
(3) *Tree KKM Theorem (Theorem* 5*), and*
(4) *Tree Fixed Point Theorem (Theorem* 7*).*

PROOF. Theorem 3 showed that (2) and (1) are equivalent. We have also seen that (1) implies (3), and (3) implies (4), so it suffices to show that (4) implies (2).

Let (V, E) be a tree, and let $f : V \to V$ be a function with finite range. We construct a finite metric tree in the following way. Let (V', E') be the finite subtree of (V, E) spanned by vertices in the range of f. We can realize this subtree as a metric tree $T = (V', E', X)$ by letting all edges have length 1. Since (V', E') is finite, the underlying space X is compact.

Now construct a continuous function $\widehat{f} : X \to X$ by extending the given f linearly across single edges, i.e., if x is a point that is some fraction of distance along an edge from vertex v to w, then we let $\widehat{f}(x)$ be the point that is the same fraction of the distance along the path from $f(v)$ to $f(w)$.

By Theorem 7, there is a point z such that $\widehat{f}(z) = z$. If $z \in V$ then we see f has a fixed vertex, as desired. Otherwise, z is on some edge $e = \{v, w\}$ and $\widehat{f}(v)$ and $\widehat{f}(w)$ must be in different components of $X \setminus \{z\}$; moreover, they must be vertices in V. But v and w are the nearest vertices to z in those two components. Hence the path from $f(v)$ to $f(w)$ in the combinatorial tree (V, E) is a path that contains e, as desired. □

6. Infinite Settings

We can extend both the Tree KKM theorem and the Tree Fixed Point Theorem to the infinite setting (for covers with infinitely many sets, and to compact maps on trees with infinitely many vertices).

The following standard lemma will be useful, so we include it here for completeness:

LEMMA 11. *Let X be a topological space. Let Λ be an infinite (not necessarily countable) index set and suppose that $\mathscr{D} = \{D_\alpha : \alpha \in \Lambda\}$ is a family of closed sets in X with the finite intersection property. Further suppose that D_α is compact for at least one $\alpha \in \Lambda$. Then*

$$\bigcap_{\alpha \in \Lambda} D_\alpha \neq \emptyset.$$

PROOF. If not, then there is no point in every D_α, hence each point is covered by some complement: D_α^c. Then $\mathscr{D}^c = \{D_\alpha^c : D_\alpha \in \mathscr{D}\}$ is an open cover of $Y = \cup\{D_\alpha : \alpha \in \Lambda\}$. Let D_β be one of the compact sets in $\mathscr{D}$. It follows that $\mathscr{D}^c$ is an open cover of D_β and thus has a finite subcover, say it is the collection $\mathscr{F}$. Since these sets cover D_β, the intersection of their complements together with D_β is empty — that is,

$$\left(\bigcap_{D_\alpha^c \in \mathscr{F}} D_\alpha \right) \cap D_\beta = \emptyset.$$

This contradicts that $\mathscr{D}$ has the finite intersection property and thus proves the lemma. □

Now we may prove a KKM theorem for infinite trees:

THEOREM 12 (KKM Theorem for Infinite Trees). *Let $T = (V, E, X)$ be a tree, let A be a (possibly infinite) subset of V. Suppose that $\mathscr{D} = \{D_a : a \in A\}$ is a KKM cover of T relative to A such that at least one set in $\mathscr{D}$ is compact. Then*

$$\bigcap_{a \in A} D_a \neq \emptyset.$$

The extra condition that one set be compact is essential. Recall again the integer tree: a line whose nodes are the integers, and whose edges are the intervals between successive integers. See Figure 3. One may construct a KKM cover by letting $D_i = [i, \infty)$, but the intersection of all such sets is empty.

FIGURE 3. The integer tree, with infinite number of vertices (at integers) and edges.

PROOF. By Lemma 11 it suffices to show that $\mathscr{D}$ has the finite intersection property. Let J be a finite subset of A. Then the family of sets $\{D_j : j \in J\}$ forms a KKM cover of T relative to J, a finite set, so Theorem 5 applies. Hence $\bigcap_{\alpha \in J} D_\alpha \neq \emptyset$. □

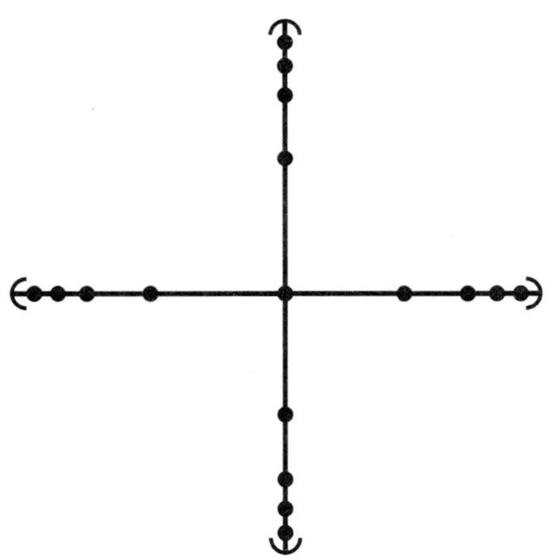

FIGURE 4. A bounded infinite tree. Vertices (not all shown) accumulate at the open endpoints of the underlying set.

Using the Tree KKM Theorem we can strengthen the Tree Fixed Point Theorem. Recall that if X and Y are metric spaces, $f : X \to Y$ is a *compact* map if $f(B)$ is contained in a compact subset of Y for every bounded set B.

THEOREM 13. *Suppose that T is a bounded tree with vertex set V. If $f : T \to T$ is a continuous compact map, then f has a fixed point.*

The compactness hypothesis of the map is reminiscent of the compactness hypothesis of the Schauder fixed point theorem [**3**].

We give a couple of (non-)examples to illustrate. Consider the integer tree of Figure 3 with the map that shifts it one unit to the right. This map does not have a fixed point; it is a compact map, but the tree is not bounded. The infinite tree of Figure 4 is bounded, but it is not compact. One may easily construct a fixed-point-free map that moves points within this tree towards the rightmost open endpoint, so that the image is not contained in any compact subset.

PROOF. Since f is compact and continuous, the image $K = \overline{f(T)}$ is compact and connected. Let $V' = V \cap K$ be the set of vertices of T that are in K. Let ∂K denote the set of points that form the boundary of K in T. Consider $V^* = V' \cup \partial K$. It is a subset of K since K is compact.

We construct a new tree T^* with vertices V^* and underlying metric space K. To start, let T' be the subtree of T spanned by the vertices in V', with edge set E'. We shall augment T' to form T^* by adding points of ∂K as leaves. So consider any $x \in \partial K$. Note that x lies in some realized edge e of T with endpoints in V.

We claim that exactly one component C of $e \setminus \{x\}$ will intersect $f(T)$. This is clear if x is an endpoint of e, else it follows from the fact that $f(T)$ is connected: if $s, t \in f(T)$ were in different components, then the path from s to t would lie in $f(T)$ and must contain x, so x could not be a boundary point of K.

Let v_x denote the unique endpoint of e that is in component C. Then either v_x is in $f(T)$ or not. We explore these cases and the tree T^* that results:

(1) If $v_x \notin f(T)$, then because $f(T)$ is connected, then $f(T)$ must lie in the interior of e, so $f(T)$ is an interval with x at one endpoint and $y \neq v_x$ at another endpoint. Then K has at most two boundary points and is an interval. Then let T^* be the tree consisting of one edge $e^* = [x, v_x]$ and two vertices x, v_x.
(2) If $v_x \in f(T)$, then $f(T)$ contains C because $x \in K = \overline{f(T)}$. Then the segment $e^*_x = [x, v_x]$ lies in K, and $v_x \in V' = V \cap K$. This construction may be done for every $x \in \partial K$. We construct T^* as the tree with vertex set $V^* = V' \cup \{x : x \in \partial K \setminus V\}$, edge set $E^* = E' \cup \{e^*_x : x \in \partial K \setminus V\}$, and underlying space K.

Since T^* has underlying space K, note that $f : T \to T$ restricts to a function $f^* : T^* \to T^*$. Then consider $\mathscr{D}_{f^*,V^*}$, the move-away KKM cover of T^* relative to V^*.

Suppose that H is a finite subset of V^*. From Theorem 8 we see that $\mathscr{D}_{f^*,H}$ is a KKM cover of T relative to H and thus, by the Tree KKM Theorem, the intersection of its sets is nonempty. Thus, $\mathscr{D}_{f^*,V^*}$ has the finite intersection property.

Because K is compact, all the sets of $\mathscr{D}_{f^*,V^*}$ are compact, so Theorem 12 shows that the intersection of sets in $\mathscr{D}_{f,V}$ is also non-empty, say it contains a point z. Then for all $v \in V^*$ we have $d(z, v) \leq d(f(z), v)$. We claim that $f(z) = z$.

Suppose that $z \neq f(z)$. If $f(z)$ were a vertex v, this would contradict that $z \in D_v$. So $f(z)$ is not a vertex and is thus a cut point. Choose $w \in V^*$ such that w and z are in different components of $T \setminus \{f(z)\}$. Then the path from w to $f(z)$ is contained in the path from w to z. Since $f(z) \neq z$, this implies that $d(z, w) > d(f(z), w)$, so that $z \notin D_w$, a contradiction. Hence $f(z) = z$.

□

7. A KKM Theorem for Cycles

Recall that a *cycle* is a finite graph with vertices $v_1, \ldots, v_n$ and edges (v_i, v_{i+1}) as well as (v_n, v_1). We define a *metric cycle* to be a triple $C = (V, E, X)$ where (V, E) is a cycle (as above) and X is an *underlying metric space* obtained from a cycle in exactly the same fashion as we obtained a metric tree from a tree in Section 3. The resulting space C is topologically a circle, partitioned into a finite set of segments (realized edges) joined at their endpoints $v_1, ..., v_n$. Between any two points of C there are exactly two paths; the metric is just the minimum length of the two paths connecting the points.

Hereafter, all metric cycles will simply be referred to as cycles.

If C_n is a cycle with n vertices and x is in C_n, let $e(x)$ be the set consisting of x and all points y that are not vertices but are on a realized edge with x. Note that by removing from C_n the set $e(x)$ as well as vertices and realized edges within, we obtain a metric tree $C_n \setminus e(x)$ with n or $(n-1)$ vertices, depending on whether or not x is a vertex of C_n. This observation will become the key to reducing KKM covers on cycles to KKM covers on trees.

Now, since there are two paths connecting any two distinct vertices in a cycle, we must slightly alter our definition of KKM cover for trees, but we want to do so in a way consistent with our definition for trees.

With this in mind, let C_n be a cycle with n vertices $V = \{v_1, v_2, \ldots, v_n\}$. A *KKM cover* of the cycle C_n is a family of closed sets $\mathscr{D} = \{D_v : v \in V\}$ such that the following conditions hold:

- each $v \in D_v$, and
- for all $v, w \in V$, *at least one of the paths* between v and w is contained in $D_v \cup D_w$.

This new path-covering property generalizes the corresponding property for trees. We can now state the main theorem of this section.

THEOREM 14. *Suppose that C_n is a cycle with vertices $V = \{v_1, v_2, \ldots, v_n\}$ and let $\mathscr{D} = \{D_v : v \in V\}$ be a KKM cover of the cycle C_n. Then there is a point x in C_n such that x is in at least $\lfloor \frac{n}{2} \rfloor + 1$ sets of $\mathscr{D}$.*

To compare this result with Theorem 5, note that KKM covers of trees have a point in *all* the sets of the cover, but KKM covers of cycles have a point in a *strict majority* of the sets.

PROOF. For each x, consider the set $\ell(x) = \{v \in V : x \in D_v\}$.

Fix x. If $|\ell(x)| \geq \lfloor \frac{n}{2} \rfloor + 1$, then we have our desired conclusion.

Else, if $|\ell(x)| \leq \lfloor \frac{n}{2} \rfloor - 1$, then let $H = V \setminus \ell(x)$; clearly $|H| \geq \lfloor \frac{n}{2} \rfloor + 1$. Note that for any pair v and w in H, the path between them covered by $D_v \cup D_w$ does not include x, so this path is still covered if we remove $e(x)$ from the graph. Then $C_n \setminus e(x)$ is a tree and and the family $\mathscr{F} = \{D_v : v \in H\}$ is a KKM cover of this tree relative to H. By Theorem 5, $\mathscr{F}$ has non-empty intersection, and it has at least $\lfloor \frac{n}{2} \rfloor + 1$ sets of $\mathscr{D}$, as desired.

The only remaining case is when $|\ell(x)| = \lfloor \frac{n}{2} \rfloor$ is constant for all x. We now show why this leads to a contradiction, but we must take some care because the sets of the cover might have several connected components.

The set of boundary points of D_v, denoted by ∂D_v, is a closed set. Moreover, ∂D_v has no interior, so the finite union of boundary points $B = \cup\{\partial D_v : v \in V\}$ is closed and has no interior. So B^c is a nonempty open set.

So choose $x \in B^c$ and since B is closed, we may find $b \in B$ that is closest to x. Then all points in $U = \{y : d(x, y) < d(x, b)\}$ are, for each $v \in V$, interior points of either D_v or D_v^c. So if $y \in U$, $\ell(x) = \ell(y)$. Since b is a limit point of U and each D_v is closed, $\ell(x) \subseteq \ell(b)$.

If for some $v \notin \ell(x)$ we have $b \in D_v$, then $\ell(x) \cup \{v\} \subseteq \ell(b)$, so the size of ℓ is not constant, a contradiction. Otherwise, for every $v \notin \ell(x)$, we have that $b \in D_v^c$, an open set, so $\ell(x) = \ell(b)$. So there is an open set W around b that contains no points of D_v for all $v \notin \ell(x)$; therefore for $w \in W$, $\ell(w) \subseteq \ell(x) = \ell(b)$. Since $b \in B$, it must be in ∂D_z for some $z \in \ell(x)$. So there is a $w \in W$ such that $w \notin D_z$, thus $\ell(w) \subseteq \ell(b) \setminus \{z\}$, so the size of ℓ is not constant, a contradiction. □

Theorem 14 has an interesting application to voting theory. In *approval voting*, each voter specifies which options she would consider acceptable, without ranking the options. Following [**1**], the set of all options available to voters is called a *(political) spectrum*; it often has a natural topology given by notions of "closeness" or "similarity" of political preferences. For instance, the political spectrum is often modeled as $\mathbb{R}$, a line with conservative positions to the right and liberal positions to the left. However, in elections over multiple issues, the spectrum might be best modeled as a subset of $\mathbb{R}^n$. Political spectra have been modeled also by a circle; often this arises by bending the linear political spectrum so that the extreme left-wing and right-wing positions are considered close; e.g., see [**8**].

For each voter, the set of options that a voter approves is called her *approval set*. We assume that approval sets are closed subsets of the spectrum, and we call the set of all voters together with their approval sets a *society*.

We call a society with a circular political spectrum *super-agreeable* if for each pair of voters i, j, one of the paths between i, j is covered by their approval sets. We remark that in many cases it is natural to assume that a voter's approval set is connected. In this situation, a super-agreeable society is simply one in which every pair of voters can find common ground, i.e., an option which they will both approve. This agrees with the definition of super-agreeable for a linear society, as in [**1**].

Then Theorem 14 then has the following corollary:

THEOREM 15. *In a super-agreeable society with a circular political spectrum, there is an option that will be approved by a strict majority of the voters.*

The value of this result is that it gives a sufficient condition for the existence of a strict majority using approval voting when the political spectrum is circular. We do not assume the approval sets have to be connected. Results for connected approval sets and weaker intersection hypotheses may be found in [**1**], who consider linear political spectra, and Hardin [**4**], who extends those results to circular political spectra.

References

[1] Deborah E. Berg, Serguei Norine, Francis Edward Su, Robin Thomas, and Paul Wollan, *Voting in agreeable societies*, Amer. Math. Monthly **117** (2010), no. 1, 27–39, DOI 10.4169/000298910X474961. MR2599465 (2011i:91065)

[2] Eli Berger, *KKM—a topological approach for trees*, Combinatorica **25** (2005), no. 1, 1–18, DOI 10.1007/s00493-005-0001-y. MR2109190 (2005h:05143)

[3] James Dugundji and Andrzej Granas. *Fixed Point Theory.* Polish Scientific Publishers, Warszawa, 1982.

[4] Christopher S. Hardin, *Agreement in circular societies*, Amer. Math. Monthly **117** (2010), no. 1, 40–49, DOI 10.4169/000298910X474970. MR2599466 (2011i:91067)

[5] Mohamed A. Khamsi, *KKM and Ky Fan theorems in hyperconvex metric spaces*, J. Math. Anal. Appl. **204** (1996), no. 1, 298–306, DOI 10.1006/jmaa.1996.0438. MR1418536 (98h:54059)

[6] B. Knaster, C. Kuratowski, and S. Mazurkiewicz. Ein beweis des fixpunktsatzes für n-dimensionale simplexe. *Fund. Math.*, 14:132–137, 1929.

[7] Kathryn L. Nyman and Francis Edward Su, *A Borsuk-Ulam equivalent that directly implies Sperner's lemma*, Amer. Math. Monthly **120** (2013), no. 4, 346–354, DOI 10.4169/amer.math.monthly.120.04.346. MR3035127

[8] Sven Gunnar Simonsen. *Nationalism and the Russian political spectrum: Locating and evaluating the extremes.* Studies on Foreign Policy Issues, 4:2000, Ministry of Foreign Affairs, Norway, 2000. Online at `http://odin.dep.no/ud/english/doc/reports/032001-220004/dok-bn.html`.

[9] Michael J. Todd, *The computation of fixed points and applications*, Springer-Verlag, Berlin-New York, 1976. Lecture Notes in Economics and Mathematical Systems, Vol. 124. MR0410732 (53 #14478)

[10] Zaifu Yang, *Computing equilibria and fixed points*, Theory and Decision Library. Series C: Game Theory, Mathematical Programming and Operations Research, vol. 21, Kluwer Academic Publishers, Boston, MA, 1999. The solution of nonlinear inequalities. MR1788059 (2001h:90003)

[11] Mark Yoseloff, *Topologic proofs of some combinatorial theorems*, J. Combinatorial Theory Ser. A **17** (1974), 95–111. MR0345836 (49 #10566)

Jane Street Capital, 1 New York Plaza, Floor 33, New York, New York 10004
E-mail address: `agnieder@gmail.com`

Department of Mathematics, University of Washington, Seattle, Washington 98195
E-mail address: `drizzolo@uw.edu`

Department of Mathematics, Harvey Mudd College, Claremont, California 91711
E-mail address: `su@math.hmc.edu`

Contemporary Mathematics
Volume **625**, 2014
http://dx.doi.org/10.1090/conm/625/12493

Cliques and cycles in distance graphs and graphs of diameters

Andrei M. Raigorodskii

Abstract. In this paper, we discuss some properties of the so-called distance graphs and graphs of diameters. The study of such graphs is well-motivated by classical problems of combinatorial geometry going back to Erdős, Hadwiger, Nelson, and Borsuk.

Contents

1. Distance graphs: definitions and motivation

Let $\mathbb{R}^d$ be the standard Euclidean space, where the distance between any two points $\mathbf{x}, \mathbf{y}$ is denoted by $|\mathbf{x} - \mathbf{y}|$. Let V be an arbitrary point set in $\mathbb{R}^d$. Let $a > 0$ be a real number. By a *complete distance graph* with the set of vertices V we mean the graph $G = (V, E)$ whose set of edges E contains *all* pairs of points from V that are at the distance a apart:

$$E = \{\{\mathbf{x}, \mathbf{y}\} : \ |\mathbf{x} - \mathbf{y}| = a\}.$$

Distance graphs are defined as arbitrary subgraphs of complete distance graphs.

It is worth noting that the two definitions are quite different in a sense. More precisely, consider the set $\mathcal{D}_n(d)$ of all (labeled) unit distance graphs in $\mathbb{R}^d$ having

---

2010 *Mathematics Subject Classification.* Primary 52C10.

Key words and phrases. Graphs of diameters, Borsuk's problem, coloring, cliques.

This work was done under the financial support of the following grants: the grant 12-01-00683 of Russian Foundation for Basic Research, the grant MD-6277.2013.1 of the Russian President, the grant NSh-2519.2012.1 supporting Leading scientific schools of Russia.

n vertices (*unit distance* means that $a = 1$). Denote by $\mathcal{CD}_n(d)$ the similar set of complete unit distance graphs. In **[66]** it is proved that

$$|\mathcal{D}_n(d)| = 2^{\left(1-\frac{1}{[d/2]}+o(1)\right)\frac{n^2}{2}}, \qquad (1)$$

which means that for $d \geq 4$, the number of distance graphs on n vertices is nearly equal to the number of subgraphs of a complete n-vertex $[d/2]$-partite graph. At the same time, as shown recently by Alon and Kupavskii (see **[2]**),

$$|\mathcal{CD}_n(d)| = 2^{(1+o(1))dn\log_2 n},$$

which is much smaller than the right hand side of (1).

Unit distance graphs are among the most studied objects of combinatorial geometry. First of all, they are at the ground of the classical Nelson–Hadwiger problem, which was proposed around 1950 (see **[41]**, **[107]**) and consists in finding the *chromatic number* $\chi(\mathbb{R}^d)$ of the space:

$$\chi(\mathbb{R}^d) = \min\left\{\chi : \ \mathbb{R}^d = V_1 \sqcup \ldots \sqcup V_\chi, \ \forall\, i\, \forall\, \mathbf{x}, \mathbf{y} \in V_i \ |\mathbf{x}-\mathbf{y}| \neq 1\right\},$$

i.e., the chromatic number of the space is the minimum number of colors needed to color all the points in $\mathbb{R}^d$ so that any two points at the distance 1 apart receive different colors. In other words, it is the chromatic number of the complete unit distance graph whose vertex set coincides with $\mathbb{R}^d$.

Due to the extremal popularity of the subject, colorings of unit distance graphs are very deeply explored. Let us just refer the reader to several books and survey articles: **[1]**, **[11]**, **[15]**, **[55]**, **[78]**, **[79]**, **[86]**, **[111]**. Here we only mention the best known bounds for the chromatic numbers:

$$\chi(\mathbb{R}^1) = 2, \ 4 \leq \chi(\mathbb{R}^2) \leq 7, \ 6 \leq \chi(\mathbb{R}^3) \leq 15 \ \textbf{[18]}, \ 7 \leq \chi(\mathbb{R}^4) \leq 54 \ \textbf{[77]},$$

dim	1	2	3	4	5	6
$\chi \geq$	2	4	6, **[73]**	7, **[14]**,**[49]**, **[56]**	9, **[14]**	11, **[16]**
dim	7	8	9	10	11	12
$\chi \geq$	15, **[78]**	16, **[67]**	21, **[62]**	23, **[62]**	25, **[57]**	27, **[56]**

$$\textbf{[85]} \ (1.239\ldots + o(1))^n \leq \chi(\mathbb{R}^n) \leq (3 + o(1))^n \ \textbf{[67]}.$$

Another well-studied question on unit distance graphs was first asked by Erdős in **[26]**: *If* $G = (V, E)$ *is a unit distance graph with* $|V| = n$*, then how many edges can it have depending on* n*?* This subject is even closer to the main line of this survey, since a natural generalization of Erdős' question concerns the maximum number of *cliques* in a unit-distance graph. One of the best sources of up-to-date references is the book **[11]**. In Section 4, we will also tell a few words about this type of questions.

2. Graphs of diameters: definitions and motivation

Let $V \subset \mathbb{R}^d$. Denote by $\operatorname{diam} V$ the value $\sup\limits_{\mathbf{x},\mathbf{y}\in V} |\mathbf{x}-\mathbf{y}|$. This quantity is called *the diameter* of V. By the *graph of diameters* of the set V we mean the graph $G = (V, E)$, where E consists of *all* possible pairs of points in V whose distance equals exactly the diameter of V. Note that here we do not consider arbitrary subgraphs of graphs of diameters as we did for distance graphs in the previous section: the graphs of diameters are necessarily complete by definition.

If V is an infinite set (say, a convex body in $\mathbb{R}^d$), then its graph of diameters can be ill-defined in a sense. Indeed, V might be open and then the corresponding graph of diameters would be empty, so that it would give no information on the structure of V. Thus, it is much more reasonable to concentrate on the study of *finite* graphs of diameters.

On the other hand, the investigation of the graphs of diameters was initiated essentially in 1933 by Borsuk (see [**9**]). Borsuk asked whether any set of diameter 1 in $\mathbb{R}^d$ can be partitioned into $d+1$ parts of smaller diameter. The positive answer to this question is called *Borsuk's conjecture.* More precisely, let $V \subset \mathbb{R}^d$ be a set of diameter 1. Denote by $f(V)$ the minimum number of parts of smaller diameter, into which V can be partitioned. Denote by $f(d)$ the maximum value of $f(V)$, where the maximum is taken over all possible sets of diameter 1 in $\mathbb{R}^d$. Borsuk's conjecture says that $f(d) = d+1$.

Of course in the case when V is finite, $f(V)$ equals the chromatic number $\chi(G)$ of the graph of diameters G of V. Thus, we come to the problem of finding colorings of graphs of diameters. In the general case, the difference between $f(V)$ and $\chi(G)$ may be great, even if V is closed. As an example, consider a sphere $V = S^{d-1} \subset \mathbb{R}^d$. Its "Borsuk number" — $f(S^{d-1})$ — is equal to $d+1$, which is due to the classical Borsuk–Ulam theorem (see [**70**]) and supports Borsuk's conjecture. However, the graph of diameters of the sphere is just a matching whose chromatic number is trivially 2.

In view of the above observations, Borsuk's problem appears to be a good motivation for the study of graphs of diameters, but the assumption that such graphs are finite still seems to be rather restrictive. Nevertheless, the history of Borsuk's problem shows that even this restriction is not so important. The point is that in 1993 Borsuk's conjecture was *disproved* in a high dimension and that *all* the counterexamples are provided by *finite* graphs of diameters. As in the previous section, we just refer the reader to the following books and surveys: [**1**], [**7**], [**11**], [**55**], [**78**], [**80**]–[**84**], [**86**]. Below, we give a list of the most relevant results:

$$f(d) = d+1, \ d \leq 3 \ [\mathbf{9}], [\mathbf{24}], [\mathbf{39}], [\mathbf{42}], [\mathbf{69}],$$

$$[\mathbf{91}] \ (1.2255\ldots + o(1))^{\sqrt{d}} \leq f(d) \leq (1.224\ldots + o(1))^d \ [\mathbf{10}], [\mathbf{102}],$$

Conjecture is false for $d \geq$	Author(s)	Year and Reference
2015	Kahn, Kalai	1993, [**53**]
946	Nilli (Alon)	1994, [**74**]
903	Grey, Weissbach	1997, [**38**]
561	Raigorodskii	1997, [**92**]
560	Weissbach	2000, [**114**]
324	Hinrichs	2002, [**45**]
321	Pikhurko	2002, [**76**]
298	Hinrichs, Richter	2003, [**46**]
65	Bondarenko	2013, [**8**]
64	Jenrich	2013, [**51**]

As in the case of distance graphs, problems on the numbers of edges, triangles, etc. in graphs of diameters are of a special interest. In Section 4, we will also discuss them.

3. What is the role of cliques and cycles in geometric graphs?

Here we want to emphasize the role that is played by cliques and cycles in a distance graph or in a graph of diameters. A theorem of Jung is well-known (see

[**52**]): *Any set of diameter 1 in* $\mathbb{R}^d$ *is contained in a ball of radius* $\sqrt{\frac{d}{2d+2}}$. Moreover, the quantity $\sqrt{\frac{d}{2d+2}}$ is just the radius of a ball circumscribed around a regular d-simplex with side length 1. But what is a regular simplex in terms of our graphs? A regular simplex with side length 1 and with k vertices is exactly a k-clique in graphs of diameters and in distance graphs. In other words, if our geometric graphs contain some k-cliques (($k-1$)-simplices), then their vertex sets cannot lie inside spheres of radii smaller than $\sqrt{\frac{k-1}{2k}}$: although the edges are certainly of length 1, the smallest spheres that might contain our graphs are far enough from the unit sphere. The same can be said about cycles.

Denote by $\omega(G)$ the *clique number* of a graph G (i.e., the maximum number of vertices in its clique). Denote by $\alpha(G)$ the *independence number* of the graph (i.e., the maximum number of pairwise non-adjacent vertices). Denote by $g(G)$ the *girth* of the graph (i.e., the minimum length of a cycle). Intuitively, the smaller is $g(G)$ or even the larger is $\omega(G)$, the bigger is the chromatic number $\chi(G)$. In 1959 Erdős broke this intuition by proving in [**27**] that for any k, l there exists a graph G such that $g(G) > k$ and $\chi(G) > l$.

Since the main problems for distance graphs and graphs of diameters are concerned with their chromatic numbers, one of the most intriguing questions here is whether it is possible to translate the results of Erdős into the language of geometric graphs. Once again, finding a geometric graph with large chromatic number and without large cliques (short cycles) would mean to get a graph that not only requires many colors in a vertex coloring, but also can be "drawn" on a small sphere (and vice versa).

First question of such type was raised by Erdős himself in 1976. He asked whether there exists a distance graph on the plane with chromatic number 4 and without triangles (see [**25**]). The question was natural, since, as we only know, $\chi(\mathbb{R}^2) \geq 4$ (cf. Section 1) and the corresponding distance graph is the well-known Moser spindle containing 4 triangles (see any of the books and surveys cited in Section 1). The answer on the question is positive (see [**115**]). Moreover, in 2000 a planar analog of Erdős' general result was given by O'Donnell (see [**21**]–[**23**]): *For every* k, *there exists a distance graph* G *in* $\mathbb{R}^2$ *with* $\chi(G) = 4$ *and* $g(G) > k$.

In small dimensions, some more results are due to Rubanov (see [**100**]).

In growing dimension, the chromatic number grows exponentially (see Section 1). Thus, the following question is well-motivated: *For which* k *there does exist a constant* $c > 1$, *a function* $\delta = o(1)$ *as* $d \to \infty$, *and a sequence* G_d *of distance graphs in* $\mathbb{R}^d$ *such that* $\chi(G_d) \geq (c+\delta)^d$ *and* $\omega(G_d) \leq k$? This question was proposed by Raigorodskii in [**89**], and we will describe a rather long history of its solution in Section 5. In Section 6, we will discuss a closely related question of finding the chromatic number of a sphere. In Section 7, we will exhibit similar (but even more complicated) results for graphs of diameters. In the next section, we will dwell on counting cliques in geometric graphs.

4. Counting cliques in distance graphs and graphs of diameters

Denote by $D_d(l, n)$ the maximum number of l-cliques in a graph of diameters with n vertices in $\mathbb{R}^d$. Let $U_d(l, n)$ be a similar quantity in the case of unit distance graphs. First, let us discuss some results for $U_d(l, n)$.

The study of the quantity $U_d(l,n)$ was started in 1946 with the paper [**26**] by Erdős. When $l=2$ the value of $U_d(l,n)=U_d(2,n)$ is just the largest number of edges in a distance graph on n vertices in $\mathbb{R}^d$. For $d=2$, the "records" are

$$[\mathbf{26}]\ c_1 n e^{\frac{c_2 \log n}{\log\log n}} \le U_2(2,n) \le c_3 n^{\frac{4}{3}}\ [\mathbf{1}],\ [\mathbf{5}],\ [\mathbf{75}],\ [\mathbf{108}],\ [\mathbf{112}],\ c_1>0,\ c_2>0,\ c_3>0.$$

The conjecture is that the lower bound is tight, but it is apparently very far from solution (see [**11**] for more details).

For $d=3$, the value of $U_3(2,n)$ is still much smaller than the number of edges in a complete graph on n vertices:

$$[\mathbf{26}]\ c_1 n^{\frac{4}{3}} \log\log n \le U_3(2,n) \le c_2 n^{\frac{3}{2}} 2^{c_3\alpha^2(n)}\ [\mathbf{17}],\ c_1>0,\ c_2>0,\ c_3>0,$$

where $\alpha(n)$ is the inverse Ackermann function.

For $d \ge 4$, any bipartite graph can be obviously realized as a complete distance graph. For $d \ge 6$, the same is true for any tripartite graph, etc. This yields the bound

$$U_d(2,n) \ge \frac{1}{2}\left(1-\frac{1}{[d/2]}\right)n^2 + o\left(n^2\right). \tag{2}$$

Now, the following asymptotics is known (see [**28**], [**29**], [**32**], [**106**]):

$$U_d(2,n) = \frac{1}{2}\left(1-\frac{1}{[d/2]}\right)n^2 + t(n),$$

where

$$t(n) = \begin{cases} n - O(1), & d \ge 4, \quad d=2k, \\ \Theta\left(n^{\frac{4}{3}}\right), & d \ge 5, \quad d=2k+1. \end{cases}$$

For $d=4$ and $n \ge 10$, the above asymptotics was changed to an exact formula by Brass and van Wamelen (see [**113**]):

$$U_4(2,n) = \left[\frac{n^2}{4}\right] + t'(n),$$

where

$$t'(n) = \begin{cases} n, & n \text{ is divisible by 8 or 10}, \\ n-1, & \text{otherwise}. \end{cases}$$

Finally, for every even $d \ge 6$, Swanepoel (see [**110**]) proved that there exists an n_0 such that for $n \ge n_0$,

$$U_d(2,n) = \frac{p-1}{2p}n^2 + t''(n) - O(1),$$

where $p = \frac{d}{2}$,

$$t''(n) = \begin{cases} n-r, & 0 \le r \le p-1, \\ n-p, & p \le r \le 3p-1, \\ n-2d+r, & 3p \le r \le 4p-1, \end{cases}$$

r is the remainder when dividing n by $4p=2d$.

For $l \ge 3$, there exist many partial solutions to the problem of finding or at least bounding $U_d(l,n)$. For almost exhaustive surveys of the corresponding results, see [**11**], [**66**].

Now, proceed to the quantity $D_d(l,n)$. Here the most studied cases are again those when $l=2$. Also, the cases $l=d$ and $l=d+1$ are of a special interest. Of course the intermediate cases are important as well, but we will not dwell on them in this paper.

The value of $D_2(2,n)$ was shown to be equal to n by Hopf and Pannwitz (see [**47**]) right after Borsuk's conjecture had been raised. From this equality, Borsuk's conjecture for finite sets in $\mathbb{R}^2$ follows immediately. In the three-dimensional case, a similar equality holds: $D_3(2,n) = 2n-2$. It was proved in 1957 independently by Grünbaum (see [**40**]), Heppes (see [**43**]), and Straszewicz (see [**109**]). Again, this equality simply yields the finite version of Borsuk's conjecture in $\mathbb{R}^3$. Another proof of this version was given a year before by Heppes and Révész (see [**44**]).

For $d \geq 4$, we know the asymptotic behaviour of $D_d(2,n)$. In fact it is the same as it was for distance graphs: $D_d(2,n) \sim U_d(2,n)$ as d is fixed and $n \to \infty$ (see [**26**], [**28**]). Swanepoel improved this result (see [**110**]). To describe his assertion let

$$t_p(n) = \frac{p-1}{2p}n^2 - O(1),$$

where $t_p(n)$ is just the maximum number of edges in a p-partite graph on n vertices (cf. the argument before equation (2)). Then, for large enough n (depending on d),

$$D_4(2,n) = \begin{cases} t_2(n) + \lceil n/2 \rceil + 1, & n \not\equiv 3 \pmod 4, \\ t_2(n) + \lceil n/2 \rceil, & n \equiv 3 \pmod 4, \end{cases}$$

$$D_5(2,n) = t_2(n) + n, \ D_d(2,n) = t_p(n) + p \text{ for even } d \geq 6, \text{ where } p = d/2,$$

$$D_d(2,n) = t_p(n) + \lceil n/p \rceil + p - 1 \text{ for odd } d \geq 7, \text{ where } p = [d/2].$$

Recently Kupavskii considered more carefully the case of $d = 4$ and proved that Swanepoel's result on $D_4(2,n)$ is true starting from $n = 52$ (see [**59**]). Moreover, he showed that for sufficiently large n,

$$D_4(3,n) = \begin{cases} (n-1)^2/4 + n, & n \equiv 1 \pmod 4, \\ (n-1)^2/4 + n - 1, & n \equiv 3 \pmod 4, \\ n(n-2)/4 + n, & n \equiv 0 \pmod 2. \end{cases}$$

Finally, he got the equality

$$D_4(4,n) = n,$$

which was also recently obtained with the help of an alternative method by Kupavskii, Bulankina, and Polyanskii (see [**12**], [**13**]).

The last equality is a special case of Schur's conjecture stating that $D_d(d,n) = n$ (see [**103**]). The conjecture is obviously true for $d = 1$ and follows from the above-mentioned result of Hopf and Pannwitz for $d = 2$. For $d = 3$, the conjecture is proved in the paper [**103**], where it is formulated. In the same paper, it is shown that $D_d(d+1,n) = 1$. Thus, the last result of Kupavskii makes a breakthrough in the study of Schur's conjecture. At the same time, in view of this result, we have an almost exhaustive list of values of the quantities $D_d(l,n)$ for $d \leq 4$ and arbitrary l, n.

Even more recently, Kupavskii and Polyanskii announced a proof of Schur's conjecture for every d.

5. Distance graphs with exponential chromatic numbers and without cliques or cycles

As we have already mentioned in Section 3, the following question was proposed by Raigorodskii in 2007 in [**89**]: *For which k there does exist a constant $c > 1$, a function $\delta = o(1)$ as $d \to \infty$, and a sequence G_d of distance graphs in $\mathbb{R}^d$ such that $\chi(G_d) \geq (c+\delta)^d$ and $\omega(G_d) < k$?* In [**89**], it was only shown that one can take k

as a function of d, which is asymptotically $o(d)$ as $d \to \infty$. Since a priori we only know that $k \le d+1$, this result was already non-trivial. However, it was clear that it could be substantially improved.

In 2008 Raigorodskii and Rubanov (see **[95]**, **[96]**) showed that any $k \ge 6$ would suffice. In view of this result, the definition of the following quantity is well-motivated:

$$\zeta_{\text{clique}}(k) = \sup\{\zeta : \exists\, \delta(d),\ \delta(d) = o(1),\ \forall\, d,\ \exists\, G \text{ in } \mathbb{R}^d,\ \omega(G) < k,\ \chi(G) \ge (\zeta + \delta(d))^d\}.$$

Roughly speaking, $\zeta_{\text{clique}}(k)$ is the maximum c in a bound $\chi(G_d) \ge (c+\delta)^d$ given in the original Raigorodskii question. In terms of this quantity, the results of **[95]**, **[96]** mean that for every $k \ge 6$, we have $\zeta_{\text{clique}}(k) > 1$.

The bounds in **[95]**, **[96]** are explicit. Moreover, they tend to $1.239\ldots$ as $k \to \infty$, which is exactly the constant in the best known lower bound for $\chi(\mathbb{R}^d)$. However, nothing was done there for $k \in \{3, 4, 5\}$, and thus the question for such values of k has been remaining open.

In some sense, the complete solution of the problem was made by Demekhin, Raigorodskii, and Rubanov in **[20]**. On the one hand, they proved that $\zeta_{\text{clique}}(k) > 1$ for every $k \in \{3, 4, 5\}$. On the other hand, they improved upon some previous bounds on $\zeta_{\text{clique}}(k)$, $k \ge 6$.

However, there is an important subtlety. The point is that, in the majority of cases, distance graphs proving that $\zeta_{\text{clique}}(k) > 1$ are not *complete*. The main idea of how to find them is as follows: take a sequence $\{G_d\}$ of complete distance graphs in $\mathbb{R}^d$, for which one knows that $\chi(G_d) \ge (c+o(1))^d$, $c > 1$; consider random subgraphs H_d of each G_d according to an appropriate probability distribution (cf. **[3]**, **[6]**, **[33]**–**[35]**, **[50]**); prove that with high probability we still have $\chi(H_d) \ge (c' + o(1))^d$, $1 < c' < c$, and we also have $\omega(H_d) < k$. Therefore, it is natural to define a quantity $\zeta'_{\text{clique}}(k)$, where

$$\zeta'_{\text{clique}}(k) = \sup\{\zeta : \exists\, \delta(d),\ \delta(d) = o(1),\ \forall\, d,\ \exists\, G \text{ complete in } \mathbb{R}^d,$$
$$\omega(G) < k,\ \chi(G) \ge (\zeta + \delta(d))^d\}.$$

In fact in **[20]** the values of $\zeta'_{\text{clique}}(k)$ were also bounded from below. Moreover, for $k \in \{3, 4, 5\}$, the bound $\zeta_{\text{clique}}(k) \ge \zeta'_{\text{clique}}(k)$ was really used. At the same time, for large k, the best results were obtained using the probabilistic method, and the bounds for $\zeta'_{\text{clique}}(k)$ tended to $1.139\ldots$, not to $1.239\ldots$ as it had been for the bounds on the initial quantity $\zeta_{\text{clique}}(k)$.

Thus, it was still of interest to understand whether $\zeta'_{\text{clique}}(k)$ can be bounded from below by something tending to a limit strictly greater than $1.139\ldots$ or even equal to $1.239\ldots$ It was also of interest to find as good bounds for $\zeta_{\text{clique}}(k)$ as possible by combining both approaches — the one with complete distance graphs and the probabilistic one.

Now, the best results in this direction are given in the papers **[60]**, **[64]**, **[65]**. However, we still only know that $\zeta'_{\text{clique}}(k) \ge c(k)$, where $c(k) \to \frac{2}{\sqrt{3}} = 1.154\ldots$, which is of course much smaller than $1.239\ldots$ (see **[60]**). An up-to-date table of bounds for both $\zeta'_{\text{clique}}(k)$ and $\zeta_{\text{clique}}(k)$ can be found in **[60]**, too. In that paper, the probabilistic approach is substantially refined. In particular, the Lovász Local Lemma is carefully used (see **[3]**, **[6]**).

Finally, in **[98]** the following result is proved: *If $k = k(d)$ is an arbitrary function such that $k(d) \to \infty$ as $d \to \infty$, then $\zeta_{\text{clique}}(k) = 1.239\ldots$*

As we see, the success is great in studying distance graphs without cliques and with large chromatic numbers. Thus, it is natural to try to study cycles instead of cliques. Define

$$\zeta_{\text{girth}}(k) = \sup\{\zeta : \exists\, \delta(d),\ \delta(d) = o(1),\ \forall\, d,\ \exists\, G \text{ in } \mathbb{R}^d,\ g(G) > k,\ \chi(G) \geq (\zeta+\delta(d))^d\},$$

$$\zeta'_{\text{girth}}(k) = \sup\{\zeta : \ \exists\ \delta(d),\ \delta(d) = o(1),\ \forall\ d,\ \exists\ G \text{ complete in } \mathbb{R}^d,\ g(G) > k,\ \chi(G) \geq (\zeta + \delta(d))^d\}.$$

Unfortunately, absolutely nothing is known about the second quantity. It is probably the most intriguing problem here. A partial solution to this problem can be given, if we agree to replace $\zeta'_{\text{girth}}(k)$ by $\zeta'_{\text{odd girth}}(k)$, in whose definition we just substitute the condition $g(G) > k$ by the condition that in G there are no cycles of *odd* length $\leq k$. In [**20**] the bounds $\zeta'_{\text{odd girth}}(k) > 1$ are proved for each k.

For $\zeta_{\text{girth}}(k)$, the probabilistic method of [**20**] and even [**60**] cannot be applied. However, Kupavskii succeeded in breaking this obstacle. In [**61**] he cleverly mixed the probabilistic tools with some results in extremal combinatorics by Frankl and Rödl (see [**36**]) and showed eventually that $\zeta_{\text{girth}}(k) > 1$ for every k. Apparently, the bounds in [**61**] are very far from being optimal, but qualitatively they give a solution to the problem in the case of non-complete distance graphs.

Some generalizations of the above-described problems have been studied. Namely, distance graphs determined by *several* distances were considered. In other words, to define such graphs one has to fix a set $\mathcal{A}$ of positive real numbers and take $V \subseteq \mathbb{R}^d$,

$$E \subseteq \{\{\mathbf{x}, \mathbf{y}\} : \ |\mathbf{x} - \mathbf{y}| \in \mathcal{A}\}.$$

The corresponding results can be found in [**20**], [**37**], [**54**], [**58**], [**71**], [**90**], [**94**], [**97**].

6. The chromatic numbers of spheres

Let S_r^{d-1} be a sphere of radius r in $\mathbb{R}^d$. Define its *chromatic number* as usual:

$$\chi(S_r^{d-1}) = \min\left\{\chi : \ S_r^{d-1} = V_1 \sqcup \ldots \sqcup V_\chi,\ \forall\, i\ \forall\, \mathbf{x}, \mathbf{y} \in V_i\ |\mathbf{x} - \mathbf{y}| \neq 1\right\}.$$

Note that varying r in the range $\left[\frac{1}{2}, \infty\right)$ is the same as fixing an r and varying the distance that is forbidden among monochromatic points. For us, it is more convenient to vary the radius and to fix the forbidden distance 1. Thus, we deal here with unit distance graphs on different spheres.

As we know from Section 3, the question of bounding the chromatic numbers of spheres is closely related to the question discussed in the previous section. However, it is even more classical, and there is an intriguing story associated with it.

Of course $\chi(S_{1/2}^{d-1}) = 2$ (cf. Section 2). In 1981 Erdős conjectured (see [**31**]) that $\chi(S_r^{d-1}) \to \infty$ as $d \to \infty$ for all $r > \frac{1}{2}$. This conjecture was proved in 1983 by Lovász who showed in [**68**] that $\chi(S_r^{d-1}) \geq d$ (see [**70**] as well). In the same paper, Lovász claimed that if $r < \sqrt{\frac{d}{2d+2}}$, then $\chi(S_r^{d-1}) \leq d+1$, and he conjectured that $\chi(S_r^{d-1})$ grows exponentially, provided $r \geq \sqrt{\frac{d}{2d+2}}$.

The claim of Lovász was completely false! We will explain it a bit later. Surprisingly, a lot of papers cite this claim *literally* (see, e.g., [**111**]) and even use it to prove some "new" results (see, e.g., [**48**]), which are consequently wrong as well.

The assertion of Lovász is as follows: *if* $r < \sqrt{\frac{d}{2d+2}} \sim \frac{1}{\sqrt{2}}$, *i.e., the length of any side of a regular d-simplex inscribed into* S_r^{d-1} *is smaller than 1, then* $\chi(S_r^{d-1}) \leq d+1$. The idea of Lovász (and his mistake) are, in some sense, natural. Take the sphere S_r^{d-1} with $r < \sqrt{\frac{d}{2d+2}}$. Inscribe a regular d-simplex into this sphere. Let $\mathbf{x}_1, \ldots, \mathbf{x}_{d+1}$ be its vertices. Let O be the center of S_r^{d-1}. Consider $d+1$ equal multidimensional angles generated by O and some d points from the set $\{\mathbf{x}_1, \ldots, \mathbf{x}_{d+1}\}$. Color the intersection of the sphere with the ith angle in the ith color. Since the side length of the simplex, *which is the diameter of any monochromatic part of the sphere*, is less than 1, we have a coloring avoiding distance 1. The mistake is just in italic. Of course, the diameters of monochromatic parts are *not* attained on the sides of the simplex. For example, if $d = 2k$, then the maximum distance is between appropriately normed $\mathbf{x}_1 + \ldots + \mathbf{x}_k$ and $\mathbf{x}_{k+1} + \ldots + \mathbf{x}_{2k}$.

The mistake of Lovász was corrected by Raigorodskii in [**87**], [**88**]. It turns out that $\chi(S_r^{d-1})$ grows exponentially, provided not only $r \geq \sqrt{\frac{d}{2d+2}}$, but $r > \frac{1}{2}$. Below, we give the exact statements.

Theorem 6.1. *For any* $r \in \left(\frac{1}{2}, \frac{1}{\sqrt{2}}\right)$, *there exists a function* $\delta_r(d) = o(1)$ *as* $d \to \infty$ *such that for every* $d \in \mathbb{N}$, *we have*

$$\chi(S_r^{d-1}) \geq \left(2\left(\frac{1}{8r^2}\right)^{\frac{1}{8r^2}}\left(1-\frac{1}{8r^2}\right)^{1-\frac{1}{8r^2}} + \delta_r(d)\right)^d.$$

One can readily see that for any $r \in \left(\frac{1}{2}, \frac{1}{\sqrt{2}}\right)$, the value

$$\gamma(r) = 2\left(\frac{1}{8r^2}\right)^{\frac{1}{8r^2}}\left(1-\frac{1}{8r^2}\right)^{1-\frac{1}{8r^2}}$$

is greater than 1. Thus, Theorem 6.1 says that the quantity $\chi(S_r^{d-1})$ grows essentially like an exponent. Moreover, if $r' \geq \frac{1}{\sqrt{2}}$, then for any $r < \frac{1}{\sqrt{2}}$, we have $S_r^{d-1} \subset S_{r'}^d$ and therefore $\chi(S_{r'}^{d-1})$ also has exponential growth. Consequently, for any r, there exists a d_0 such that for all $d \geq d_0$, we have $\chi(S_r^{d-1}) \gg d+1$ contradicting the assertion of Lovász.

The gap between exponents from Theorem 6.1 and the linear function from Lovász' assertion is quite large. Thus, one may expect that superlinear lower bounds for $\chi(S_r^{d-1})$ hold not only for a constant $r > \frac{1}{2}$, but also for some sequences $r_d \to \frac{1}{2}$ as $d \to \infty$. The most general statement of this kind is given in Theorem 6.2.

Theorem 6.2. *Let* $\mathbb{P}$ *be the set of prime numbers. Let* $f(x)$ *be such a function that for any* $x \in \mathbb{R}$, $x \geq 0$,

$$f(x) = \min\{p \in \mathbb{P} : \ p > x\} - x.$$

Let

$$m(x) = \max\{m < x : \ m \equiv 0 \pmod 4\}.$$

Consider a sequence $\{r_d\}_{d=1}^{\infty}$, *where* $r_d > \frac{1}{2}$ *for each* $d \in \mathbb{N}$. *Set*

$$p(d) = \frac{m(d)}{8r_d^2} + f\left(\frac{m(d)}{8r_d^2}\right).$$

If

$$\frac{m(d)}{4} < p(d) \le \frac{m(d)}{2}, \ d \in \mathbb{N},$$

then,

$$\chi(S_{r_d}^{d-1}) \ge \frac{\binom{m(d)}{m(d)/2}}{\binom{m(d)}{p(d)}}.$$

Comparing a rather implicit result of Theorem 6.2 with the erroneous Lovász claim Raigorodskii gets the following theorem (see [**88**]).

Theorem 6.3. *Consider a sequence* $\{r_d\}_{d=1}^{\infty}$, *where* $r_d > \frac{1}{2}$ *for each* $d \in \mathbb{N}$. *Let* $\kappa < 2$, *and let* $p(d)$ *be the same as in Theorem 6.2. If*

$$\frac{m(d)}{4} < p(d) < \frac{m(d)}{2} - \sqrt{\frac{m(d)\log(m(d))}{\kappa}}, \ d \in \mathbb{N},$$

then,

$$\chi(S_{r_d}^{d-1}) > d+1, \ \forall\, d \ge d_0.$$

The bound given by Theorem 6.3 depends on estimates for the function $f(x)$. Determining the exact asymptotic behaviour of $f(x)$ is a very hard problem of analytical number theory (see [**30**]). Now, the best upper bound is $f(x) = O\left(x^{0.525-\varepsilon}\right)$ with a very small $\varepsilon > 0$ (see [**4**]). However, it is conjectured that $f(x) = O\left(\log^2 x\right)$ (see [**19**]). The tightest lower bound is given in [**99**] and [**101**], but it is sublogarithmic and conjectured to be far from the truth. Using this information, Raigorodskii derives in [**88**]

Theorem 6.4. *Assume that* $c_0 > 0$ *is such that* $f(x) \le c_0 x^{0.525}$ *for every* x. *Then, there exists a constant* $c_0' > 0$ *such that for any sequence of radii* r_d *satisfying the inequality*

$$r_d \ge \frac{1}{2} + \frac{c_0'}{d^{0.475}},$$

we have the bound

$$\chi(S_{r_d}^{d-1}) > d+1, \ \forall\, d \ge d_0.$$

Theorem 6.5. *Assume that* $c_1 > 0$ *is such that* $f(x) \le c_1 \log^2 x$ *for every* x. *Then, there exists a constant* $c_1' > 0$ *such that for any sequence of radii* r_d *satisfying the inequality*

$$r_d \ge \frac{1}{2} + c_1'\sqrt{\frac{\log d}{d}},$$

we have the bound

$$\chi(S_{r_d}^{d-1}) > d+1, \ \forall\, d \ge d_0.$$

So $r_d > \frac{1}{2}$ may be quite close to the value $\frac{1}{2}$, and, nevertheless, the chromatic numbers will exceed the Lovász "upper bound". Finally, it is of interest for which sequences of r_d, we do really have the bound $\chi(S_{r_d}^{d-1}) \le d+1$. In [**88**], we find

Theorem 6.6. *There exists a constant $c_2 > 0$ such that for any sequence of radii r_d satisfying the inequality*

$$r_d \leq \frac{1}{2} + \frac{c_2}{d},$$

we have the bound

$$\chi(S_{r_d}^{d-1}) \leq d+1, \ \forall \ d \geq d_0.$$

This theorem is already proved using the idea of inscribing a regular simplex into a given sphere. However, here the radius is not just $< \sqrt{\frac{d}{2d+2}} \sim \frac{1}{\sqrt{2}}$, but it is like $\frac{1}{2} + O\left(\frac{1}{d}\right)$!

Now, the only remaining problem, which seems to be hard enough, is to understand what happens when r_d lies between, say, $\frac{1}{2} + O\left(\frac{1}{d}\right)$ and $\frac{1}{2} + O\left(\sqrt{\frac{\log d}{d}}\right)$.

For small-dimensional cases of the sphere coloring problem see, e.g., [**104**], [**105**].

7. Counterexamples to Borsuk's conjecture on spheres of small radii

All the counterexamples to Borsuk's conjecture listed in Section 2 do contain many cliques of any size (which is not surprising) and thus lie on spheres whose radii are asymptotically close to $\frac{1}{\sqrt{2}}$ (cf. Section 3). Therefore, a natural question is whether it is possible to find counterexamples on spheres of other radii $> \frac{1}{2}$. This question was proposed and partially solved by Raigorodskii in [**93**], where counterexamples for $r > \sqrt{\frac{3}{8}}$ were constructed. A complete solution was given by Kupavskii and Raigorodskii in [**63**]. Let us describe those results.

Theorem 7.1. *For any $r > \frac{1}{2}$, there exists a $d_0 = d_0(r)$ such that for every $d \geq d_0$, one can find a finite set $V \subset S_r^{d-1}$ which has diameter 1 and does not admit a partition into $d+1$ parts of smaller diameter, i.e., the chromatic number of its graph of diameters is greater than $d+1$.*

Theorem 7.1 means that there exist finite graphs of diameters without any prescribed cliques and cycles, but with large chromatic number (contradicting Borsuk's conjecture). How large can be bounds for the chromatic numbers? To formulate an even stronger result than the one of Theorem 7.1, introduce the quantity

$$f_r(d) = \max_{V \subset S_r^{d-1}, \ \mathrm{diam}\, V = 1} f(V).$$

In these terms, Theorem 7.1 says that for any $r > \frac{1}{2}$, there exists a $d_0 = d_0(r)$ such that for every $d \geq d_0$, $f_r(d) > d+1$. Moreover,

Theorem 7.2. *For any $r > \frac{1}{2}$, there exist numbers $k = k(r) \in \mathbb{N}$, $c = c(r) > 1$ and a function $\delta(d) = o(1)$ as $d \to \infty$ such that*

$$f_r(d) \geq (c + \delta(d))^{\sqrt[2k]{d}}.$$

Theorem 7.2 means that if r is fixed and exceeds $\frac{1}{2}$, then the order of magnitude of the value $f_r(d)$ is at least $e^{g(d)}$, where $g(d)$ is just a fixed positive power of d. So not only the quantity $f_r(d)$ is greater than $d+1$ starting from some d_0, but also it is *substantially* greater than the conjectured value. This fact allows one to use some optimization and to prove eventually the following theorem (see [**63**]).

Theorem 7.3. *Let $r = r(d) = \frac{1}{2} + \varphi(d)$, where $\varphi = o(1)$ and $\varphi(d) \geq c\frac{\log\log d}{\log d}$ for all d and a large enough $c > 0$. Then, there exists a d_0 such that for $d \geq d_0$, $f_{r(d)}(d) > d + 1$.*

In other words, Borsuk's conjecture can be disproved by constructing sets of diameter 1 that lie on spheres with radii tending to $\frac{1}{2}$ as $d \to \infty$. Here, in the case of $\varphi(d) = \Theta\left(\frac{\log\log d}{\log d}\right)$, there is already no room to spare in the bound $f_r(d) > d+1$. So using this approach, one only may discuss the value of a constant c in Theorem 7.3, which is of course not significant.

Now, a natural question arises: perhaps $f_r(d) \leq d + 1$, provided $r = r(d) = O\left(\frac{\log\log d}{\log d}\right)$? Unfortunately, the best result given in [**63**] is

Theorem 7.4. *Let $r = r(d) = \frac{1}{2} + \varphi(d)$, where $\varphi = O(1/d)$. Then, $f_r(d) \leq d + 1$.*

Thus, a gap remains between the two functions $\frac{1}{d}$ and $\frac{\log\log d}{\log d}$. To reduce this gap would be of a great interest.

References

[1] János Pach and Pankaj K. Agarwal, *Combinatorial geometry*, Wiley-Interscience Series in Discrete Mathematics and Optimization, John Wiley & Sons, Inc., New York, 1995. A Wiley-Interscience Publication. MR1354145 (96j:52001)

[2] Noga Alon, Andrey Kupavskii, *Two notions of unit distance graphs*, J. Comb. Theory, Ser. A **125** (2014), 1–17 [Contents]

[3] Noga Alon and Joel H. Spencer, *The probabilistic method*, 2nd ed., Wiley-Interscience Series in Discrete Mathematics and Optimization, Wiley-Interscience [John Wiley & Sons], New York, 2000. With an appendix on the life and work of Paul Erdős. MR1885388 (2003f:60003)

[4] R. C. Baker, G. Harman, and J. Pintz, *The difference between consecutive primes. II*, Proc. London Math. Soc. (3) **83** (2001), no. 3, 532–562, DOI 10.1112/plms/83.3.532. MR1851081 (2002f:11125)

[5] József Beck and Joel Spencer, *Unit distances*, J. Combin. Theory Ser. A **37** (1984), no. 3, 231–238, DOI 10.1016/0097-3165(84)90047-5. MR769215 (86a:52015)

[6] Béla Bollobás, *Random graphs*, 2nd ed., Cambridge Studies in Advanced Mathematics, vol. 73, Cambridge University Press, Cambridge, 2001. MR1864966 (2002j:05132)

[7] Vladimir Boltyanski, Horst Martini, and Petru S. Soltan, *Excursions into combinatorial geometry*, Universitext, Springer-Verlag, Berlin, 1997. MR1439963 (98b:52001)

[8] Andriy Bondarenko, *On Borsuk's conjecture for two-distance sets*, Discrete and Comput. Geom. **3** (2014), no. 51, 509–515.

[9] K. Borsuk, *Drei Sätze über die n-dimensionale euklidische Sphäre*, Fundamenta Math., **20** (1933), 177–190.

[10] J. Bourgain and J. Lindenstrauss, *On covering a set in* $\mathbf{R}^N$ *by balls of the same diameter*, Geometric aspects of functional analysis (1989–90), Lecture Notes in Math., vol. 1469, Springer, Berlin, 1991, pp. 138–144, DOI 10.1007/BFb0089220. MR1122618 (92g:52018)

[11] Peter Brass, William Moser, and János Pach, *Research problems in discrete geometry*, Springer, New York, 2005. MR2163782 (2006i:52001)

[12] V.V. Bulankina, A.B. Kupavskii, A.A. Polyanskii, *On Schur's conjecture in* $\mathbb{R}^4$, to appear in Math. Notes.

[13] V.V. Bulankina, A.B. Kupavskii, A.A. Polyanskii, *A note on Schur's conjecture in* $\mathbb{R}^4$, Dokl. Math. **89** (2014), no. 1, 88–91.

[14] Kent Cantwell, *Finite Euclidean Ramsey theory*, J. Combin. Theory Ser. A **73** (1996), no. 2, 273–285. MR1370133 (97g:05164)

[15] Kiran B. Chilakamarri, *The unit-distance graph problem: a brief survey and some new results*, Bull. Inst. Combin. Appl. **8** (1993), 39–60. MR1217358 (94e:05003)
[16] Josef Cibulka, *On the chromatic number of real and rational spaces*, Geombinatorics **18** (2008), no. 2, 53–65. MR2456223
[17] Kenneth L. Clarkson, Herbert Edelsbrunner, Leonidas J. Guibas, Micha Sharir, and Emo Welzl, *Combinatorial complexity bounds for arrangements of curves and spheres*, Discrete Comput. Geom. **5** (1990), no. 2, 99–160, DOI 10.1007/BF02187783. MR1032370 (91f:52021)
[18] D. Coulson, *A 15-colouring of 3-space omitting distance one*, Discrete Math. **256** (2002), no. 1-2, 83–90, DOI 10.1016/S0012-365X(01)00183-2. MR1927057 (2003c:05073)
[19] H. Cramér, *On the order of magnitude of the difference between consecutive prime numbers*, Acta Arithmetica, 2 (1936), 23–46.
[20] E. E. Demëkhin, A. M. Raĭgorodskiĭ, and O. I. Rubanov, *Distance graphs that have a large chromatic number and contain no cliques or cycles of a given size* (Russian, with Russian summary), Mat. Sb. **204** (2013), no. 4, 49–78, DOI 10.1070/SM2013v204n04ABEH004310; English transl., Sb. Math. **204** (2013), no. 3-4, 508–538. MR3097579
[21] Paul O'Donnell, *Arbitrary girth, 4-chromatic unit distance graphs in the plane. I. Graph description*, Geombinatorics **9** (2000), no. 3, 145–152. MR1746081
[22] Paul O'Donnell, *Arbitrary girth, 4-chromatic unit distance graphs in the plane. II. Graph embedding*, Geombinatorics **9** (2000), no. 4, 180–193. MR1763978
[23] Robert Hochberg and Paul O'Donnell, *Some 4-chromatic unit-distance graphs without small cycles*, Geombinatorics **5** (1996), no. 4, 137–141. MR1380143
[24] H. G. Eggleston, *Covering a three-dimensional set with sets of smaller diameter*, J. London Math. Soc. **30** (1955), 11–24. MR0067473 (16,734b)
[25] P. Erdős, *Unsolved Problems*, Congress Numerantium XV — Proceedings of the 5th British Comb. Conf. 1975, (1976), 681.
[26] P. Erdös, *On sets of distances of n points*, Amer. Math. Monthly **53** (1946), 248–250. MR0015796 (7,471c)
[27] P. Erdős, *Graph theory and probability*, Canad. J. Math. **11** (1959), 34–38. MR0102081 (21 #876)
[28] P. Erdős, *On sets of distances of n points in Euclidean space* (English, with Russian summary), Magyar Tud. Akad. Mat. Kutató Int. Közl. **5** (1960), 165–169. MR0141007 (25 #4420)
[29] P. Erdős, *On some applications of graph theory to geometry*, Canad. J. Math. **19** (1967), 968–971. MR0219438 (36 #2520)
[30] Paul Erdős, *Some unsolved problems*, Magyar Tud. Akad. Mat. Kutató Int. Közl. **6** (1961), 221–254. MR0177846 (31 #2106)
[31] P. Erdős, R.L. Graham, *Problem proposed at the 6th Hungarian combinatorial conference*, Eger, July 1981.
[32] P. Erdős and J. Pach, *Variations on the theme of repeated distances*, Combinatorica **10** (1990), no. 3, 261–269, DOI 10.1007/BF02122780. MR1092543 (92b:52037)
[33] P. Erdős and A. Rényi, *On random graphs. I*, Publ. Math. Debrecen **6** (1959), 290–297. MR0120167 (22 #10924)
[34] P. Erdős and A. Rényi, *On the evolution of random graphs* (English, with Russian summary), Magyar Tud. Akad. Mat. Kutató Int. Közl. **5** (1960), 17–61. MR0125031 (23 #A2338)
[35] P. Erdős and A. Rényi, *On the evolution of random graphs* (English, with French summary), Bull. Inst. Internat. Statist. **38** (1961), 343–347. MR0148055 (26 #5564)
[36] Peter Frankl and Vojtěch Rödl, *Forbidden intersections*, Trans. Amer. Math. Soc. **300** (1987), no. 1, 259–286, DOI 10.2307/2000598. MR871675 (88m:05003)
[37] E. S. Gorskaya, I. M. Mitricheva, V. Yu. Protasov, and A. M. Raĭgorodskiĭ, *Estimation of the chromatic numbers of Euclidean space by convex minimization methods* (Russian, with Russian summary), Mat. Sb. **200** (2009), no. 6, 3–22, DOI 10.1070/SM2009v200n06ABEH004019; English transl., Sb. Math. **200** (2009), no. 5-6, 783–801. MR2553072 (2010k:05089)
[38] J. Grey, B. Weissbach, *Ein weiteres Gegenbeispiel zur Borsukschen Vermutung*, Univ. Magdeburg, Fakultät für Mathematik, 1997, Preprint 25.
[39] B. Grünbaum, *A simple proof of Borsuk's conjecture in three dimensions*, Proc. Cambridge Philos. Soc. **53** (1957), 776–778. MR0090072 (19,763d)

[40] B. Gruenbaum, *A proof of Vazonyi's conjecture*, Bull. Res. Council Israel. Sect. A. **6** (1956), 77–78. MR0087115 (19,304d)

[41] H. Hadwiger, *Ein Ueberdeckungssatz für den Euklidischen Raum* (German), Portugaliae Math. **4** (1944), 140–144. MR0011108 (6,120d)

[42] Aladár Heppes, *On the partitioning of three-dimensional point-sets into sets of smaller diameter* (Hungarian), Magyar Tud. Akad. Mat. Fiz. Oszt. Közl. **7** (1957), 413–416. MR0095450 (20 #1952)

[43] A. Heppes, *Beweis einer Vermutung von A. Vázsonyi* (German, with Russian summary), Acta Math. Acad. Sci. Hungar. **7** (1956), 463–466. MR0087116 (19,304e)

[44] A. Heppes and P. Révész, *Zum Borsukschen Zerteilungsproblem* (German, with Russian summary), Acta Math. Acad. Sci. Hungar. **7** (1956), 159–162. MR0080897 (18,323d)

[45] Aicke Hinrichs, *Spherical codes and Borsuk's conjecture*, Discrete Math. **243** (2002), no. 1-3, 253–256, DOI 10.1016/S0012-365X(01)00202-3. MR1874745 (2002k:52020)

[46] Aicke Hinrichs and Christian Richter, *New sets with large Borsuk numbers*, Discrete Math. **270** (2003), no. 1-3, 137–147, DOI 10.1016/S0012-365X(02)00833-6. MR1997894 (2004e:52003)

[47] H. Hopf, E. Pannwitz, Aufgabe Nr. 167, Jahresbericht Deutsch. Math.-Verein., 43 (1934), p. 114.

[48] Boris Horvat, Jan Kratochvíl, and Tomaž Pisanski, *On the computational complexity of degenerate unit distance representations of graphs*, Combinatorial algorithms, Lecture Notes in Comput. Sci., vol. 6460, Springer, Heidelberg, 2011, pp. 274–285, DOI 10.1007/978-3-642-19222-7_28. MR2788834

[49] L. L. Ivanov, *An estimate for the chromatic number of the space* $\mathbb{R}^4$ (Russian), Uspekhi Mat. Nauk **61** (2006), no. 5(371), 181–182, DOI 10.1070/RM2006v061n05ABEH004361; English transl., Russian Math. Surveys **61** (2006), no. 5, 984–986. MR2328262 (2008g:05075)

[50] Svante Janson, Tomasz Łuczak, and Andrzej Rucinski, *Random graphs*, Wiley-Interscience Series in Discrete Mathematics and Optimization, Wiley-Interscience, New York, 2000. MR1782847 (2001k:05180)

[51] T. Jenrich, *A 64-dimensional two-distance counterexample to Borsuk's conjecture*, arXiv:1308.0206.

[52] H.W.E. Jung, *Über die kleinste Kugel, die eine räumliche Figur einschliesst*, J. reine und angew. Math., 123 (1901), 241–257.

[53] Jeff Kahn and Gil Kalai, *A counterexample to Borsuk's conjecture*, Bull. Amer. Math. Soc. (N.S.) **29** (1993), no. 1, 60–62, DOI 10.1090/S0273-0979-1993-00398-7. MR1193538 (94a:52007)

[54] Y. Katznelson, *Chromatic numbers of Cayley graphs on* $\mathbb{Z}$ *and recurrence*, Combinatorica **21** (2001), no. 2, 211–219, DOI 10.1007/s004930100019. Paul Erdős and his mathematics (Budapest, 1999). MR1832446 (2002h:05065)

[55] Victor Klee and Stan Wagon, *Old and new unsolved problems in plane geometry and number theory*, The Dolciani Mathematical Expositions, vol. 11, Mathematical Association of America, Washington, DC, 1991. MR1133201 (92k:00014)

[56] A. B. Kupavskiĭ, *Colorings of spheres embedded in* $\mathbb{R}^n$ (Russian, with Russian summary), Mat. Sb. **202** (2011), no. 6, 83–110, DOI 10.1070/SM2011v202n06ABEH004169; English transl., Sb. Math. **202** (2011), no. 5-6, 859–886. MR2849314 (2012e:52044)

[57] Andrey Kupavskiy, *On the chromatic number of* $\mathbb{R}^n$ *with an arbitrary norm*, Discrete Math. **311** (2011), no. 6, 437–440, DOI 10.1016/j.disc.2010.12.005. MR2799896 (2012d:52028)

[58] A. B. Kupavskiĭ, *On the chromatic number of* $\mathbb{R}^n$ *with a set of forbidden distances* (Russian), Dokl. Akad. Nauk **435** (2010), no. 6, 740–743, DOI 10.1134/S1064562410060323; English transl., Dokl. Math. **82** (2010), no. 3, 963–966. MR2814571

[59] A.B. Kupavskii, *Diameter graphs in* $\mathbb{R}^4$, arXiv:1306.3910.

[60] A.B. Kupavskii, *Explicit and probabilistic constructions of distance graphs with small clique numbers and large chromatic numbers*, Ivestiya: Mathematics **78** (2014), no. 1, 59–89.

[61] Andrey B. Kupavskii, *Distance graphs with large chromatic number and arbitrary girth*, Mosc. J. Comb. Number Theory **2** (2012), no. 2, 52–62. MR2988526

[62] A. B. Kupavskiĭ and A. M. Raĭgorodskiĭ, *On the chromatic number of* $\mathbb{R}^9$ (Russian, with English and Russian summaries), Fundam. Prikl. Mat. **14** (2008), no. 5, 139–154, DOI 10.1007/s10958-009-9708-4; English transl., J. Math. Sci. (N. Y.) **163** (2009), no. 6, 720–731. MR2533583 (2010i:05125)

[63] Andrey Kupavskii and Andrey Raigorodskii, *Counterexamples to Borsuk's conjecture on spheres of small radii*, Mosc. J. Comb. Number Theory **2** (2012), no. 4, 27–48. MR3065279

[64] A. B. Kupavskiĭ and A. M. Raĭgorodskiĭ, *On distance graphs with large chromatic numbers and small clique numbers* (Russian), Dokl. Akad. Nauk **444** (2012), no. 5, 483–487, DOI 10.1134/S1064562412030295; English transl., Dokl. Math. **85** (2012), no. 3, 394–398. MR3027131

[65] A.B. Kupavskiy, A.M. Raigorodskii, *On obstacles for realizing distance graphs with large chromatic numbers on spheres of small radii*, Mat. Sbornik, 204 (2013), N10, 47–90; English transl., Sbornik Math., 204 (2013), N10, 1435–1479.

[66] Andrey B. Kupavskii, Andrei M. Raigorodskii, and Maria V. Titova, *New bounds for the distance Ramsey number*, Discrete Math. **313** (2013), no. 22, 2566–2574, DOI 10.1016/j.disc.2013.07.022. MR3095431

[67] D. G. Larman and C. A. Rogers, *The realization of distances within sets in Euclidean space*, Mathematika **19** (1972), 1–24. MR0319055 (47 #7601)

[68] L. Lovász, *Self-dual polytopes and the chromatic number of distance graphs on the sphere*, Acta Sci. Math. (Szeged) **45** (1983), no. 1-4, 317–323. MR708798 (84i:05051)

[69] V. V. Makeev, *On affine images of a rhombo-dodecahedron circumscribed about a three-dimensional convex body* (Russian, with English and Russian summaries), Zap. Nauchn. Sem. S.-Peterburg. Otdel. Mat. Inst. Steklov. (POMI) **246** (1997), no. Geom. i Topol. 2, 191–195, 200; English transl., J. Math. Sci. (New York) **100** (2000), no. 3, 2307–2309. MR1631812 (99e:52005)

[70] Jiří Matoušek, *Using the Borsuk-Ulam theorem*, Universitext, Springer-Verlag, Berlin, 2003. Lectures on topological methods in combinatorics and geometry; Written in cooperation with Anders Björner and Günter M. Ziegler. MR1988723 (2004i:55001)

[71] N. G. Moshchevitin and A. M. Raĭgorodskiĭ, *Colorings of the space* $\mathbb{R}^n$ *with several forbidden distances* (Russian, with Russian summary), Mat. Zametki **81** (2007), no. 5, 733–743, DOI 10.1134/S0001434607050112; English transl., Math. Notes **81** (2007), no. 5-6, 656–664. MR2348824 (2008j:52030)

[72] V. F. Moskva and A. M. Raĭgorodskiĭ, *New lower bounds for the independence numbers of distance graphs with vertices in* $\{-1,0,1\}^n$ (Russian), Mat. Zametki **89** (2011), no. 2, 319–320, DOI 10.1134/S0001434611010366; English transl., Math. Notes **89** (2011), no. 1-2, 307–308. MR2840429

[73] Oren Nechushtan, *On the space chromatic number*, Discrete Math. **256** (2002), no. 1-2, 499–507, DOI 10.1016/S0012-365X(00)00406-4. MR1927575 (2003c:05084)

[74] A. Nilli, *On Borsuk's problem*, Jerusalem combinatorics '93, Contemp. Math., vol. 178, Amer. Math. Soc., Providence, RI, 1994, pp. 209–210, DOI 10.1090/conm/178/01901. MR1310585 (95j:52002)

[75] János Pach and Gábor Tardos, *Forbidden paths and cycles in ordered graphs and matrices*, Israel J. Math. **155** (2006), 359–380, DOI 10.1007/BF02773960. MR2269435 (2007g:05097)

[76] O. Pikhurko, *Borsuk's conjecture fails in dimensions 321 and 322*, arXiv: CO/0202112, 2002.

[77] Radoš Radoičić and Géza Tóth, *Note on the chromatic number of the space*, Discrete and computational geometry, Algorithms Combin., vol. 25, Springer, Berlin, 2003, pp. 695–698, DOI 10.1007/978-3-642-55566-4_32. MR2038498 (2004m:05266)

[78] A. M. Raĭgorodskiĭ, *The Borsuk problem and the chromatic numbers of some metric spaces* (Russian, with Russian summary), Uspekhi Mat. Nauk **56** (2001), no. 1(337), 107–146, DOI 10.1070/rm2001v056n01ABEH000358; English transl., Russian Math. Surveys **56** (2001), no. 1, 103–139. MR1845644 (2002m:54033)

[79] A.M. Raigorodskii, *The chromatic numbers*, Moscow Centre for Continuous Mathematical Education (MCCME), Moscow, Russia, 2003 (book in Russian).

[80] A.M. Raigorodskii, *Borsuk's problem*, Moscow Centre for Continuous Mathematical Education (MCCME), Moscow, Russia, 2006 (book in Russian).

[81] A. M. Raĭgorodskiĭ, *Around the Borsuk conjecture* (Russian, with Russian summary), Sovrem. Mat. Fundam. Napravl. **23** (2007), 147–164, DOI 10.1007/s10958-008-9196-y; English transl., J. Math. Sci. (N. Y.) **154** (2008), no. 4, 604–623. MR2342528 (2008j:52035)

[82] Andreii M. Raigorodskii, *Three lectures on the Borsuk partition problem*, Surveys in contemporary mathematics, London Math. Soc. Lecture Note Ser., vol. 347, Cambridge Univ. Press, Cambridge, 2008, pp. 202–247. MR2388494 (2009b:52020)

[83] Andreĭ M. Raĭgorodskiĭ, *The Borsuk partition problem: the seventieth anniversary*, Math. Intelligencer **26** (2004), no. 3, 4–12, DOI 10.1007/BF02986745. MR2088008 (2005e:52001)

[84] A.M. Raigorodskii, *Coloring Distance Graphs and Graphs of Diameters*, Thirty Essays on Geometric Graph Theory, J. Pach ed., Springer, 2013, 429–460.

[85] A. M. Raĭgorodskiĭ, *On the chromatic number of a space* (Russian), Uspekhi Mat. Nauk **55** (2000), no. 2(332), 147–148, DOI 10.1070/rm2000v055n02ABEH000281; English transl., Russian Math. Surveys **55** (2000), no. 2, 351–352. MR1781075 (2001i:05072)

[86] A.M. Raigorodskii, *The linear algebra method in combinatorics*, Moscow Centre for Continuous Mathematical Education (MCCME), Moscow, Russia, 2007 (book in Russian).

[87] A. M. Raĭgorodskiĭ, *On the chromatic numbers of spheres in Euclidean spaces* (Russian), Dokl. Akad. Nauk **432** (2010), no. 2, 174–177, DOI 10.1134/S1064562410030117; English transl., Dokl. Math. **81** (2010), no. 3, 379–382. MR2723588 (2011j:05122)

[88] A. M. Raigorodskii, *On the chromatic numbers of spheres in* $\mathbb{R}^n$, Combinatorica **32** (2012), no. 1, 111–123, DOI 10.1007/s00493-012-2709-9. MR2927634

[89] A. M. Raĭgorodskiĭ, *On distance graphs that have a large chromatic number but do not contain large simplices* (Russian), Uspekhi Mat. Nauk **62** (2007), no. 6(378), 187–188, DOI 10.1070/RM2007v062n06ABEH004493; English transl., Russian Math. Surveys **62** (2007), no. 6, 1224–1225. MR2382811

[90] A. M. Raĭgorodskiĭ, *The Erdős-Hadwiger problem and the chromatic numbers of finite geometric graphs* (Russian, with Russian summary), Mat. Sb. **196** (2005), no. 1, 123–156, DOI 10.1070/SM2005v196n01ABEH000874; English transl., Sb. Math. **196** (2005), no. 1-2, 115–146. MR2141326 (2006b:05054)

[91] A. M. Raĭgorodskiĭ, *On a bound in the Borsuk problem* (Russian), Uspekhi Mat. Nauk **54** (1999), no. 2(326), 185–186, DOI 10.1070/rm1999v054n02ABEH000146; English transl., Russian Math. Surveys **54** (1999), no. 2, 453–454. MR1711207 (2000i:54041)

[92] A. M. Raĭgorodskiĭ, *On dimensionality in the Borsuk problem* (Russian), Uspekhi Mat. Nauk **52** (1997), no. 6(318), 181–182, DOI 10.1070/RM1997v052n06ABEH002184; English transl., Russian Math. Surveys **52** (1997), no. 6, 1324–1325. MR1611334 (99c:52005)

[93] A. M. Raĭgorodskiĭ, *Counterexamples to Borsuk's conjecture on spheres of small radius* (Russian), Dokl. Akad. Nauk **434** (2010), no. 2, 161–163, DOI 10.1134/S1064562410050108; English transl., Dokl. Math. **82** (2010), no. 2, 719–721. MR2759271

[94] A. M. Raĭgorodskiĭ and M. I. Absalyamova, *A lower bound for the chromatic number of the space* $\mathbb{R}^n$ *with forbidden distances and the metric* l_1 (Russian), Chebyshevskiĭ Sb. **7** (2006), no. 4(20), 105–112. MR2455629 (2009j:05091)

[95] A. M. Raigorodskii and O. I. Rubanov, *On the clique and the chromatic numbers of high-dimensional distance graphs*, Number theory and applications, Hindustan Book Agency, New Delhi, 2009, pp. 149–155. MR2547497 (2011a:05117)

[96] A. M. Raĭgorodskiĭ and O. I. Rubanov, *Distance graphs with large chromatic number and without large cliques* (Russian, with Russian summary), Mat. Zametki **87** (2010), no. 3, 417–428, DOI 10.1134/S0001434610030119; English transl., Math. Notes **87** (2010), no. 3-4, 392–402. MR2761598 (2012g:05061)

[97] A. M. Raĭgorodskiĭ and I. M. Shitova, *On the chromatic numbers of real and rational spaces with real or rational forbidden distances* (Russian, with Russian summary), Mat. Sb. **199** (2008), no. 4, 107–142, DOI 10.1070/SM2008v199n04ABEH003934; English transl., Sb. Math. **199** (2008), no. 3-4, 579–612. MR2410142 (2009f:05101)

[98] A.M. Raigorodskii, A.E. Zvonarev, *On distance graphs with large chromatic numbers and small clique numbers*, Proceedings of Moscow Institute of Physics and Technology, 4 (2012), N1, 122–126 (in Russian).

[99] R. A. Rankin, *The difference between consecutive prime numbers. V*, Proc. Edinburgh Math. Soc. (2) **13** (1962/1963), 331–332. MR0160767 (28 #3978)

[100] O. I. Rubanov, *Chromatic numbers of three-dimensional distance graphs without tetrahedra* (Russian), Mat. Zametki **82** (2007), no. 5, 797–800, DOI 10.1134/S0001434607110156; English transl., Math. Notes **82** (2007), no. 5-6, 718–721. MR2399959 (2009c:05091)

[101] Arnold Schönhage, *Eine Bemerkung zur Konstruktion grosser Primzahllücken* (German), Arch. Math. (Basel) **14** (1963), 29–30. MR0146154 (26 #3680)

[102] Oded Schramm, *Illuminating sets of constant width*, Mathematika **35** (1988), no. 2, 180–189, DOI 10.1112/S0025579300015175. MR986627 (89m:52013)

[103] Zvi Schur, Micha A. Perles, Horst Martini, and Yaakov S. Kupitz, *On the number of maximal regular simplices determined by n points in* $\mathbb{R}^d$, Discrete and computational geometry, Algorithms Combin., vol. 25, Springer, Berlin, 2003, pp. 767–787, DOI 10.1007/978-3-642-55566-4_36. MR2038502 (2004m:52017)

[104] G. J. Simmons, *On a problem of Erdös concerning a 3-coloring of the unit sphere*, Discrete Math. **8** (1974), 81–84. MR0329952 (48 #8291)

[105] Gustavus J. Simmons, *The chromatic number of the sphere*, J. Austral. Math. Soc. Ser. A **21** (1976), no. 4, 473–480. MR0419280 (54 #7303)

[106] M. Simonovits, *A method for solving extremal problems in graph theory, stability problems*, Theory of Graphs (Proc. Colloq., Tihany, 1966), Academic Press, New York, 1968, pp. 279–319. MR0233735 (38 #2056)

[107] Alexander Soifer, *The mathematical coloring book*, Springer, New York, 2009. Mathematics of coloring and the colorful life of its creators; With forewords by Branko Grünbaum, Peter D. Johnson, Jr. and Cecil Rousseau. MR2458293 (2010a:05005)

[108] J. Spencer, E. Szemerédi, and W. Trotter Jr., *Unit distances in the Euclidean plane*, Graph theory and combinatorics (Cambridge, 1983), Academic Press, London, 1984, pp. 293–303. MR777185 (86m:52015)

[109] S. Straszewicz, *Sur un problème géométrique de P. Erdös* (French, with Russian summary), Bull. Acad. Polon. Sci. Cl. III. **5** (1957), 39–40, IV–V. MR0087117 (19,304f)

[110] Konrad J. Swanepoel, *Unit distances and diameters in Euclidean spaces*, Discrete Comput. Geom. **41** (2009), no. 1, 1–27, DOI 10.1007/s00454-008-9082-x. MR2470067 (2010f:52031)

[111] L. A. Székely, *Erdős on unit distances and the Szemerédi-Trotter theorems*, Paul Erdős and his mathematics, II (Budapest, 1999), Bolyai Soc. Math. Stud., vol. 11, János Bolyai Math. Soc., Budapest, 2002, pp. 649–666. MR1954746 (2004a:52028)

[112] László A. Székely, *Crossing numbers and hard Erdős problems in discrete geometry*, Combin. Probab. Comput. **6** (1997), no. 3, 353–358, DOI 10.1017/S0963548397002976. MR1464571 (98h:52030)

[113] Paul van Wamelen, *The maximum number of unit distances among n points in dimension four*, Beiträge Algebra Geom. **40** (1999), no. 2, 475–477. MR1720118 (2000i:52031)

[114] Bernulf Weißbach, *Sets with large Borsuk number*, Beiträge Algebra Geom. **41** (2000), no. 2, 417–423. MR1801432 (2001k:52022)

[115] Nicholas Wormald, *A 4-chromatic graph with a special plane drawing*, J. Austral. Math. Soc. Ser. A **28** (1979), no. 1, 1–8. MR541161 (80k:05060)

Moscow State University, Mechanics and Mathematics Faculty, Department of Mathematical Statistics and Random Processes, Leninskie gory, 119991, Moscow, Russia; Moscow Institute of Physics and Technology, Faculty of Innovations and High Technology, Department of Discrete Mathematics, Institutskiy per. 9, 141700, Moscow Region, Dolgoprudny, Russia; Yandex division of Theoretical and Applied Research, 119021, Leo Tolstoi st. 16, Moscow, Russia.

E-mail address: mraigor@yandex.ru

Contemporary Mathematics
Volume **625**, 2014
http://dx.doi.org/10.1090/conm/625/12494

New bounds for equiangular lines

Alexander Barg and Wei-Hsuan Yu

To Ilya Dumer, on the occasion of his 60th birthday

ABSTRACT. A set of lines in $\mathbb{R}^n$ is called equiangular if the angle between each pair of lines is the same. We address the question of determining the maximum size of equiangular line sets in $\mathbb{R}^n$, using semidefinite programming to improve the upper bounds on this quantity. Improvements are obtained in dimensions $24 \le n \le 136$. In particular, we show that the maximum number of equiangular lines in $\mathbb{R}^n$ is 276 for all $24 \le n \le 41$ and is 344 for $n = 43$. This provides a partial resolution of the conjecture set forth by Lemmens and Seidel (1973).

1. Introduction

A set of lines in a metric space is called equiangular if the angle between each pair of lines is the same. We are interested in upper bounds on the number of equiangular lines in $\mathbb{R}^n$. In other words, if we have a set of unit vectors $S = \{x_i\}_{i=1}^M$ and there is a constant $c > 0$ such that $|\langle x_i, x_j\rangle| = c$ for all $1 \le i \ne j \le M$, what is the maximum cardinality of S? Denote this quantity by $M(n)$. The problem of determining $M(n)$ looks elementary but a general answer has so far proved elusive: Until recently, the maximum number of equiangular lines in $\mathbb{R}^n$ was known only for 16 values of the dimension n. The history of this problem started with Hanntjes [**8**] who found $M(n)$ for $n = 2$ and 3 in 1948. Van Lint and Seidel [**11**] found the largest number of equiangular lines for $4 \le n \le 7$. In 1973, Lemmens and Seidel [**10**] used linear-algebraic methods to determine $M(n)$ for most values of n in the region $8 \le n \le 23$. Gerzon (see [**10**]) gave the following upper on $M(n)$.

THEOREM 1.1 (Gerzon). *If there are M equiangular lines in $\mathbb{R}^n$, then*

$$M \le \frac{n(n+1)}{2} \tag{1.1}$$

Gerzon's upper bound can be attained only for a very small number of values of n. Currently, such constructions are known only for $n = 2, 3, 7$, and 23. Neumann (see [**10**], Theorem 3.2) proved a fundamental result in this area:

THEOREM 1.2 (Neumann). *If there are M equiangular lines in $\mathbb{R}^n$ with angle $\arccos \alpha$ and $M > 2n$, then $1/\alpha$ is an odd integer.*

2010 *Mathematics Subject Classification.* Primary 52C35; Secondary 94B75.
Key words and phrases. Semidefinite Programming, Two-Distance Sets, Tight Designs.

n	$M(n)$	$1/\alpha$		n	$M(n)$	$1/\alpha$
2	3	2		17	48-50	5
3	6	$\sqrt{5}$		18	48-61	5
4	6	3; $\sqrt{5}$		19	72-76	5
5	10	3		20	90-96	5
6	16	3		21	126	5
$7 \le n \le 13$	28	3		22	176	5
14	28-30	$3; 5$		23	276	5
15	36	5		$24 \le n \le 42$	≥ 276	5
16	40-42	5		43	≥ 344	7

TABLE 1. Known bounds on $M(n)$ in small dimensions

n	$M(n)$	SDP bound		n	$M(n)$	SDP bound
3	6	6		18	48-61	61
4	6	6		19	72-76	76
5	10	10		20	90-96	96
6	16	16		21	126	126
$7 \le n \le 13$	28	28		22	176	176
14	28-30	30		23	276	276
15	36	36		$24 \le n \le 41$	276	276
16	40-42	42		42	≥ 276	288
17	48-50	51		43	344	344

TABLE 2. Bounds on $M(n)$ including new results

Note that if M attains the Gerzon bound, then $(n+2)\alpha^2 = 1$ [**10**, Thm.3.5]. Therefore, if the cardinality of an equiangular line set attains the Gerzon bound, then n has to be 2 or 3 or an odd square minus two and the angle between pairs of lines is $\arccos 1/(\sqrt{n+2})$.

A set of unit vectors $S = \{x_1, x_2, \dots\} \subset \mathbb{R}^n$ is called *two-distance* if $\langle x_i, x_j\rangle \in \{a, b\}$ for some a, b and all $i \neq j$.

THEOREM 1.3 (Larman, Rogers, and Seidel [**9**]). *Let S be a spherical two-distance set in $\mathbb{R}^n$. If $|S| > 2n+3$ and $a > b$, then $b = \frac{ka-1}{k-1}$ for some integer k such that $2 \le k \le (1+\sqrt{2n})/2$.*

The condition $|S| > 2n+3$ was improved to $|S| > 2n+1$ by Neumaier [**13**]. He also gave an example of a two-distance set with cardinality $2n+1$ that violates the integerality condition of k. This example is obtained from the spherical embedding of the conference graph.

If the spherical two-distance set gives rise to equiangular lines, then $a = -b$, so Theorem 1.3 implies that $a = 1/(2k-1)$, which is the statement of the Neumann theorem. The assumption of Theorem 1.3 is more restrictive than of Theorem 1.2, but in return we obtain an upper bound on k. For instance, if $n = 40$, then k can be only 2 or 3, so the angle has to be $\arccos\alpha$, where $\alpha = 1/3$ or $1/5$. The assumption of Theorem 1.3 is satisfied since there exist equiangular line sets with $M \ge 2n+4$ for all $n \ge 15$.

The known bounds on $M(n)$ for small dimensions are summarized in Table 1 [**10**], [**15**] (the latter for the upper bound on $M(17)$); in particular, $M(n)$ was known exactly only if $2 \leq n \leq 13; n = 15, 21, 22, 23$. In the unsettled cases the best known upper bound on $M(n)$ is usually the Gerzon bound. Lemmens and Seidel [**10**, Thm. 4.5] further showed that

$$M_{1/3}(n) \leq 2(n-1), \quad n \geq 16, \tag{1.2}$$

where $M_\alpha(n)$ is the maximum size of an equiangular line set when the value of the angle is $\arccos\alpha$. They also conjectured that $M_{1/5}(n) = 276$ for $23 \leq n \leq 185$, observing that if this conjecture is true, then $M(n) = 276$ for $24 \leq n \leq 41$ and $M(43) = 344$. Note that generally we have [**10**]:

$$M_\alpha(n) \leq \frac{n(1-\alpha^2)}{1-n\alpha^2} \tag{1.3}$$

valid for all α such that the denominator is positive. This inequality is sometimes called the *relative bound* as opposed to the "absolute bound" of (1.1).

In this paper we use the semidefinite programming (SDP) method to derive some new bounds on $M(n)$. Our main results are summarized in Table 2. In particular, exact values of $M(n)$ are obtained for $24 \leq n \leq 41$ and for $n = 43$ where previous results gave divergent bounds: we show that $M(n) = 276$ for $24 \leq n \leq 41$ and $M(43) = 344$. These results are established by performing computations with SDP. We also show that $M_{1/5}(n) = 276$ for $23 \leq n \leq 60$. These results resolve a part of the Lemmens-Seidel conjecture and enable us to obtain the results in Table 2. For $44 \leq n \leq 136$, we also obtain new upper bounds on $M(n)$, improving upon the Gerzon bound, although no new exact values are found in this range. Below in the paper we give a more complete table of the computation results.

An interesting question relates to the asymptotic behavior of $M(n)$ for $n \to \infty$. For a long time the best known constructions were able to attain the growth order of $M(n) = \Omega(n)$, until D. de Caen [**6**] constructed a family of $\frac{2}{9}(n+1)^2$ equiangular lines in $\mathbb{R}^n$ for $n = 3 \cdot 2^{2t-1}, t \in \mathbb{N}$. Thus, currently the best asymptotic results are summarized as follows:

$$\frac{2}{9} \leq \limsup_{n\to\infty} \frac{M(n)}{n^2} \leq \frac{1}{2}, \tag{1.4}$$

where the upper bound is from (1.1). The question of the correct order of growth represents a difficult unresolved problem. Contributing to the study of the asymptotic bounds, we show that for $n = 3(2k-1)^2 - 4$ and $\alpha = \frac{1}{2k-1}$, for all integer $k \geq 2$,

$$M_\alpha(n) \leq \frac{(n+1)(n+2)}{6}. \tag{1.5}$$

Added in proof: After this paper was accepted, C Greaves, et al. posted a preprint [16] in which the upper bounds for $n = 14, 16$ were improved to $M(14) \leq 29$ and $M(16) \leq 41$, respectively.

2. SDP bounds for equiangular lines

Many problems in operations research, combinatorial optimization, control theory, and discrete geometry can be modelled or approximated as semidefinite programming. SDP optimization problems are usually stated in the following form:

$$\min c^T x$$

$$\text{subject to} \quad F_0 + \sum_{i=1}^{m} F_i x_i \succeq 0, \quad x \in \mathbb{R}^m,$$

where $c \in \mathbb{R}^m$ is a given vector of coefficients, $F_i, i = 0, 1, \dots$ are $n \times n$ symmetric matrices, and "$\succeq$" means that the matrix is positive semidefinite. SDP problems fall in the class of convex optimization problems since the domain of feasible solutions is a convex subset of $\mathbb{R}^m$. For the case of diagonal matrices F_i, SDP turns into a linear programming (LP) problem. Properties of SDP problems and algorithms for their solution are discussed, for instance, in [**5**]. Most SDP solvers such as CSDP, Sedumi, SDPT3 use interior point methods originating with Karmarkar's celebrated algorithm (we used CVX toolbox in Matlab).

Let $\mathcal{C} \subset S^{n-1}$ be a set of unit vectors in $\mathbb{R}^n$ such that $\langle x, x' \rangle \le a$ for all $x, x' \in \mathcal{C}, x \ne x'$ (a *spherical code*). As shown by Bachoc and Vallentin [**1**], the problem of estimating the maximum size of $\mathcal{C}$ can be stated as an SDP problem. In particular, for $a = 1/2$, this is the famous "kissing number problem", i.e., the question about the maximum number of nonoverlapping unit spheres that can touch a given unit sphere. A particular case of the main result in [**1**] was used in [**4**] to find new bounds on the maximum cardinality of spherical two-distance sets.

Let us introduce some notation. Let $G_k^{(n)}(t), k = 0, 1, \dots$ denote the Gegenbauer polynomials of degree k, i.e., a family of polynomials defined recursively as follows: $G_0^{(n)} \equiv 1, G_1^{(n)}(t) = t$, and

$$G_k^{(n)}(t) = \frac{(2k+n-4)tG_{k-1}^{(n)}(t) - (k-1)G_{k-2}^{(n)}(t)}{k+n-3}, \quad k \ge 2.$$

Following [**1**], define a $(p-k+1) \times (p-k+1)$ matrix $Y_k^n(u, v, t), k \ge 0$,

$$(Y_k^n(u,v,t))_{ij} = u^i v^j ((1-u^2)(1-v^2))^{k/2} G_k^{(n-1)}\Big(\frac{t-uv}{\sqrt{(1-u^2)(1-v^2)}}\Big),$$

where $p \in \mathbb{N}$, and a matrix $S_k^n(u, v, t)$ by setting

$$S_k^n(u,v,t) = \frac{1}{6}\sum_{\sigma} Y_k^n(\sigma(u,v,t)),$$

where the sum is over all permutations of 3 elements. Note that $(S_k^n(1,1,1))_{ij} = 0$ for all i, j and all $k \ge 1$. Let $\mathcal{C}$ be a spherical code. As shown in [**7**],

$$\sum_{(x,y)\in\mathcal{C}^2} G_k^{(n)}(\langle x, y\rangle) \ge 0, \tag{2.1}$$

and as shown in [**1**],

$$\sum_{(x,y,z)\in\mathcal{C}^3} S_k^n(\langle x,y\rangle, \langle x,z\rangle, \langle y,z\rangle) \succeq 0. \tag{2.2}$$

Inequalities (2.1) and (2.2) can be used to formulate a general SDP problem for upper bounds on the cardinality of spherical codes in $\mathbb{R}^n$ [**1**].

Using the approach of [**4**], we obtain the following SDP bound on $M(n)$.

THEOREM 2.1. *Let $\mathcal{C}$ be a set of equiangular lines with inner product values either a or $-a$. Let p be the positive integer. The cardinality $|\mathcal{C}|$ is bounded above by the solution of the following semi-definite programming problem :*

$$1 + \frac{1}{3}\max(x_1 + x_2) \tag{2.3}$$

subject to

$$\begin{pmatrix} 1 & 0 \\ 0 & 1 \end{pmatrix} + \frac{1}{3}\begin{pmatrix} 0 & 1 \\ 1 & 1 \end{pmatrix}(x_1 + x_2) + \begin{pmatrix} 0 & 0 \\ 0 & 1 \end{pmatrix}(x_3 + x_4 + x_5 + x_6) \succeq 0 \tag{2.4}$$

$$\begin{aligned} S_k^n(1,1,1) + S_k^n(a,a,1)x_1 + S_k^n(-a,-a,1)x_2 + S_k^n(a,a,a)x_3 \\ + S_k^n(a,a,-a)x_4 + S_k^n(a,-a,-a)x_5 + S_k^n(-a,-a,-a)x_6 \succeq 0 \end{aligned} \tag{2.5}$$

$$3 + G_k^{(n)}(a)x_1 + G_k^{(n)}(-a)x_2 \geq 0, \tag{2.6}$$

where $k = 0, 1, \cdots, p$ *and* $x_j \geq 0$, $j = 1, \cdots, 6$.

To compute bounds on $M(n)$, we found solutions of the SDP problem (2.3)-(2.6), restricting calculations to the case $p = 5$. In Table 3 we list the values of SDP bounds for all possible angles except the angle $\arccos \frac{1}{3}$ which is not included because of (1.2) (note that the SDP bounds for other angles are much greater than $2(n-1)$). The column labelled 'max' refers to the maximum of the SDP bounds among all possible angles. The last column in the table gives the value of the angle for which the maximum is attained.

Some comments on the tables are in order. Observe that $M_{1/5} = 276$ for $23 \leq n \leq 60$. Combined with the results of [**10**], this implies that $M(n) = 276$ for $23 \leq n \leq 41$ and $M(43) = 344$. The case $n = 42$ remains open since we only obtain that $276 \leq M(42) \leq 288$ for the angle $\arccos 1/7$.

Improvements of the Gerzon upper bound (1.1) are obtained for $n \leq 136$. The last 3 entries in Table 3 produced no improvements, and are marked by an asterisk because of that. Similarly, the SDP problem yielded no improvements for higher dimensions.

An interesting, unexplained observation regarding this table is that the SDP bound for $M_\alpha(n)$ has long stable ranges for dimensions starting with the value $n = d^2 - 2$, where d is an odd integer and $\alpha = 1/d$. For instance, one such region begins with $d = 5$, another with $d = 7$. The same phenomenon can observed for $d = 9$ where the SDP value $M_\alpha(n) \leq 3160$ is obtained for all values of n satisfying $79 \leq n \leq 227$ and for $d = 11$ where the value 7140 appears for all $n, 119 \leq n \leq 347$.

Note that the SDP bound gives the same value as the Gerzon bound for $n = 47, 79$ and 119, and that these three dimensions are of the form $n = (2k-1)^2 - 2$, where $k \geq 2$ is a positive integer. Bannai, Munemasa, and Venkov [**2**] showed that for $n = 47, 79$ the maximum possible size $M(n)$ cannot attain this value while the case $n = 119$ is still open. The result of [**2**] relies on the fact that an equiangular line set in $\mathbb{R}^n$ with cardinality $\frac{n(n+1)}{2}$ gives rise to a spherical two-distance set of size $(n-1)(n+2)/2$ in $\mathbb{R}^{n-1}$, and such sets are related to tight spherical 4-designs whose existence can be sometimes ruled out.

Based on the earlier results and our calculations, we make the following

CONJECTURE: *There exist* 1128 *equiangular lines in* $\mathbb{R}^{48}$ *with angle arccos*$(1/7)$ *and* 3160 *equiangular lines in* $\mathbb{R}^{80}$ *with angle arccos*$(1/9)$.

If this conjecture is true, then $M(n) = 1128$ for $48 \leq n \leq 75$ and $M(n) = 3160$ for $80 \leq n \leq 116$.

3. Tight spherical designs of harmonic index 4 and equiangular lines

DEFINITION 3.1. Let t be a natural number. A finite subset X of the unit sphere S^{n-1} is called a *spherical t-design* if, for any polynomial $f(x) = f(x_1, x_2, \ldots, x_n)$

n	1/5	1/7	1/9	1/11	1/13	1/15	max	Gerzon	angle
22	176	39	29	26	25	24	176	253	1/5
23	276	42	31	28	26	25	276	276	1/5
24	276	46	33	29	27	26	276	300	1/5
25	276	50	35	31	29	28	276	325	1/5
26	276	54	37	32	30	29	276	351	1/5
27	276	58	40	34	31	30	276	378	1/5
28	276	64	42	36	33	31	276	406	1/5
29	276	69	44	37	34	33	276	435	1/5
30	276	75	47	39	36	34	276	465	1/5
31	276	82	49	41	37	35	276	496	1/5
32	276	90	52	43	39	37	276	528	1/5
33	276	99	55	45	40	38	276	561	1/5
34	276	108	57	46	42	39	276	595	1/5
35	276	120	60	48	43	41	276	630	1/5
36	276	132	64	50	45	42	276	666	1/5
37	276	148	67	52	47	44	276	703	1/5
38	276	165	70	54	48	45	276	741	1/5
39	276	187	74	57	50	46	276	780	1/5
40	276	213	78	59	52	48	276	820	1/5
41	276	246	82	61	53	49	276	861	1/5
42	276	288	86	63	55	51	288	903	1/7
43	276	344	90	66	57	52	344	946	1/7
44	276	422	95	68	59	54	422	990	1/7
45	276	540	100	71	60	56	540	1035	1/7
46	276	736	105	73	62	57	736	1081	1/7
47	276	1128	110	76	64	59	1128	1128	1/7
48	276	1128	116	78	66	60	1128	1176	1/7
49	276	1128	122	81	68	62	1128	1225	1/7
50	276	1128	129	84	70	64	1128	1275	1/7
51	276	1128	136	87	72	65	1128	1326	1/7
52	276	1128	143	90	74	67	1128	1378	1/7
53	276	1128	151	93	76	69	1128	1431	1/7
54	276	1128	160	96	78	70	1128	1485	1/7
55	276	1128	169	100	81	72	1128	1540	1/7
56	276	1128	179	103	83	74	1128	1596	1/7
57	276	1128	190	106	85	76	1128	1653	1/7
58	276	1128	201	110	87	77	1128	1711	1/7
59	276	1128	214	114	90	79	1128	1770	1/7
60	276	1128	228	118	92	81	1128	1830	1/7
61	279	1128	244	122	94	83	1128	1891	1/7
62	290	1128	261	126	97	85	1128	1953	1/7
63	301	1128	280	130	99	87	1128	2016	1/7
64	313	1128	301	134	102	89	1128	2080	1/7
65	326	1128	325	139	105	91	1128	2145	1/7
66	339	1128	352	144	107	92	1128	2211	1/7
67	353	1128	382	148	110	94	1128	2278	1/7
68	367	1128	418	153	113	97	1128	2346	1/7
69	382	1128	460	159	115	99	1128	2415	1/7
70	398	1128	509	164	118	101	1128	2485	1/7
71	416	1128	568	170	121	103	1128	2556	1/7
72	434	1128	640	176	124	105	1128	2628	1/7
73	453	1128	730	182	127	107	1128	2701	1/7
74	473	1128	845	188	130	109	1128	2775	1/7
75	494	1128	1000	195	134	112	1128	2850	1/7
76	517	1128	1216	202	137	114	1216	2926	1/9
77	542	1128	1540	210	140	116	1540	3003	1/9
78	568	1128	2080	217	144	118	2080	3081	1/9
79	596	1128	3160	225	147	121	3160	3160	1/9
80	626	1128	3160	234	151	123	3160	3240	1/9
81	658	1128	3160	243	154	126	3160	3321	1/9
82	693	1128	3160	252	158	128	3160	3403	1/9

n	1/5	1/7	1/9	1/11	1/13	1/15	max	Gerzon	angle
82	693	1128	3160	252	158	128	3160	3403	1/9
83	731	1128	3160	262	162	130	3160	3486	1/9
84	772	1128	3160	272	166	133	3160	3570	1/9
85	816	1128	3160	283	170	136	3160	3655	1/9
86	866	1128	3160	294	174	138	3160	3741	1/9
87	920	1128	3160	307	178	141	3160	3828	1/9
88	979	1128	3160	320	182	143	3160	3916	1/9
89	1046	1128	3160	333	186	146	3160	4005	1/9
90	1120	1128	3160	348	191	149	3160	4095	1/9
91	1203	1128	3160	364	196	152	3160	4186	1/9
92	1298	1128	3160	380	200	154	3160	4278	1/9
93	1406	1128	3160	398	205	157	3160	4371	1/9
94	1515	1128	3160	417	210	160	3160	4465	1/9
95	1556	1128	3160	438	215	163	3160	4560	1/9
96	1599	1128	3160	460	220	166	3160	4656	1/9
97	1644	1128	3160	485	226	169	3160	4753	1/9
98	1691	1128	3160	511	231	172	3160	4851	1/9
99	1739	1128	3160	540	237	176	3160	4950	1/9
100	1790	1128	3160	571	243	179	3160	5050	1/9
101	1842	1128	3160	606	249	182	3160	5151	1/9
102	1897	1128	3160	644	255	185	3160	5253	1/9
103	1954	1128	3160	686	262	189	3160	5356	1/9
104	2014	1128	3160	734	268	192	3160	5460	1/9
105	2077	1128	3160	787	275	196	3160	5565	1/9
106	2142	1128	3160	848	282	199	3160	5671	1/9
107	2211	1128	3160	917	289	203	3160	5778	1/9
108	2282	1128	3160	997	297	206	3160	5886	1/9
109	2358	1128	3160	1090	305	210	3160	5995	1/9
110	2437	1128	3160	1200	313	214	3160	6105	1/9
111	2521	1128	3160	1332	321	218	3160	6216	1/9
112	2609	1128	3160	1493	330	222	3160	6328	1/9
113	2702	1128	3160	1695	339	226	3160	6441	1/9
114	2800	1128	3160	1954	348	230	3160	6555	1/9
115	2904	1128	3160	2300	357	234	3160	6670	1/9
116	3015	1128	3160	2784	367	238	3160	6786	1/9
117	3132	1128	3160	3510	378	242	3510	6903	1/11
118	3257	1128	3160	4720	388	247	4720	7021	1/11
119	3390	1128	3160	7140	399	251	7140	7140	1/11
120	3532	1128	3160	7140	411	256	7140	7260	1/11
121	3684	1128	3160	7140	423	260	7140	7381	1/11
122	3848	1128	3160	7140	436	265	7140	7503	1/11
123	4024	1128	3160	7140	449	270	7140	7626	1/11
124	4214	1128	3160	7140	462	275	7140	7750	1/11
125	4419	1128	3160	7140	477	280	7140	7875	1/11
126	4643	1128	3160	7140	492	285	7140	8001	1/11
127	4887	1128	3160	7140	508	290	7140	8128	1/11
128	5153	1128	3160	7140	524	295	7140	8256	1/11
129	5447	1128	3160	7140	541	301	7140	8385	1/11
130	5770	1128	3160	7140	560	306	7140	8515	1/11
131	6130	1128	3160	7140	579	312	7140	8646	1/11
132	6531	1130	3160	7140	599	317	7140	8778	1/11
133	6982	1158	3160	7140	620	323	7140	8911	1/11
134	7493	1187	3160	7140	643	329	7493	9045	1/5
135	8075	1218	3160	7140	667	336	8075	9180	1/5
136	8747	1249	3160	7140	692	342	8747	9316	1/5
*137	9528	1282	3160	7140	719	348	9528	9453	1/5
*138	10450	1315	3160	7140	747	355	10450	9591	1/5
*139	11553	1350	3160	7140	778	362	11553	9730	1/5

TABLE 3. Values of the SDP bound on $M(n), 22 \leq n \leq 139$

of degree at most t, the following equality holds :

$$\frac{1}{|S^{n-1}|}\int_{S^{n-1}} f(x)d\sigma(x) = \frac{1}{|X|}\sum_{x\in X} f(x).$$

A spherical t-design is called tight if it attains the LP bound of [**7**], also called the *absolute bound.*

An equivalent definition of spherical designs can be given in terms of harmonic polynomials. Let $\mathrm{Harm}_t(\mathbb{R}^n)$ be the set of homogeneous harmonic polynomials of degree t on $\mathbb{R}^n$. Then the set X is a spherical design [**7**] if

$$\sum_{x\in X} f(x) = 0 \quad \forall f(x) \in \mathrm{Harm}_j(\mathbb{R}^n), 1 \le j \le t.$$

The following definition was recently proposed by Bannai, Okuda, and Tagami [**3**]: A *spherical design of harmonic index* t is a finite subset $X \subset S^{n-1}$ such that

$$\sum_{x\in X} f(x) = 0 \quad \forall f(x) \in \mathrm{Harm}_t(\mathbb{R}^n).$$

An LP bound for spherical designs of harmonic index t was derived in [**3**]. Similarly, if this bound is attained, then the design is called tight. Our interest in tight spherical designs of a fixed harmonic index is motivated by a result in [**3**] which shows that a tight design of index 4 gives rise to an equiangular line set in $\mathbb{R}^n$ with angle $\arccos a = \sqrt{3/(n+4)}$. Since $a = \frac{1}{2k-1}$ for some integer $k \ge 2$, we find that $n = 3(2k-1)^2 - 4$. These considerations motivate the following result.

THEOREM 3.2. *Let $n = 3(2k-1)^2 - 4, k \ge 2$. The cardinality N of any equiangular line set in $\mathbb{R}^n$ with inner product $a = 1/(2k-1)$ satisfies the inequality*

$$|S| \le \frac{(n+1)(n+2)}{6}. \tag{3.1}$$

PROOF. To prove this result we use the LP bound of [**7**] that has the following form: *Let $T \subset [-1,1]$. Let $S = \{x_1, x_2, \dots, x_N\}$ be a set of unit vectors in $\mathbb{R}^n$ such $\langle x_i, x_j\rangle \in T \cup \{1\}$. Let $f(t) = \sum_k f_k G_k^n(t)$ be a polynomial such that $f_0 > 0, f_k \ge 0, k \ge 1$ and that $f(t) \le 0$ for all $t \in T$. Then*

$$|S| \le \Big\lfloor \frac{f(1)}{f_0} \Big\rfloor. \tag{3.2}$$

Consider the polynomial

$$f(t) = (t^2 - a^2)\Big(t^2 + \frac{a^2 n + 4a^2 - 6}{n+4}\Big).$$

Let $X \subset \mathbb{R}^n$ be an equiangular line set with inner product a. Then $T = \{\pm a\}$, and $f(t) = 0$ for $t \in T$. Computing the Gegenbauer expansion of $f(t)$, we obtain

$$f_0 = -\frac{a^4 n^2 + 6a^2 n(a^2-1) + 8a^4 - 6a^2(n+2) + 3}{n^2 + 6n + 8}$$

$$f_1 = f_2 = f_3 = 0$$

$$f_4 = \frac{n^2 - 1}{(n+2)(n+4)}$$

We need to check that $f_0 > 0$. Substituting the values of n and a, we obtain

$$f_0 = \frac{8k(k-1)}{(2k-1)^4(12k^2 - 12k + 1)} \ge 0 \quad \text{for } k \ge 2.$$

Thus, $f(t)$ satisfies the conditions of the LP bound, and we obtain

$$|S| \leq \frac{f(1)}{f_0} = \frac{(a^2-1)(n+2)(n+a^2n+4a^2-2)}{a^4n^2+6a^4n+8a^4-6a^2n-12a^2+3}.$$

In particular, putting $a = \frac{1}{2k-1}$ and $n = 3(2k-1)^2 - 4 = 12k^2 - 12k - 1$, we obtain

$$\frac{f(1)}{f_0} = \frac{(n+1)(n+2)}{6}. \quad \square$$

This theorem gives infinitely many values of n for which the upper bound $M_a(n)$ is strictly less than the Gerzon bound, yielding the asymptotic constant $1/6$ for the growth rate of the quantity $M_\alpha(n)$ (cf. (1.4)-(1.5)).

REMARK 3.3. Observe that the relative bound (1.3) is an instance of the LP bound (3.2); see [**7**]. Thus, the SDP bound (2.3)-(2.6) is as strong or stronger than the bound (1.3).

REMARK 3.4. Using SDP, we further show that for some dimensions the LP bound (3.1) cannot be attained. Indeed, for $k = 3, 4, 5$ we obtain the values of the dimension $n = 71, 143, 239$, respectively, and the SDP bound implies that

$$M_{1/5}(71) \leq 416, \quad M_{1/7}(143) \leq 1506, \quad M_{1/9}(239) \leq 3902,$$

which is much smaller than the values $876, 3480, 9640$ obtained from (3.1). Extending these calculations, we have shown that for $k \leq 54$ and $n = 3(2k-1)^2 - 4 \leq 34343$ the SDP bound improves upon the LP bound (3.1).

In conclusion, we note that the value of the maximum in the LP problem for the maximum cardinality of equiangular line sets with a given angle can be explicitly characterized. The LP problem has the following form:

$$M_a(n) \leq \max\{1 + x_1 + x_2, x_1 \geq 0, x_2 \geq 0\} \tag{3.3}$$

subject to

$$1 + G_k^n(a)x_1 + G_k^n(-a)x_2 \geq 0 \quad \text{for } k = 1, 2, \ldots. \tag{3.4}$$

THEOREM 3.5. *Let $a \in (0, 1)$,*

$$g_n = \min_{k \geq 0} \frac{1}{|G_k^n(a)|} \tag{3.5}$$

where k is even and such that $G_k^n(a) < 0$. Then

$$M_a(n) \leq g_n + 1,$$

where the value $g_n + 1$ is the solution of the LP problem (3.3),(3.4).

PROOF. Let k be even, then $G_k^n(t)$ is an even function, so inequalities (3.4) take the form

$$1 + G_k^n(a)(x_1 + x_2) \geq 0, \quad k = 2m, m \in \mathbb{N}. \tag{3.6}$$

These inequalities define a set of half-planes whose boundaries are parallel to the objective function. The inequalities for odd k are bounded by lines that are perpendicular to the boundaries of the even-indexed constraints, and therefore can be disregarded. We conclude that the maximum is attained on the line $1+G_k^n(a)(x_1+x_2) = 0$ for some even k. The inequalities with k such that $G_k^n(a) \geq 0$

are trivially satisfied, therefore, we consider only those values of k when $G_k^n(a) < 0$. Eq. (3.6) implies that, for all even k,

$$x_1 + x_2 \leq -\frac{1}{G_k^n(a)} = \frac{1}{|G_k^n(a)|}.$$

This completes the proof. □

To give an example of using this theorem, take $n = 71$ and $a = \frac{1}{5}$. To find a bound on $M_a(n)$, we estimate the quantity g_n in (3.5) by computing

$$\min_{0 \leq k \leq 100} \frac{1}{|G_k^{(71)}(1/5)|}$$

for all even k such that $G_k^{(71)}(1/5) < 0$. The smallest value is obtained for $k = 4$, and $G_4^{(71)}(1/5) = -1/875$. Thus, we obtain $M_{1/5}(71) \leq 876$. Of course, it could be possible that for greater k we obtain a smaller value of the bound, but this is not supported by our experiments (although we do not have a proof that $k = 4$ is the optimal choice).

Experiments also suggest that $k = 4$ may be the universal optimal choice for infinitely many values of n and a. Indeed, we have

$$G_4^n(x) = \frac{(n+2)(n+4)x^4 - 6(n+2)x^2 + 3}{n^2 - 1}.$$

Taking $n = 3(2t-1)^2 - 4$ and $a = 1/(2t-1)$, where $t \geq 2$, we obtain the expression

$$\frac{1}{G_4^n(a)} + 1 = 2t(t-1)(12t^2 - 12t + 1) = \frac{(n+1)(n+2)}{6}$$

which coincides with the LP bound (3.1).

Acknowledgement: We thank Peter Casazza, Eiichi Bannai, and Ferenc Szöllősi for insightful discussions on the topics of this paper.

Alexander Barg was supported in part by NSF grants DMS1101697, CCF1217245, and CCF1217894, and by NSA grant H98230-12-1-0260. Wei-Hsuan Yu was supported in part by NSF grant CCF CCF1217245.

References

[1] C. Bachoc and F. Vallentin, *New upper bounds for kissing numbers from semidefinite programming*, J. Amer. Math. Soc. **21** (2008), no. 3, 909–924, DOI 10.1090/S0894-0347-07-00589-9. MR2393433 (2009c:52029)

[2] E. Bannai, A. Munemasa, and B. Venkov, *The nonexistence of certain tight spherical designs*, Algebra i Analiz **16** (2004), no. 4, 1–23, DOI 10.1090/S1061-0022-05-00868-X; English transl., St. Petersburg Math. J. **16** (2005), no. 4, 609–625. MR2090848 (2005e:05022)

[3] E. Bannai, T. Okuda, and M. Tagami, *Spherical designs of harmonic index t*, http://arxiv.org/pdf/1308.5101v1.pdf (August 2013).

[4] Alexander Barg and Wei-Hsuan Yu, *New bounds for spherical two-distance sets*, Exp. Math. **22** (2013), no. 2, 187–194, DOI 10.1080/10586458.2013.767725. MR3047911

[5] L. Vandenberghe and S. Boyd, *Semidefinite programming*, SIAM Rev. **38** (1996), no. 1, 49–95, DOI 10.1137/1038003. MR1379041 (96m:90005)

[6] D. de Caen, *Large equiangular sets of lines in Euclidean space*, Electron. J. Combin. **7** (2000), Research Paper 55, 3 pp. (electronic). MR1795615 (2001h:51026)

[7] P. Delsarte, J. M. Goethals, and J. J. Seidel, *Spherical codes and designs*, Geometriae Dedicata **6** (1977), no. 3, 363–388. MR0485471 (58 #5302)

[8] J. Haantjes, *Equilateral point-sets in elliptic two- and three-dimensional spaces*, Nieuw Arch. Wiskunde (2) **22** (1948), 355–362. MR0023530 (9,369c)

[9] D. G. Larman, C. A. Rogers, and J. J. Seidel, *On two-distance sets in Euclidean space*, Bull. London Math. Soc. **9** (1977), no. 3, 261–267. MR0458308 (56 #16511)

[10] P. W. H. Lemmens and J. J. Seidel, *Equiangular lines*, J. Algebra **24** (1973), 494–512. MR0307969 (46 #7084)

[11] J. H. van Lint and J. J. Seidel *Equiangular point sets in elliptic geometry*, Proc. Nederl Akad. Wetensch Series A **69** (1966), 335-348.

[12] G. Nebe and B. Venkov, *On tight spherical designs*, Algebra i Analiz **24** (2012), no. 3, 163–171, DOI 10.1090/S1061-0022-2013-01249-0; English transl., St. Petersburg Math. J. **24** (2013), no. 3, 485–491. MR3014131

[13] A. Neumaier, *Distance matrices, dimension, and conference graphs*, Nederl. Akad. Wetensch. Indag. Math. **43** (1981), no. 4, 385–391. MR639855 (82m:05071)

[14] I. J. Schoenberg, *Positive definite functions on spheres*, Duke Math. J. **9** (1942), 96–108. MR0005922 (3,232c)

[15] M. A. Sustik, J. A. Tropp, I. S. Dhillon, and R. W. Heath Jr., *On the existence of equiangular tight frames*, Linear Algebra Appl. **426** (2007), no. 2-3, 619–635, DOI 10.1016/j.laa.2007.05.043. MR2350682 (2008f:15066)

[16] G. Greaves, J. H. Koolen, A. Munemaza, and F. Szöllősi, *Equiangular lines in Euclidean spaces*, preprint arXiv:1403:2155 (http://arxiv.org/abs/1403.2155).

Dept. of ECE and Inst. for Systems Research, University of Maryland, College Park, Maryland 20742 — and — Inst. for Problems of Inform. Trans., RAS, Moscow, Russia

E-mail address: abarg@umd.edu

Department of Math. and Inst. for Systems Research, University of Maryland, College Park, Maryland 20742

E-mail address: mathyu@math.umd.edu

Contemporary Mathematics
Volume **625**, 2014
http://dx.doi.org/10.1090/conm/625/12495

Formal duality and generalizations of the Poisson summation formula

Henry Cohn, Abhinav Kumar, Christian Reiher, and Achill Schürmann

Abstract. We study the notion of formal duality introduced by Cohn, Kumar, and Schürmann in their computational study of energy-minimizing particle configurations in Euclidean space. In particular, using the Poisson summation formula we reformulate formal duality as a combinatorial phenomenon in finite abelian groups. We give new examples related to Gauss sums and make some progress towards classifying formally dual configurations.

1. Introduction

The Poisson summation formula connects the sum of a function over a lattice $\Lambda \subset \mathbb{R}^n$ with the sum of its Fourier transform over the dual lattice Λ^*; recall that Λ^* is spanned by the dual basis (with respect to the inner product) to any basis of Λ. In fact, Poisson summation completely characterizes the notion of duality for lattices. In a computational study of energy minimization for particle configurations, Cohn, Kumar, and Schürmann [**CKS**] found several examples of non-lattice configurations exhibiting a similar *formal duality* with respect to a version of Poisson summation. In this paper, we place these examples in a broader context, produce new examples using the theory of Gauss sums, and take the first steps towards a classification of formally dual configurations.

Energy minimization is a natural problem in geometric optimization, which generalizes the sphere packing problem of arranging congruent, non-overlapping spheres as densely as possible in $\mathbb{R}^n$. The *energy* $E_f(\mathcal{C})$ of a configuration $\mathcal{C} \subset \mathbb{R}^n$ with respect to a radial potential function $f\colon \mathbb{R}_{>0} \to \mathbb{R}$ is defined to be the average over $x \in \mathcal{C}$ of the energy of x, which is

$$E_f(x, \mathcal{C}) = \sum_{y \in \mathcal{C} \setminus \{x\}} f(|x-y|).$$

Of course these sums might diverge or the average over x might not be well defined. We therefore restrict $\mathcal{C}$ to be a periodic configuration, i.e., the union of finitely many translates of a lattice in $\mathbb{R}^n$, and we consider only potential functions that decrease rapidly enough at infinity to ensure convergence. See Section 9 of [**CK1**] for more details.

2010 *Mathematics Subject Classification.* Primary 05B40, 11H31; Secondary 52C17.
The first author was supported in part by National Science Foundation grants DMS-0757765 and DMS-0952486 and by a grant from the Solomon Buchsbaum Research Fund.

For each potential function f, the energy minimization problem asks for the configuration $\mathcal{C}$ that minimizes $E_f(\mathcal{C})$ subject to fixing the point density $\delta(\mathcal{C})$ (i.e., the number of points per unit volume).

In [**CKS**], the authors undertook an experimental study of energy minima in low dimensions for Gaussian potential functions. This is the Gaussian core model from mathematical physics [**S**], and Gaussian potential functions also play a key role in the mathematical theory of universal optimality [**CK1**], because they span the cone of completely monotonic functions of squared distance. (If a configuration minimizes all Gaussian potentials simultaneously, then it minimizes many others as well, such as inverse power laws.) For the potential function $G_c(r) := \exp(-\pi c r^2)$, as $c \to \infty$ the potential energy for each point is dominated by the contribution from its nearest neighbors. In the limit, minimizing the energy requires maximizing the distance between the nearest neighbors and thus maximizing the density of the corresponding sphere packing. We can therefore view energy minimization with c large as a "soft-matter" version of sphere packing, in which small distances between particles are allowed but heavily penalized, and we recover the hard sphere model in the limit as $c \to \infty$.

Maximizing density is a necessary condition for optimality as $c \to \infty$, but it is not sufficient, since two optimal sphere packings needn't have the same energy. For example, one may contain fewer pairs of nearest neighbors, in which case it will have lower energy when c is large. As shown in [**CK2**], the densest lattice packing in $\mathbb{R}^n$ fails to minimize energy for large c when $n = 5$ or $n = 7$. Further results were obtained in [**CKS**], which reported on numerical searches for energy minima among periodic configurations with $1 \le n \le 9$ and a range of values of c. (The results in [**CKS**] are formulated in terms of a fixed potential function and varying particle density, but that is equivalent to our perspective here under rescaling to fix the density.)

The most noteworthy finding from [**CKS**] was that in each dimension, the energy-minimizing structures for the potential functions G_c and $G_{1/c}$ seem to be *formally dual* (except in certain narrow ranges of phase coexistence). Formal duality generalizes the more familiar notion of duality for lattices. We will recall the definition in Section 2; the key property is that if $\mathcal{P}$ and $\mathcal{Q}$ are formal duals, then formal duality relates the f-potential energy of $\mathcal{P}$ to the $\widehat{f}$-potential energy of $\mathcal{Q}$ for all potential functions f, where $\widehat{f}$ is the Fourier transform of f. Note that G_c and $G_{1/c}$ are Fourier transforms of each other, up to scalar multiplication.

To describe the simulation results from [**CKS**], we will need some notation. Let D_n^+ be the periodic configuration consisting of the union of the checkerboard lattice

$$D_n = \{(x_1, \dots, x_n) \in \mathbb{Z}^n \mid x_1 + \dots + x_n \equiv 0 \pmod{2}\}$$

and its translate by the all-halves vector (note that D_n^+ is actually a lattice if n is even), and for $\alpha > 0$ let

$$D_n^+(\alpha) = \{(x_1, \dots, x_{n-1}, \alpha x_n) \mid (x_1, \dots, x_n) \in D_n^+\}$$

be obtained by scaling the last coordinate.

The numerical experiments in [**CKS**] indicate that in dimension 5, the family of configurations $D_5^+(\alpha)$ minimize the G_c-energy, with α some function of the parameter c, except in a small interval around $c = 1$ (in this interval there is phase coexistence and the optimal configuration is probably not periodic). For instance, as $c \to \infty$, the minima seem to approach $D_5^+(2)$, which is the tight packing Λ_5^2 in the notation of [**CS1**]. Similarly, in dimension 7 the $D_7^+(\alpha)$ family seems to be optimal. In three cases there are single configurations that seem to minimize potential energy for the entire family of Gaussian potential functions: D_4 in dimensional 4, E_8 in dimension 8 (consistent with the conjecture of universal optimality from [**CK1**]), and D_9^+ in dimension 9. In dimension 6, the energy minima are experimentally seen to be E_6 and its dual for $c \to \infty$ and $c \to 0$, respectively; around the central point $c = 1$ experiments yield the following periodic configuration $\mathcal{P}_6(\alpha)$, where α depends on c. Let $\mathcal{P}_6$ be the lattice $D_3 \oplus D_3$, along with its translates by the three vectors $v_1 = (-1/2, -1/2, -1/2, 1, 1, 1)$, $v_2 = (1, 1, 1, -1/2, -1/2, -1/2)$, and $v_3 = v_1 + v_2$. Then $\mathcal{P}_6(\alpha)$ is obtained from $\mathcal{P}_6$ by scaling the first three coordinates by α and the last three by $1/\alpha$.

Whether or not these families are the true global minima, they certainly exhibit the phenomenon of formal duality. Namely, $D_n^+(\alpha)$ is formally dual to an isometric copy of $D_n^+(1/\alpha)$, and $\mathcal{P}_6(\alpha)$ is formally dual to an isometric copy of itself. See Section VI of [**CKS**] for a proof for $D_n^+(\alpha)$ and a sketch of the analogous proof for $\mathcal{P}_6(\alpha)$. Formal duality comes as a surprise, because most configurations do not have formal duals at all. The experimental findings lead to a natural question: do the global minima for Gaussian potential energy in Euclidean space always appear in families exhibiting formal duality? Outside of certain narrow ranges for the parameter c, where one observes phase coexistence leading to aperiodic minima, all the numerical data from [**CKS**] is consistent with formal duality.

The structures found in [**CKS**] have been the only known examples of formally dual pairs other than lattices. In this paper, we present a new family of examples based on Gauss sums, we analyze structural properties of formally dual configurations, and we take the first steps towards a classification.

2. Poisson summation formulas and duality

We first recall the Poisson summation formula. Given a well-behaved function $f\colon \mathbb{R}^n \to \mathbb{R}$ (for example, a Schwartz function, though much weaker hypotheses will suffice), define its Fourier transform $\widehat{f}\colon \mathbb{R}^n \to \mathbb{R}$ by

$$\widehat{f}(y) = \int_{\mathbb{R}^n} f(x) e^{-2\pi i \langle x, y\rangle}\, dx.$$

Then the Poisson summation formula states that for a lattice $\Lambda \subset \mathbb{R}^n$,

$$\sum_{x\in\Lambda} f(x) = \frac{1}{\operatorname{covol}(\Lambda)} \sum_{y \in \Lambda^*} \widehat{f}(y),$$

where

$$\Lambda^* = \{y \in \mathbb{R}^n \mid \langle x, y\rangle \in \mathbb{Z} \text{ for all } x \in \Lambda\}$$

is the dual lattice and $\operatorname{covol}(\Lambda) = \operatorname{vol}(\mathbb{R}^n/\Lambda)$ is the volume of a fundamental domain of Λ. The Poisson summation formula is a useful identity in many areas of mathematics. For instance, it can be used to prove analytic continuation and the functional equation for the Riemann zeta function.

As a consequence of Poisson summation,

$$f(0) + E_f(\Lambda) = \frac{1}{\operatorname{covol}(\Lambda)}\big(\widehat{f}(0) + E_{\widehat{f}}(\Lambda^*)\big)$$

for every lattice Λ. Here $\widehat{f}$ is an abuse of notation, in which we treat the potential function $f\colon \mathbb{R}_{>0} \to \mathbb{R}$ as a radial function on $\mathbb{R}^n$.

It follows that a lattice Λ minimizes E_f among lattices with a fixed covolume if and only if Λ^* minimizes $E_{\widehat{f}}$. The most important special case is the Gaussian potential function $G_c(r) = \exp(-\pi c r^2)$, which has n-dimensional Fourier transform $\widehat{G_c}(r) = c^{-n/2}\exp(-\pi r^2/c)$. In this case Poisson summation relates $E_{G_c}(\Lambda)$ to $E_{G_{1/c}}(\Lambda^*)$.

One could ask if there is a reasonable analogue of the Poisson summation formula for non-lattices. The obvious generalization would be to ask for periodic configurations $\mathcal{P}$ and $\mathcal{Q}$ with

$$\sum_{x\in\mathcal{P}} f(x) = \delta(\mathcal{P}) \sum_{y\in\mathcal{Q}} \widehat{f}(y)$$

for all well-behaved f. Here $\delta(\mathcal{P})$ is the point density of $\mathcal{P}$: if $\mathcal{P}$ consists of N translates of a lattice Λ, then $\delta(\mathcal{P}) = N/\operatorname{covol}(\Lambda)$. However, the requirement above is too stringent, for it forces $\mathcal{P}$ and $\mathcal{Q}$ to be lattices, by Theorem 1 in [**C**]. Instead, we are really interested in the differences between points in $\mathcal{P}$, at least for the purposes of potential energy, so we modify the notion of duality as follows. For a Schwartz function $f\colon \mathbb{R}^n \to \mathbb{R}$ and a periodic configuration $\mathcal{P} = \bigcup_{j=1}^N (\Lambda + v_j)$ (where Λ is a lattice), we let

$$\Sigma_f(\mathcal{P}) = \frac{1}{N}\sum_{j,k=1}^{N}\sum_{x\in\Lambda} f(x+v_j-v_k)$$

be the *average pair sum* of f over $\mathcal{P}$. It is also the average over all points $x\in\mathcal{P}$ of $\Sigma_f(x,\mathcal{P}) = \sum_{y\in\mathcal{P}} f(y-x)$, and this interpretation shows that it is independent of the decomposition of $\mathcal{P}$ as $\bigcup_{j=1}^N(\Lambda+v_j)$. Note that when f is a radial function, this sum is related to the potential energy by $\Sigma_f(\mathcal{P}) = E_f(\mathcal{P}) + f(0)$, but we do not require f to be radial.

DEFINITION 2.1. We say two periodic configurations $\mathcal{P}$ and $\mathcal{Q}$ in $\mathbb{R}^n$ are *formally dual* to each other if $\Sigma_f(\mathcal{P}) = \delta(\mathcal{P})\Sigma_{\widehat{f}}(\mathcal{Q})$ for every Schwartz function $f\colon \mathbb{R}^n \to \mathbb{R}$.

For a lattice, pair sums reduce to sums over the lattice itself. Thus, two lattices are formally dual if and only if they are actually dual.

We define formal duality only for periodic configurations, although there may be interesting extensions to the aperiodic case. Note also that the formal dual of a configuration needn't be unique. One form of non-uniqueness is obvious: if $\mathcal{Q}$ is a formal dual of $\mathcal{P}$, then so are $\mathcal{Q}+t$ and $-\mathcal{Q}+t$ for all vectors t. However, formal duals are not unique even modulo these transformations. See Remark 3.3 for an example.

REMARK 2.2. If $\mathcal{P}$ and $\mathcal{Q}$ are formally dual as above, then we can prove $\delta(\mathcal{P})\delta(\mathcal{Q}) = 1$ by considering a steep Gaussian $f(x) = \exp(-\pi c|x|^2)$ and letting $c\to\infty$. Therefore the relation of being formally dual is symmetric.

Our notion of formal duality is stronger than another version in the literature (see, for example, the question on p. 185 of [**CS2**]). The other version asks for equality only for radial functions, which is equivalent to a statement about the average theta series. For clarity we call that version radial formal duality:

DEFINITION 2.3. We say two periodic configurations $\mathcal{P}$ and $\mathcal{Q}$ in $\mathbb{R}^n$ are *radially formally dual* to each other if $\Sigma_f(\mathcal{P}) = \delta(\mathcal{P})\Sigma_{\widehat{f}}(\mathcal{Q})$ for every *radial* Schwartz function $f\colon \mathbb{R}^n \to \mathbb{R}$.

If Λ_1 and Λ_2 are distinct lattices in $\mathbb{R}^n$ with the same theta series, then Λ_1 and Λ_2^* are radially formally dual but not dual and hence not formally dual. The most interesting case is when Λ_1 and Λ_2 are not isometric (for example, D_{16}^+ and $E_8 \oplus E_8$), but the simplest case is when Λ_2 is a rotation of Λ_1.

The discrete analogue of radial formal duality has been investigated in the coding theory literature, with several striking examples such as Kerdock and Preparata codes [**HKCSS**].

Radial formal duality is all one needs for studying energy under radial potential functions, but the stronger definition arose in the examples from [**CKS**] and possesses a richer structure theory. For example, Lemma 2.4 below fails for radial formal duality (let $\mathcal{P}$ be $\mathbb{Z}^2$, let $\mathcal{Q}$ be $\mathbb{Z}^2$ rotated by an angle of $\pi/4$, and let ϕ be the diagonal matrix with entries 2 and 1).

We will now transform the notion of formal duality into a more combinatorial definition about subsets of abelian groups, rather than the continuous setting of periodic configurations and potential functions. The first step is the following easy result, which is Lemma 2 in [**CKS**].

LEMMA 2.4. *Let $\mathcal{P}$ and $\mathcal{Q}$ be periodic configurations of $\mathbb{R}^n$ which are formally dual to each other, and let $\phi \in \mathrm{GL}_n(\mathbb{R})$ be an invertible linear transformation of the space. Then $\phi(\mathcal{P})$ and $(\phi^t)^{-1}(\mathcal{Q})$ are formally dual to each other.*

Here ϕ^t is the adjoint of ϕ with respect to the inner product $\langle \cdot, \cdot \rangle$ (i.e., its matrix is the transpose of that of ϕ).

PROOF. If f is any Schwartz function, then so is $g = f \circ \phi$, and

$$\widehat{g} = \frac{1}{\det(\phi)} \widehat{f} \circ (\phi^t)^{-1}.$$

Therefore,

$$\begin{aligned} \Sigma_f\bigl(\phi(\mathcal{P})\bigr) = \Sigma_{f\circ\phi}(\mathcal{P}) &= \Sigma_g(\mathcal{P}) \\ &= \delta(\mathcal{P})\,\Sigma_{\widehat{g}}(\mathcal{Q}) \\ &= \delta(\mathcal{P}) \cdot \frac{1}{\det(\phi)} \cdot \Sigma_{\widehat{f}\circ(\phi^t)^{-1}}(\mathcal{Q}) \\ &= \delta\bigl(\phi(\mathcal{P})\bigr)\,\Sigma_{\widehat{f}}\bigl((\phi^t)^{-1}(\mathcal{Q})\bigr), \end{aligned}$$

which shows that $\phi(\mathcal{P})$ and $(\phi^t)^{-1}(\mathcal{Q})$ are formally dual. □

This lemma shows that, for a periodic configuration, the property of having a formal dual depends only on the underlying abelian group and coset structure, rather than how the configuration is embedded into $\mathbb{R}^n$. In other words, having a formal dual is not a metric property.

For further progress in making formal duality more combinatorial, we will need to remove the Fourier transform from the definition. We can do so using Poisson summation, as follows. The statement looks complicated, but it will be an essential tool for simplifying the duality theory.

LEMMA 2.5. *Let* $\mathcal{P} = \bigcup_{j=1}^{N}(\Lambda + v_j)$ *and* $\mathcal{Q} = \bigcup_{j=1}^{M}(\Gamma + w_j)$ *be periodic configurations with underlying lattices* Λ *and* Γ, *respectively. Then* $\mathcal{P}$ *and* $\mathcal{Q}$ *are formally dual if and only if for all Schwartz functions* $f\colon \mathbb{R}^n \to \mathbb{R}$,

$$\sum_{y\in\Lambda^*} \widehat{f}(y) \left| \frac{1}{N}\sum_{j=1}^{N} e^{2\pi i\langle v_j, y\rangle} \right|^2 = \frac{1}{M}\sum_{j,k=1}^{M}\sum_{z\in\Gamma} \widehat{f}(z + w_j - w_k).$$

PROOF. Let $v \in \mathbb{R}^n$. By Poisson summation for the function $x \mapsto f(x+v)$,

$$\sum_{x\in\Lambda} f(x+v) = \frac{1}{\operatorname{covol}(\Lambda)} \sum_{y\in\Lambda^*} e^{2\pi i\langle v, y\rangle}\widehat{f}(y).$$

Using this, if $\mathcal{P} = \bigcup_{j=1}^{N}(\Lambda + v_j)$, then

$$\begin{aligned}\Sigma_f(\mathcal{P}) &= \frac{1}{N}\sum_{j,k=1}^{N}\sum_{x\in\Lambda} f(x + v_j - v_k)\\ &= \frac{1}{N\operatorname{covol}(\Lambda)} \sum_{y\in\Lambda^*} \widehat{f}(y) \left| \sum_{j=1}^{N} e^{2\pi i\langle v_j, y\rangle} \right|^2\\ &= \delta(\mathcal{P}) \sum_{y\in\Lambda^*} \widehat{f}(y) \left| \frac{1}{N}\sum_{j=1}^{N} e^{2\pi i\langle v_j, y\rangle} \right|^2.\end{aligned}$$

Formal duality holds if and only if this quantity equals

$$\delta(\mathcal{P})\Sigma_{\widehat{f}}(\mathcal{Q}) = \frac{\delta(\mathcal{P})}{M}\sum_{j,k=1}^{M}\sum_{z\in\Gamma} \widehat{f}(z + w_j - w_k),$$

as desired. □

This lemma has powerful consequences for the cosets of $\mathcal{P}$ in Λ. Recall that for a set A in an abelian group G, we define $A - A = \{x - y \mid x, y \in A\}$.

COROLLARY 2.6. *Let* Λ *and* Γ *be underlying lattices of formally dual configurations* $\mathcal{P}$ *and* $\mathcal{Q}$, *respectively. Then* $\mathcal{P} - \mathcal{P} \subseteq \Gamma^*$ *and* $\mathcal{Q} - \mathcal{Q} \subseteq \Lambda^*$.

PROOF. It is enough to show the latter statement, since the former follows by symmetry. By Lemma 2.5,

$$\sum_{y\in\Lambda^*} \widehat{f}(y) \left| \frac{1}{N}\sum_{j=1}^{N} e^{2\pi i\langle v_j, y\rangle} \right|^2 = \frac{1}{M}\sum_{j,k=1}^{M}\sum_{z\in\Gamma} \widehat{f}(z + w_j - w_k)$$

for every Schwartz function f. Since $\widehat{f}$ is an arbitrary Schwartz function, this forces the set $\{z + w_j - w_k \mid 1 \le j, k \le M \text{ and } z \in \Gamma\}$, which is exactly $\mathcal{Q} - \mathcal{Q}$, to be contained in Λ^*. □

The following corollary holds for exactly the same reason.

COROLLARY 2.7. *Let $\mathcal{P} = \bigcup_{j=1}^{N}(\Lambda + v_j)$ and $\mathcal{Q} = \bigcup_{j=1}^{M}(\Gamma + w_j)$ be periodic configurations, such that $\mathcal{P} - \mathcal{P} \subseteq \Gamma^*$ and $\mathcal{Q} - \mathcal{Q} \subseteq \Lambda^*$. Then $\mathcal{P}$ is formally dual to $\mathcal{Q}$ if and only if for every $y \in \Lambda^*$,*

$$\left|\frac{1}{N}\sum_{j=1}^{N} e^{2\pi i \langle v_j, y\rangle}\right|^2 = \frac{1}{M}\cdot \#\{(z,j,k) \mid 1 \le j,k \le M,\ z \in \Gamma, \text{ and } y = z + w_j - w_k\},$$

i.e., $1/M$ times the number of ways the coset $y + \Gamma$ can be written as a difference of two of the M cosets of Γ in $\mathcal{Q}$.

From now on, we will assume without loss of generality that $0 \in \mathcal{P}$ (and therefore $\Lambda \subseteq \mathcal{P}$), and similarly $0 \in \mathcal{Q}$. We may do so because formal duality is clearly translation-invariant.

Now $\mathcal{P} = \mathcal{P} - 0 \subseteq \mathcal{P} - \mathcal{P} \subseteq \Gamma^*$, so $\mathcal{P}$ can be represented as a subset S of size N in the finite abelian group Γ^*/Λ. Similarly, $\mathcal{Q}$ corresponds to a subset T of M points in Λ^*/Γ. The natural pairing

$$(\Gamma^*/\Lambda) \times (\Lambda^*/\Gamma) \to S^1 \subset \mathbb{C}^*$$

given by

$$\langle x + \Lambda, y + \Gamma\rangle = e^{2\pi i \langle x, y\rangle} \tag{2.1}$$

identifies the two groups as duals. In other words, we view Λ^*/Γ as the group $\widehat{G}$ of characters on $G := \Gamma^*/\Lambda$, with $\chi \in \widehat{G}$ acting on $g \in G$ via $\chi(g) = \langle g, \chi\rangle$. Note that in (2.1), $\langle \cdot, \cdot \rangle$ denotes both the pairing between G and $\widehat{G}$ and the Euclidean inner product, but the type of the inputs makes the usage unambiguous. We will also canonically identify G with the dual of $\widehat{G}$ and treat the pairing between them as symmetric.

Because $v_1, \dots, v_N \in \mathcal{P} \subseteq \Gamma^*$, the quantity

$$\left|\frac{1}{N}\sum_{j=1}^{N} e^{2\pi i \langle v_j, y\rangle}\right|^2$$

from Corollary 2.7 only depends on y modulo Γ.

We can now reformulate formal duality as follows. Let the Fourier transform of a function $f \colon G \to \mathbb{C}$ be $\widehat{f} \colon \widehat{G} \to \mathbb{C}$, defined by

$$\widehat{f}(y) = \frac{1}{\sqrt{|G|}} \sum_{x \in G} f(x)\overline{\langle x, y\rangle} = \frac{1}{\sqrt{|G|}} \sum_{x \in G} f(x) y(-x).$$

THEOREM 2.8. *With notation as above, let $\mathcal{P}$ correspond to the translates of Λ by elements of $S = \{v_1, \dots, v_N\} \subseteq G = \Gamma^*/\Lambda$, and $\mathcal{Q}$ correspond to the translates of Γ by $T = \{w_1, \dots, w_M\} \subseteq \widehat{G} = \Lambda^*/\Gamma$. Then $\mathcal{P}$ and $\mathcal{Q}$ are formally dual if and only if the following equivalent conditions hold.*

(1) *For every $y \in \widehat{G}$,*

$$\left|\frac{1}{N}\sum_{i=1}^{N} \langle v_i, y\rangle\right|^2 = \frac{1}{M}\cdot \#\{(j,k) \mid 1 \le j,k \le M \text{ and } y = w_j - w_k\}.$$

(2) *For every function* $f\colon G \to \mathbb{C}$,

$$\frac{1}{N^{3/2}} \sum_{j,k=1}^{N} f(v_j - v_k) = \frac{1}{M^{3/2}} \sum_{j,k=1}^{M} \widehat{f}(w_j - w_k).$$

PROOF. The equivalence of statement (1) and formal duality is a mild rephrasing of Corollary 2.7. To see why (1) is equivalent to (2), we first note that

$$f(x) = \frac{1}{\sqrt{|G|}} \sum_{y \in \widehat{G}} \widehat{f}(y) \langle x, y \rangle.$$

We now have

$$\begin{aligned}
\frac{1}{N^{3/2}} \sum_{j,k=1}^{N} f(v_j - v_k) &= \frac{1}{N^{3/2}} \cdot \frac{1}{\sqrt{|G|}} \sum_{j,k=1}^{N} \sum_{y \in \widehat{G}} \widehat{f}(y) \langle v_j - v_k, y \rangle \\
&= \frac{1}{N^{3/2}} \cdot \frac{1}{\sqrt{|G|}} \sum_{j,k} \sum_{y} \widehat{f}(y) \langle v_j, y \rangle \overline{\langle v_k, y \rangle} \\
&= \frac{1}{N^{3/2}} \cdot \frac{1}{\sqrt{|G|}} \sum_{y} \widehat{f}(y) \sum_{j,k} \langle v_j, y \rangle \overline{\langle v_k, y \rangle} \\
&= \frac{1}{N^{3/2}} \cdot \frac{1}{\sqrt{|G|}} \sum_{y} \widehat{f}(y) \left| \sum_{j} \langle v_j, y \rangle \right|^2 \\
&= \sqrt{\frac{N}{|G|}} \cdot \sum_{y} \widehat{f}(y) \left| \frac{1}{N} \sum_{j} \langle v_j, y \rangle \right|^2 .
\end{aligned}$$

The last expression equals

$$\frac{1}{M} \sqrt{\frac{N}{|G|}} \sum_{j,k=1}^{M} \widehat{f}(w_j - w_k)$$

for every f if and only if (1) holds. Thus, we have shown that (1) is equivalent to

$$\frac{1}{N^{3/2}} \sum_{j,k=1}^{N} f(v_j - v_k) = \frac{1}{M} \sqrt{\frac{N}{|G|}} \sum_{j,k=1}^{M} \widehat{f}(w_j - w_k). \tag{2.2}$$

To complete the proof of equivalence, we will show that (1) and (2) each imply $|G| = MN$ (in which case (2.2) is equivalent to (2)).

First, assume (1). If we sum over all $y \in \widehat{G}$ and apply orthogonality of distinct characters on $\widehat{G}$, we find that

$$\sum_{y \in \widehat{G}} \left| \frac{1}{N} \sum_{i=1}^{N} \langle v_i, y \rangle \right|^2 = \sum_{y \in \widehat{G}} \frac{1}{N^2} \sum_{i=1}^{N} |\langle v_i, y \rangle|^2 = \frac{1}{N^2} N |\widehat{G}|.$$

Thus, (1) yields

$$\frac{1}{N^2} N |\widehat{G}| = \frac{1}{M} \cdot M^2,$$

implying $|G| = |\widehat{G}| = MN$. Assuming (2), we can apply it with f being the characteristic function of the identity in G to obtain

$$\frac{1}{N^{3/2}} \cdot N = \frac{1}{M^{3/2}} \cdot \frac{1}{\sqrt{|G|}} \cdot M^2,$$

which again implies $|G| = MN$. □

DEFINITION 2.9. We say that subsets S of a finite abelian group G and T of $\widehat{G}$ are *formally dual* if the following equivalent conditions hold.

(1) For every $y \in \widehat{G}$,

$$\left| \frac{1}{|S|} \sum_{v \in S} \langle v, y \rangle \right|^2 = \frac{1}{|T|} \cdot \#\{(w, w') \in T \times T \mid y = w - w'\}.$$

(2) For every function $f \colon G \to \mathbb{C}$,

$$\frac{1}{|S|^{3/2}} \sum_{v, v' \in S} f(v - v') = \frac{1}{|T|^{3/2}} \sum_{w, w' \in T} \widehat{f}(w - w').$$

Thus, Theorem 2.8 reduces formal duality in Euclidean space to the setting of finite abelian groups.

REMARK 2.10. The second criterion in the definition immediately implies that the relation of formal duality is symmetric. However, the first criterion seems to be more useful for concrete calculations, and it is the one we will use in our examples.

3. Examples

The simplest examples of formally dual configurations in $\mathbb{R}^n$ are of course lattices and their duals. These correspond to taking the trivial abelian group $G = \{0\}$, with $S = G$ and $T = \widehat{G} = \{0\}$.

3.1. The TITO configuration. The simplest non-trivial example of a pair of formally dual configurations is the following. Consider the abelian group $G = \mathbb{Z}/4\mathbb{Z}$, and identify $\widehat{G} = \mathbb{Z}/4\mathbb{Z}$ via the pairing $\langle x, y \rangle = e^{2\pi i x y/4}$. Let $S = T = \{0, 1\}$. We check condition (1) of Definition 2.9 as follows for each value of y:

$$\begin{aligned}
y = 0&: \quad \left|\frac{1}{2}(1+1)\right|^2 = 1 = \frac{1}{2}\#\{(0,0),(1,1)\},\\
y = 1&: \quad \left|\frac{1}{2}(1+i)\right|^2 = \frac{1}{2} = \frac{1}{2}\#\{(1,0)\},\\
y = 2&: \quad \left|\frac{1}{2}(1-1)\right|^2 = 0 = \frac{1}{2}\#\{\},\\
y = 3&: \quad \left|\frac{1}{2}(1-i)\right|^2 = \frac{1}{2} = \frac{1}{2}\#\{(0,1)\}.
\end{aligned}$$

Thus, S and T are formally dual to each other. We call this configuration TITO, which stands for "two-in two-out":

$$\cdots \quad \bullet \quad \bullet \quad \circ \quad \circ \quad \bullet \quad \bullet \quad \circ \quad \circ \quad \cdots$$

TITO yields the following formally self-dual configuration in one-dimensional Euclidean space $\mathbb{R}$:

$$\mathcal{P} = \mathcal{Q} = 2\mathbb{Z} \cup (2\mathbb{Z} + 1/2).$$

All of the examples from [**CKS**] described in the introduction are products of copies of $\mathbb{Z}$ and the TITO configuration $\mathcal{P}$, up to linear transformations. For example, it is not hard to check that for odd n we can obtain D_n^+ from the product $\mathcal{P} \times \mathbb{Z}^{n-1}$. (Recall that for even n, D_n^+ is a lattice.) Similarly, the putative optimum $\mathcal{P}_6$ in six dimensions can be obtained from $\mathcal{P}^2 \times \mathbb{Z}^4$. These product decompositions imply formal duality, by the following lemma.

LEMMA 3.1. *Let $S_1 \subseteq G_1$ and $T_1 \subseteq \widehat{G_1}$ be formal duals, and let $S_2 \subseteq G_2$ and $T_2 \subseteq \widehat{G_2}$ be formal duals. Then $S_1 \times S_2 \subseteq G_1 \times G_2$ is formally dual to $T_1 \times T_2 \subseteq \widehat{G_1} \times \widehat{G_2}$.*

PROOF. This follows directly from the second criterion in Definition 2.9. (Of course the first criterion also leads to a simple proof.) Setting $G = G_1 \times G_2$, $S = S_1 \times S_2$, and $T = T_1 \times T_2$ and identifying $\widehat{G}$ with $\widehat{G_1} \times \widehat{G_2}$, we must show that every function $f\colon G \to \mathbb{C}$ satisfies

$$\frac{1}{|S|^{3/2}} \sum_{v,v' \in S} f(v - v') = \frac{1}{|T|^{3/2}} \sum_{w,w' \in T} \widehat{f}(w - w').$$

This identity follows immediately from taking the product of the corresponding identities for G_1 and G_2 if there are functions $f_i\colon G_i \to \mathbb{C}$ such that $f(x_1, x_2) = f_1(x_1)f_2(x_2)$ for all $(x_1, x_2) \in G_1 \times G_2$. Such functions span all the functions on $G_1 \times G_2$, which completes the proof. ☐

3.2. The Gauss sum configurations. We now consider the case $G = (\mathbb{Z}/p\mathbb{Z})^2$ and $\widehat{G} = (\mathbb{Z}/p\mathbb{Z})^2$, with p an odd prime. The pairing is given by

$$\langle (a, b), (c, d) \rangle = \zeta_p^{ac+bd},$$

where $\zeta_p = e^{2\pi i/p}$.

THEOREM 3.2. *For all nonzero elements α and β of $\mathbb{Z}/p\mathbb{Z}$, the subsets $S = \{(\alpha n^2, \beta n) \mid n \in \mathbb{Z}/p\mathbb{Z}\}$ and $T = \{(n, n^2) \mid n \in \mathbb{Z}/p\mathbb{Z}\}$ are formally dual to each other.*

PROOF. Recall that the absolute value squared of the classical Gauss sum $\sum_{n=1}^{p} \zeta_p^{n^2}$ is p. It follows by completing the square that

$$\left| \sum_{n=1}^{p} \zeta_p^{c\alpha n^2 + d\beta n} \right|^2 = \begin{cases} p^2 & \text{if } p \text{ divides } c \text{ and } d, \\ 0 & \text{if } p \text{ divides } c \text{ but not } d, \text{ and} \\ p & \text{if } p \text{ does not divide } c. \end{cases}$$

Thus, to check formal duality using criterion (1) from Definition 2.9, we just need to verify that the system of equations

$$(c, d) = (j - k, j^2 - k^2)$$

has p solutions if $c = d = 0$, no solution if $c = 0, d \neq 0$ and exactly one solution if $c, d \neq 0$. The first two of these statements are obvious. For the last one, note that we may solve $j + k = d/c$, which leads to a unique solution $(j, k) = \left(\frac{1}{2}(\frac{d}{c} + c), \frac{1}{2}(\frac{d}{c} - c)\right)$, since 2 is invertible modulo p. ☐

REMARK 3.3. Because α and β can vary, the formal dual of a subset is not unique, even modulo translation and automorphisms.

4. Structure theory in the cyclic case

4.1. Basic structure theory. We begin with a few observations on the structure of formally dual sets.

The first basic observation is that if $S \subseteq G$ and $T \subseteq \widehat{G}$ are formally dual, and $x \in G$, $y \in \widehat{G}$, then $S + x$ and $T + y$ are also formally dual (since formal duality only cares about differences of elements).

Let G be a finite abelian group, and H a subgroup of G. Viewing $\widehat{G} = \mathrm{Hom}(G, S^1)$ and $\widehat{H} = \mathrm{Hom}(H, S^1)$, we have a natural restriction map $\phi\colon \widehat{G} \to \widehat{H}$, with kernel the annihilator of H, i.e.,

$$H^\perp := \{y \in \widehat{G} \mid \langle x, y\rangle = 1 \text{ for all } x \in H\}.$$

Now, if $S \subseteq H$ and $T \subseteq \widehat{H}$ are formally dual subsets, we may regard S as a subset of G and lift T to $\widehat{G}$ using ϕ^{-1}.

LEMMA 4.1. *The subsets $S \subseteq H$ and $T \subseteq \widehat{H}$ are formally dual if and only if $S \subseteq G$ and $\phi^{-1}(T) \subseteq \widehat{G}$ are formally dual.*

PROOF. The easiest way to see this is to use condition (1) of Definition 2.9, with the roles of G and $\widehat{G}$ reversed. It says that $S \subseteq H$ and $T \subseteq \widehat{H}$ are formally dual iff for all $x \in H$,

$$\left|\frac{1}{|T|}\sum_{w\in T}\langle w, x\rangle\right|^2 = \frac{1}{|S|}\cdot \#\{(v, v') \in S \times S \mid x = v - v'\}.$$

Under the above transformation from $(H, \widehat{H}, S, T)$ to $(G, \widehat{G}, S, \phi^{-1}(T))$, the right side remains unchanged if $x \in H$, while the left side becomes

$$\left|\frac{1}{(G:H)|T|}\sum_{z\in\phi^{-1}(T)}\langle z, x\rangle\right|^2 = \left|\frac{1}{|T|}\sum_{w\in T}\langle w, x\rangle\right|^2$$

since for every $z \in \widehat{G}$ mapping to $w \in \widehat{H}$ under ϕ, we have $\langle z, x\rangle = \langle w, x\rangle$, and there are exactly $(G : H)$ such z for any w. Thus, for $x \in H$ condition (1) holds for $(G, \widehat{G}, S, \phi^{-1}(T))$ iff it holds for $(H, \widehat{H}, S, T)$.

On the other hand, if $x \notin H$, then

$$\#\{(v, v') \in S \times S \mid x = v - v'\} = 0,$$

since $S - S \subseteq H - H = H$. The sum

$$\sum_{z\in\phi^{-1}(T)}\langle z, x\rangle$$

also vanishes: for each $t \in T$, let t_0 be any element of $\phi^{-1}(t)$, and then

$$\sum_{z\in\phi^{-1}(\{t\})}\langle z, x\rangle = \sum_{y\in H^\perp}\langle y + t_0, x\rangle = \langle t_0, x\rangle\sum_{y\in H^\perp}\langle y, x\rangle = 0,$$

because $y \mapsto \langle y, x\rangle$ is a non-trivial character of $H^\perp$, which sums to zero over $H^\perp$. This completes the proof of equivalence. □

In fact, this construction is reversible.

LEMMA 4.2. *Let $S \subseteq H \leq G$ be formally dual to $T \subseteq \widehat{G}$. Then T is invariant under addition by any element of $H^{\perp}$, and the image of T under the restriction map ϕ is a formal dual to $S \subseteq H$ in $\widehat{H}$.*

Here $H \leq G$ means H is a subgroup of G.

PROOF. For $y \in \widehat{H}$, define its multiplicity by

$$m(y) = \#\left(\phi^{-1}(y) \cap T\right).$$

Evidently $0 \leq m(y) \leq (G : H)$ for all $y \in \widehat{H}$. We will begin by refining this to $m(y) \in \{0, (G : H)\}$. Recall that for each $x \in G$,

$$\left|\sum_{w \in T} \langle x, w\rangle\right|^2 = \frac{|T|^2}{|S|} \cdot \#\{(v, v') \in S \times S \mid x = v - v'\}. \tag{4.1}$$

Summing this over all $x \in H$, the left side becomes

$$\sum_{x \in H}\left|\sum_{y \in \widehat{H}} m(y)\langle x, y\rangle\right|^2 = \sum_{x \in H}\sum_{y, y' \in \widehat{H}} m(y)m(y')\langle x, y - y'\rangle.$$

Interchanging the order of summation, we see that this equals

$$|H| \sum_{y, y' \in \widehat{H}} m(y)m(y')\delta_{y,y'} = |H| \sum_{y \in \widehat{H}} m(y)^2.$$

To simplify the right side of (4.1) after summing over $x \in H$, we observe that all differences of the form $v - v'$ are automatically in H. We thus get

$$|H| \sum_{y \in \widehat{H}} m(y)^2 = |T|^2 |S|.$$

Using $|S||T| = |G|$ and $\sum_{y \in \widehat{H}} m(y) = |T|$ and canceling $|H|$, we may rewrite this as

$$\sum_{y \in \widehat{H}} m(y)^2 = (G : H) \sum_{y \in \widehat{H}} m(y).$$

Now for each individual $y \in \widehat{H}$ we have $m(y)^2 \leq (G : H)m(y)$, with equality if and only if $m(y) \in \{0, (G : H)\}$. Hence the previous equation is only possible if this is indeed the case for all $y \in \widehat{H}$.

It follows that T is invariant under translation by $H^{\perp}$, because for each y, $\phi^{-1}(y)$ consists of an $H^{\perp}$-orbit of size $(G : H)$. Thus, we are in the situation covered by Lemma 4.1, and we conclude that $S \subseteq H$ and $\phi(T) \subseteq \widehat{H}$ are formally dual. □

The above results correspond to producing new formally dual configurations in Euclidean space by taking a smaller underlying lattice. Let us say that S and T are a *primitive* pair of formally dual configurations if S is not contained in a coset of a proper subgroup of G and T is not contained in a coset of a proper subgroup of $\widehat{G}$. In the classification of formal duals, we may restrict to the primitive case.

4.2. The 1-dimensional case. When G is cyclic, we conjecture that there are no primitive formally dual configurations except the trivial example and TITO. We are able to prove the conjecture when $G = \mathbb{Z}/p^2\mathbb{Z}$, with p an odd prime. The same is obviously true for $\mathbb{Z}/p\mathbb{Z}$, since the product of the sizes of the dual configurations would be p. By contrast, Theorem 3.2 shows that there are nontrivial examples in $(\mathbb{Z}/p\mathbb{Z})^2$.

PROPOSITION 4.3. *Let p be an odd prime. Then there are no primitive formally dual configurations in $G = \mathbb{Z}/p^2\mathbb{Z}$ and its dual.*

PROOF. If such configurations exist, then they must both have size p. Let $S = \{v_1, \dots, v_p\}$ and $T = \{w_1, \dots, w_p\}$ be formally dual, where we have identified $\widehat{G}$ with $\mathbb{Z}/p^2\mathbb{Z}$ via the pairing $\langle x, y\rangle = \zeta^{xy}$ with $\zeta = e^{2\pi i/p^2}$. We assume without loss of generality that $v_1 = w_1 = 0$.

From the first condition of Definition 2.9, we obtain

$$\left|\sum_{i=1}^{p} \zeta^{yv_i}\right|^2 = p \cdot n_y,$$

where we set $n_y = \#\{(j,k) \mid w_j - w_k = y\}$. That is,

$$p + \sum_{i \neq j} \zeta^{y(v_i - v_j)} = p \cdot n_y.$$

So $Z_y := \sum_{i\neq j} \zeta^{y(v_i-v_j)}$ is the rational integer $p(n_y - 1)$. Now, note that as y ranges over all the numbers modulo p^2 that are coprime to p, the algebraic numbers Z_y are all conjugates of each other. Since they are integers, they are all equal, and so are the numbers n_y. Furthermore, we cannot have $n_y = 0$ for all y coprime to p; otherwise all of $w_1, \dots, w_p$ would be multiples of p (since $w_1 = 0$) and T would be contained in a subgroup. Thus $n_y \geq 1$, and $Z_y \geq 0$. But their sum

$$\sum_{\gcd(y,p)=1} \sum_{i\neq j} \zeta^{y(v_i-v_j)} = \sum_{i\neq j} \sum_{\gcd(y,p)=1} \zeta^{y(v_i-v_j)}$$

equals zero, because the inner sum is zero for every pair $i \neq j$. (This follows from $\sum_{j=1}^{p^2} \zeta^j = 0$ and $\sum_{j=1}^{p} \zeta^{pj} = 0$.) Therefore $n_y = 1$ for all y, which means the differences $w_i - w_j$ for $i \neq j$ cover all the $p(p-1)$ elements modulo p^2 that are coprime to p exactly once. This is impossible by the following lemma, so we get a contradiction. □

LEMMA 4.4. *Let p be an odd prime. Then there is no subset S of $\mathbb{Z}/p^2\mathbb{Z}$ whose difference set $\{x - y \mid x, y \in S,\ x \neq y\}$ is the set of elements coprime to p.*

PROOF. Assume there is such a set S. Then the elements of S must be distinct modulo p, since otherwise some difference would be a multiple of p. Without loss of generality $0 \in S$, since we can translate S arbitrarily. We list the elements as

$$x_0 = 0, \quad x_1 = 1 + a_1 p, \quad x_2 = 2 + a_2 p, \quad \dots, \quad x_{p-1} = (p-1) + a_{p-1} p,$$

where the integers a_i are well defined modulo p. Now, among the differences, the numbers congruent to 1 modulo p are

$$\begin{aligned} x_1 - x_0 &= 1 + a_1 p, \\ x_2 - x_1 &= 1 + (a_2 - a_1)p, \\ &\vdots \\ x_{p-1} - x_{p-2} &= 1 + (a_{p-1} - a_{p-2})p, \\ x_0 - x_{p-1} &= p^2 - (p-1) - a_{p-1}p = 1 + (p - 1 - a_{p-1})p. \end{aligned}$$

Since these differences are all distinct modulo p^2, we need $a_1, a_2 - a_1, \dots, p-1-a_{p-1}$ to be distinct modulo p. Taking their (telescoping) sum, we get

$$p - 1 \equiv 0 + 1 + \cdots + (p-1) = \frac{p(p-1)}{2} \pmod{p},$$

which is impossible for odd p. □

We thank Gregory Minton for providing the above short proof of the lemma.

5. Non-existence of some formal duals

In this section, we show that some well-known packings do not have formal duals.

5.1. Barlow packings. Recall that the Kepler conjecture was settled by Hales, [**H**] based partially on his work with Ferguson [**HF**]. As a result, the face-centered cubic lattice A_3 gives a densest sphere packing in $\mathbb{R}^3$. It has uncountably many equally dense competitors, the Barlow packings, obtained by layering the densest planar arrangement (i.e., the hexagonal lattice A_2) in different ways. The periodic packings among them are the only periodic packings of maximal density in $\mathbb{R}^3$. The face-centered cubic lattice has a formal dual, namely its dual lattice, and it is natural to ask whether the other periodic Barlow packings have formal duals. Proposition 5.1 shows that they do not.

The periodic Barlow packings can be constructed as follows (see [**CS1**] for more details and a geometric description). Let k be the number of hexagonal layers in a period. The A_2 lattice is spanned by two unit vectors v_1 and v_2, making an angle of $\pi/3$, and the underlying lattice Λ of the Barlow packing is spanned by v_1, v_2, and v_3, where v_3 is a vector of length $k\sqrt{2/3}$ that is orthogonal to v_1 and v_2. In addition to this lattice, we need to specify how much each layer is offset. Let $a_0, \dots, a_{k-1}$ be elements of $\{0, 1, 2\}$ with $a_j \neq a_{j+1}$ for all j (where we interpret indices modulo k). Then the entire periodic configuration consists of the cosets of Λ by the translation vectors $a_j(v_1 + v_2)/3 + jv_3/k$.

PROPOSITION 5.1. *The only periodic Barlow packing that has a formal dual is the face-centered cubic lattice.*

The face-centered cubic case is when k is a multiple of 3, the sequence $a_0, \dots, a_{k-1}$ is periodic modulo 3, and $\{a_0, a_1, a_2\} = \{0, 1, 2\}$.

For the proof of Proposition 5.1, consider any periodic Barlow packing, with the notation established above. Transforming to the setting of abelian groups and applying Lemma 4.2, we can let G be the group $\mathbb{Z}/3\mathbb{Z} \times \mathbb{Z}/k\mathbb{Z}$ (generated by $(v_1 + v_2)/3$ and v_3/k modulo Λ). The question becomes whether the subset

$S = \{(a_j, j) \mid 0 \le j < k\}$ of G has a formal dual. We identify the dual group $\widehat{G}$ with $\mathbb{Z}/3\mathbb{Z} \times \mathbb{Z}/k\mathbb{Z}$ via the pairing

$$\langle (a,b),(c,d)\rangle = \omega^{ac}\zeta^{bd},$$

where $\omega = e^{2\pi i/3}$ and $\zeta = e^{2\pi i/k}$.

LEMMA 5.2. *If S has a formal dual, then k is a multiple of* 3.

PROOF. Let T be a formal dual of S, which must have $|T| = 3$. If we take $y = (1,0)$ in

$$\frac{|T|}{|S|^2}\left|\sum_{v\in S}\langle y,v\rangle\right|^2 = \#\{(w,w')\in T : y = w - w'\},$$

we find that

$$\frac{3}{k^2}\left|\sum_{j=0}^{k-1}\omega^{a_j}\right|^2$$

is an integer, which must be 0, 1, 2, or 3. It cannot be 3 because $a_0,\dots,a_{k-1}$ are not all equal. If it is 1 or 2, then $\left|\sum_{j=0}^{k-1}\omega^{a_j}\right|^2$ is $k^2/3$ or $2k^2/3$ and is also an algebraic integer, so k must be divisible by 3. Finally, if $\sum_{j=0}^{k-1}\omega^{a_j} = 0$, then the polynomial $\sum_{j=0}^{k-1}x^{a_j}$ is divisible by $1+x+x^2$ in $\mathbb{Z}[x]$, and setting $x=1$ shows that k is a multiple of 3. □

For the remainder of the proof of Proposition 5.1, suppose S does have a formal dual T. Because k is a multiple of 3, we can replace ω with $\zeta^{k/3}$ and write the condition for formal duality with $y = (r,s)$ as

$$\frac{3}{k^2}\left|\sum_{j=0}^{k-1}\zeta^{ra_jk/3+sj}\right|^2 = \#\{(w,w')\in T\times T : (r,s) = w - w'\}.$$

Without loss of generality, we can let $T = \{0, t_1, t_2\}$. There are at most six nonzero differences of elements of T, namely $\pm t_1$, $\pm t_2$, and $\pm(t_1 - t_2)$. Thus, there can be at most six nonzero vectors (r,s) for which

$$\sum_{j=0}^{k-1}\zeta^{ra_jk/3+sj} \ne 0.$$

Our next step is to show that whenever (r,s) satisfies $\sum_j \zeta^{ra_jk/3+sj} \ne 0$, its second coordinate s must have a large factor in common with k. For example, in the face-centered cubic case s is always divisible by $k/3$, and this divisibility corresponds to the periodicity of $a_0,\dots,a_{k-1}$ modulo 3.

Let $m = k/3$, and write $\sum_j \zeta^{ra_jm+sj}$ in terms of $\zeta' := \zeta^{\gcd(m,s)}$, which is a primitive root of unity of order $k/\gcd(m,s)$. The automorphisms of $\mathbb{Q}(\zeta')$ are given by $\zeta' \mapsto (\zeta')^u$ with u a unit modulo $k/\gcd(m,s)$. These maps preserve whether $\sum_j \zeta^{ra_jm+sj}$ vanishes, and they amount to multiplying $y = (r,s)$ by u. Note that $us = u's$ in $\mathbb{Z}/k\mathbb{Z}$ iff $u \equiv u' \pmod{k/\gcd(k,s)}$, and $us \ne 0$ if $s \ne 0$.

LEMMA 5.3. *Given positive integers a and b with a dividing b, there exist $\varphi(a)$ units modulo b that are distinct modulo a.*

Here φ denotes the Euler totient function. Lemma 5.3 amounts to the standard fact that the restriction map from $(\mathbb{Z}/b\mathbb{Z})^\times$ to $(\mathbb{Z}/a\mathbb{Z})^\times$ is surjective; we will provide a proof for completeness.

PROOF. Factor b/a as $b'a'$, where a' contains all the prime factors that also divide a and b' contains all those that do not. Then units modulo a are also units modulo aa', and we can use the Chinese remainder theorem to lift them to values that are 1 modulo b' and the same modulo aa'. The result is $\varphi(a)$ units modulo b that are distinct modulo a. □

Given a nonzero element (r,s) for which $\sum_j \zeta^{ra_jm+sj} \neq 0$, we can now apply Lemma 5.3 with $a = k/\gcd(k,s)$ and $b = k/\gcd(m,s)$ to find at least $\varphi\big(k/\gcd(k,s)\big)$ distinct, nonzero elements (ur, us) of $\mathbb{Z}/3\mathbb{Z} \times \mathbb{Z}/k\mathbb{Z}$ such that $\sum_j \zeta^{ura_jm+usj} \neq 0$. Thus, $\varphi\big(k/\gcd(k,s)\big) \leq 6$, which implies $k/\gcd(k,s) \leq 18$.

If $\varphi\big(k/\gcd(k,s)\big) \geq 3$ for some nonzero $(r,s) \in T - T$, then at least three elements of $\{\pm t_1, \pm t_2, \pm(t_1 - t_2)\}$ have the same value of $\gcd(k,s)$ for their second coordinate s. Call this common value g. It follows from the pigeonhole principle that g divides the second coordinate of at least two of t_1, t_2, and $t_1 - t_2$, and hence all three of them. Therefore every element of T has second coordinate a multiple of g, and $k/g \leq 18$.

The other possibility is that $\varphi\big(k/\gcd(k,s)\big) \leq 2$ for all nonzero $(r,s) \in T - T$. Then $k/\gcd(k,s) \in \{1,2,3,4,6\}$ for all such (r,s), and the least common multiple of these numbers is 12. Letting g be the greatest common divisor of $\gcd(k,s)$ for all nonzero $(r,s) \in T - T$, we find that every element of T has second coordinate a multiple of g, with $k/g \leq 12$.

Thus, in every case T is contained in the subgroup of $\mathbb{Z}/3\mathbb{Z} \times \mathbb{Z}/k\mathbb{Z}$ generated by $(1,0)$ and $(0,g)$, for some g with $k/g \leq 18$. By Lemma 4.2, S must be invariant under the annihilator of this subgroup, which is generated by $(0, k/g)$. In other words, the layers in the Barlow packing are periodic modulo k/g, where $k/g \leq 18$.

This means we can assume without loss of generality that there are at most 18 layers (i.e., $k \leq 18$). Furthermore, we can assume $a_0 = 0$ and $a_1 = 1$. Then there are few enough possibilities to enumerate them by computer, and one can check that the integrality conditions

$$\frac{3}{k^2}\left|\sum_{j=0}^{k-1} \zeta^{ra_jk/3+sj}\right|^2 \in \mathbb{Z}$$

rule out all cases except the face-centered cubic lattice. This completes the proof of Proposition 5.1.

5.2. The Best packing in $\mathbb{R}^{10}$. The Best packing is the densest known packing in $\mathbb{R}^{10}$. It is a periodic configuration, consisting of 40 translates of a lattice. It can be constructed as the subset of $\mathbb{Z}^{10}$ that reduces modulo 2 to the nonlinear $(10, 40, 4)$ Best binary code (see [**CS2**, p. 140]).

PROPOSITION 5.4. *The Best configuration does not have a formal dual.*

PROOF. Again applying Lemma 4.2, we can assume $G = (\mathbb{Z}/2\mathbb{Z})^{10}$ and $S \subseteq G$ is the Best code. Since $|S| = 40$ does not divide $|G| = 1024$, there cannot be a formal dual. □

It remains an open question whether the Best packing has a radial formal dual [**CS2**, p. 185]. It seems unlikely that it has one, but radial formal duality does not support the sort of structural analysis we have used to prove Proposition 5.4.

6. Open questions

We conclude with some open questions about formal duality. Formal duality initially arose in the simulations described in [**CKS**], and its occurrence there remains unexplained: although our results in this paper substantially clarify the algebraic foundations of this duality theory, they give no conceptual explanation of why periodic energy minimization ground states in low dimensions seem to exhibit formal duality. That is the most puzzling aspect of the theory.

It would be interesting to classify all formally dual pairs. Is every example derived from the trivial construction, TITO, and the Gauss sum construction by taking products and inflating the group (as in Lemma 4.1)?

TITO feels like a characteristic two relative of the Gauss sum construction, but it occurs in $\mathbb{Z}/4\mathbb{Z}$ rather than $(\mathbb{Z}/2\mathbb{Z})^2$. Is there a unified construction that subsumes TITO and the Gauss sum cases?

Conway and Sloane have given a conjectural list of all the "tight" packings in up to nine dimensions [**CS1**]. Their list is believed to include all the densest periodic packings in these dimensions. Can one analyze which ones have formal duals, perhaps by adapting the proof of Proposition 5.1? Note that the list contains at least a few non-lattice packings with formal duals, namely Λ_5^2, Λ_6^2, and Λ_7^3, as shown in [**CKS**].

Acknowledgments

We thank Noam Elkies, Gregory Minton, and Peter Sarnak for helpful conversations.

References

[CK1] Henry Cohn and Abhinav Kumar, *Universally optimal distribution of points on spheres*, J. Amer. Math. Soc. **20** (2007), no. 1, 99–148, DOI 10.1090/S0894-0347-06-00546-7. MR2257398 (2007h:52009)

[CK2] Henry Cohn and Abhinav Kumar, *Counterintuitive ground states in soft-core models*, Phys. Rev. E (3) **78** (2008), no. 6, 061113, 7pp, DOI 10.1103/PhysRevE.78.061113. MR2546058 (2010i:82096)

[CKS] H. Cohn, A. Kumar, and A. Schürmann, *Ground states and formal duality relations in the Gaussian core model*, Phys. Rev. E (3) **80** (2009), no. 6, 061116, 7pp. DOI 10.1103/PhysRevE.80.061116.

[CS1] J. H. Conway and N. J. A. Sloane, *What are all the best sphere packings in low dimensions?*, Discrete Comput. Geom. **13** (1995), no. 3–4, 383–403, DOI 10.1007/BF02574051. MR1318784 (96c:52031)

[CS2] J. H. Conway and N. J. A. Sloane, *Sphere packings, lattices and groups*, 3rd ed., Grundlehren der Mathematischen Wissenschaften [Fundamental Principles of Mathematical Sciences], vol. 290, Springer-Verlag, New York, 1999. With additional contributions by E. Bannai, R. E. Borcherds, J. Leech, S. P. Norton, A. M. Odlyzko, R. A. Parker, L. Queen and B. B. Venkov. MR1662447 (2000b:11077)

[C] Antonio Córdoba, *La formule sommatoire de Poisson* (French, with English summary), C. R. Acad. Sci. Paris Sér. I Math. **306** (1988), no. 8, 373–376. MR934622 (89c:46059)

[H] Thomas C. Hales, *A proof of the Kepler conjecture*, Ann. of Math. (2) **162** (2005), no. 3, 1065–1185, DOI 10.4007/annals.2005.162.1065. MR2179728 (2006g:52029)

[HF] Thomas C. Hales and Samuel P. Ferguson, *A formulation of the Kepler conjecture*, Discrete Comput. Geom. **36** (2006), no. 1, 21–69, DOI 10.1007/s00454-005-1211-1. MR2229658 (2007d:52022)

[HKCSS] A. Roger Hammons Jr., P. Vijay Kumar, A. R. Calderbank, N. J. A. Sloane, and Patrick Solé, *The* $\mathbf{Z}_4$*-linearity of Kerdock, Preparata, Goethals, and related codes*, IEEE Trans. Inform. Theory **40** (1994), no. 2, 301–319, DOI 10.1109/18.312154. MR1294046 (95k:94030)

[S] F. H. Stillinger, *Phase transitions in the Gaussian core system*, J. Chem. Phys. **65** (1976), 3968–3974, DOI 10.1063/1.432891.

Microsoft Research New England, One Memorial Drive, Cambridge, Massachusetts 02142

E-mail address: `cohn@microsoft.com`

Department of Mathematics, Massachusetts Institute of Technology, Cambridge, Massachusetts 02139

E-mail address: `abhinav@math.mit.edu`

Mathematisches Seminar der Universität Hamburg, Bundesstr. 55, D-20146 Hamburg, Germany

E-mail address: `Christian.Reiher@uni-hamburg.de`

Institute for Mathematics, University of Rostock, 18051 Rostock, Germany

E-mail address: `achill.schuermann@uni-rostock.de`

Contemporary Mathematics
Volume **625**, 2014
http://dx.doi.org/10.1090/conm/625/12496

On constructions of semi-bent functions from bent functions

Gérard Cohen and Sihem Mesnager

In honour of our friend Ilya Dumer for his 60th birthday

ABSTRACT. Plateaued functions are significant in cryptography as they possess various desirable cryptographic properties. Two important classes of plateaued functions are those of bent functions and semi-bent functions, due to their combinatorial and algebraic properties. Constructions of bent functions have been extensively investigated. However only few constructions of semi-bent functions have been proposed in the literature. In general, finding new constructions of bent and semi-bent functions is not a simple task. The paper is devoted to the construction of semi-bent functions with even number of variables. We show that bent functions give rise to primary and secondary-like constructions of semi-bent functions.

1. Introduction

A Boolean function over the Galois field $\mathbb{F}_{2^n}$ is said r-plateaued if the values of its Walsh transform belong to the set $\{0, \pm 2^{\frac{n+r}{2}}\}$, $r = 0, 1, \cdots n$. Plateaued functions [**29, 30**] are significant in cryptography as they possess desirable various cryptographic characteristics such as high nonlinearity, resiliency, propagation criteria, low additive autocorrelation and high algebraic degree. Two important classes of plateaued functions are those of bent functions and of semi-bent functions, due to their algebraic and combinatorial properties.

Bent functions introduced in 1974 [**9**] ,[**27**] are extremal objects in combinatorics and Boolean function theory. Bent functions exist only with even number of variables. They have been studied for about 35 years (even more, under the name of difference sets in elementary Abelian 2-groups). The motivation for the study of these particular difference sets is mainly cryptographic but bent functions play also a role in sequence theory, as difference sets and especially in coding theory, as elements of Reed-Muller (RM) codes. Indeed, bent functions realize the maximal possible distance from first-order RM codes; RM codes are quite popular, particularly in view of their recursive structure, exploited for their decoding in a few influencial papers by Ilya Dumer (see [**10**],[**1**] for recent advances). By achieving optimum nonlinearity, bent functions permit to resist linear attacks in the best possible way. Bent functions also satisfy the propagation criterion with respect to the non-zero vector. But, being neither balanced nor correlation immune, they are

2010 *Mathematics Subject Classification.* Primary 12Y05.
Key words and phrases. Boolean functions, Walsh transform, Bent functions, Semi-bent functions.

improper for direct cryptographic use. Thanks to the well known Parseval identity, the maximum nonlinearity they attain implies that the Hadamard Walsh transform of an n-variable (n even) bent function takes only the two values $\pm 2^{n/2}$. A good survey of bent functions can be found in the book chapter of Carlet [**4**].

Semi-bent functions have been introduced by Chee, Lee and Kim [**8**] and previously investigated under the name of three-valued almost optimal Boolean functions [**2**]. Semi-bent functions exist in even or odd dimension. In both cases, they are defined in terms of Walsh Hadamard transform. In even dimension, an n-variable Boolean function is said to semi -bent if its Hadamard Walsh transform takes three values 0 and $\pm 2^{\frac{n+2}{2}}$. Very recently, the development of the theory of semi-bent functions has increased. The motivation for their study is firstly related to their use in cryptography (we recall that in the design of cryptographic functions, various characteristics need be considered simultaneously). Indeed, unlike bent functions, semi-bent functions can also be balanced and resilient. They also possess various desirable characteristics such as a low Hadamard transform (which provides protection against fast correlation attacks [**19**] and linear cryptanalysis [**18**]), have low autocorrelation, satisfy the propagation criteria and high algebraic degree. Secondly, beside their practical use in cryptography, they are also widely used in code division multiple access (CDMA) communication systems for sequence design [**11**], [**26**], [**12**], [**13**], [**14**], [**15**], [**16**].

A lot of research has been devoted to designing constructions of bent functions. The reader can see [**4**] for general constructions of bent functions and the paper [**20**] for a complete state of the art on bent functions over the Galois field $\mathbb{F}_{2^n}$. However, only few constructions have been proposed for semi-bent functions. In even dimension, there exist some constructions of quadratic semi-bent functions ([**7**] and in [**28**]) and infinite classes of semi-bent functions with maximal algebraic degree obtained very recently in [**6**]. The reader can also see the reference [**21**] for recent results dealing with the constructions of semi-bent functions via Dillon and Niho exponents under some conditions (on the coefficients of the Boolean functions defined on $\mathbb{F}_{2^n}$) directly related to the Kloosterman sums; in particular, it was shown in [**24**] and [**21**] that the zeros and the value four of binary Kloosterman sums give rise to semi-bent functions in even dimension with maximum degree, as well as to constructions of semi-bent functions of multiples traces terms under some conditions (on the coefficients of the Boolean functions defined on $\mathbb{F}_{2^n}$) involving the Dickson polynomials. Very recently, it was shown in [**6**] and [**23**] that the oval polymomials from finite projective geometry give rise to several constructions of semi-bent functions. In these references, several constructions of semi-bent functions in bivariate representation obtained from bent functions have been provided.

In this paper, we focus on the constructions of semi-bent functions. The idea is to exploit the known constructions of bent functions to design new semi-bent functions and therefore extend the list of the known primary constructions of semi-bent functions in even dimension. We organize this paper as follows. Section 2 is an introductory part providing some preliminaries including definitions and background related to Boolean functions. Section 3 is devoted to the constructions of semi-bent functions from bent functions. Firstly, we revisited a part of a joint work of the second author with Carlet [**6**] by studying more in details (and providing a direct proof of) those constructions on the Galois field $\mathbb{F}_{2^n}$ (n even) by considering $\frac{n}{2}$-spreads of $\mathbb{F}_{2^n}$. Secondly, we treat the case of a kind of recursive construction of semi-bent functions (we shall call it a "secondary-like construction"). We prove that an indirect sum involving both bent and semi-bent functions leads to semi-bent functions. Finally, we construct semi-bent functions on $\mathbb{F}_{2^{n+2}}$ coming from bent functions on $\mathbb{F}_{2^n}$.

2. Notation and preliminaries

For any set E, $E^\star = E \setminus \{0\}$ and $\#E$ will denote the cardinality of E.

• *Boolean functions and polynomial forms*:

Let n be a positive integer. A Boolean function f in n variables is a mapping from $\mathbb{F}_2^n$ to $\mathbb{F}_2$. In cryptography, the most usual representation of these functions is the *algebraic Normal Form* (ANF) :

$$f(x_1, \cdots, x_n) = \sum_{I \subseteq \{1,\ldots,n\}} a_I \left(\prod_{i \in I} x_i \right)$$

where the a_I's are in $\mathbb{F}_2$. The terms $\prod_{i\in I} x_i$ are called monomials. The *algebraic degree* of a Boolean function f equals the global degree of its (unique) ANF, that is, the maximum degree of those monomials whose coefficients are nonzero.

Another possible representation of Boolean functions uses the identification between the vector-space $\mathbb{F}_2^n$ and the finite field $\mathbb{F}_{2^n}$. It represents any Boolean function as a polynomial in one variable $x \in \mathbb{F}_{2^n}$ of the form $f(x) = \sum_{j=0}^{2^n-1} a_j x^j$ where the a_j's are elements of the field. This representation exists for every function from $\mathbb{F}_{2^n}$ to $\mathbb{F}_{2^n}$ and such function f is Boolean if and only if a_0 and a_{2^n-1} belong to $\mathbb{F}_2$ and $a_{2j} = a_j^2$ for every $j \neq 0,\ 2^n - 1$, where $2j$ is taken modulo $2^n - 1$. This allows representing $f(x)$ in a (unique) trace expansion of the form called its *polynomial form*. First, recall that, for any positive integer k and r dividing k, the trace function from $\mathbb{F}_{2^k}$ to $\mathbb{F}_{2^r}$, denoted by Tr_r^k, is the mapping defined as:

$$Tr_r^k(x) := \sum_{i=0}^{\frac{k}{r}-1} x^{2^{ir}} = x + x^{2^r} + x^{2^{2r}} + \cdots + x^{2^{k-r}}.$$

In particular, we denote the *absolute trace* over $\mathbb{F}_2$ of an element $x \in \mathbb{F}_{2^n}$ by $Tr_1^n(x) = \sum_{i=0}^{n-1} x^{2^i}$. We make use of the following known property of the trace function $Tr_1^n(x) = Tr_1^n(x^2)$ and for every integer r dividing k, the transitivity property of Tr_r^k, that is, $Tr_1^k = Tr_1^r \circ Tr_r^k$.

Now, the polynomial form of a Boolean function defined on $\mathbb{F}_{2^n}$ f is given by :

$$f(x) = \sum_{j\in\Gamma_n} Tr_1^{o(j)}(a_j x^j) + \epsilon(1 + x^{2^n-1})$$

where

-Γ_n is the set of integers obtained by choosing one element in each cyclotomic class of 2 modulo $2^n - 1$ (the most usual choice for j is the smallest element in its cyclotomic class, called the coset leader of the class),

- $o(j)$ is the size of the cyclotomic coset of 2 modulo $2^n - 1$ containing j,

- $a_j \in \mathbb{F}_{2^{o(j)}}$,

- $\epsilon = wt(f)$ modulo 2 where $wt(f)$, is the *Hamming weight* of the image vector of f, that is, the cardinality of its support $supp(f) := \{x \in \mathbb{F}_{2^n} \mid f(x) = 1\}$.

The algebraic degree of f is then equal to the maximum 2-weight of an exponent j for which $a_j \neq 0$ if $\epsilon = 0$ and to n if $\epsilon = 1$. Recall that the 2-weight $w_2(j)$ of an integer j equals by definition the number of 1's in its binary expansion. In particular an affine function is a Boolean function whose algebraic degree is at most 1.

- *Walsh-Hadamard transform*

Let $\chi : \mathbb{F}_2 \mapsto \mathbb{Z}$ denote the nontrivial additive character of $\mathbb{F}_2$. The *"sign" function* of a Boolean function f is the integer-valued function $\chi_f = (-1)^f$.

Let f be a Boolean function defined on $\mathbb{F}_2^n$. Then the *Walsh Hadamard transform* of f is the discrete Fourier transform of χ_f, whose value at $\omega \in \mathbb{F}_2^n$ is defined as follows:

$$\forall \omega \in \mathbb{F}_2^n, \quad \widehat{\chi_f}(\omega) = \sum_{x \in \mathbb{F}_2^n} (-1)^{f(x)+\omega \cdot x}$$

where "$\cdot$" is the scalar product in $\mathbb{F}_2^n$ defined as $x \cdot y = \sum_{i=1}^n x_i y_i$.

The notion of Walsh transform refers to a scalar product (note that in the definition of the Walsh transform, we can take any inner product; the cryptographic properties are not related to a particular choice, therefore the issue of the choice of the isomorphism does not arise). When $\mathbb{F}_2^n$ is identified with the field $\mathbb{F}_{2^n}$ by an isomorphism between these two n-dimensional vector spaces over $\mathbb{F}_2$, it is convenient to choose the isomorphism such that the canonical scalar product "$\cdot$" in $\mathbb{F}_2^n$ coincides with the canonical scalar product in $\mathbb{F}_{2^n}$, which is the trace of the product : $x \cdot y = \sum_{i=1}^n x_i y_i = Tr_1^n(xy)$ for $x, y \in \mathbb{F}_{2^n}$. Thus if f is a Boolean function defined on $\mathbb{F}_{2^n}$ then, the Walsh Hadamard transform of f is the discrete Fourier transform of χ_f, whose value at $\omega \in \mathbb{F}_{2^n}$ is defined as follows:

$$\forall \omega \in \mathbb{F}_{2^n}, \quad \widehat{\chi_f}(\omega) = \sum_{x \in \mathbb{F}_{2^n}} (-1)^{f(x)+Tr_1^n(\omega x)}.$$

The Walsh transform satisfies the well-known Parseval's relation

$$\sum_{\omega \in \mathbb{F}_{2^n}} \widehat{\chi_f}^2(\omega) = 2^{2n}$$

and also the inverse Fourier formula

$$\sum_{\omega \in \mathbb{F}_{2^n}} \widehat{\chi_f}(\omega) = 2^n(-1)^{f(0)}.$$

Thanks to Parseval's relation and the inverse Fourier formula, one can prove the following well-known statement.

LEMMA 2.1. *Let f be a function on $\mathbb{F}_{2^n}$ such that for all $\omega \in \mathbb{F}_{2^n}$, $\widehat{\chi_f}(\omega) \geq 0$, then f is linear.*

- *Bent functions and semi-bent functions*:

Bent functions can be defined in terms of the Walsh transform as follows.

DEFINITION 2.2. A Boolean function $f : \mathbb{F}_{2^n} \to \mathbb{F}_2$ (n even) is said to be bent if $\widehat{\chi_f}(\omega) = \pm 2^{\frac{n}{2}}$, for all $\omega \in \mathbb{F}_{2^n}$.

Using Parseval's identity, one can prove (see for instance [**22**]) the following useful criterion of bentness in terms of congruence.

LEMMA 2.3. *Let g be a function on $\mathbb{F}_{2^n}$ with $n = 2m$. Then g is bent if and only if* $\forall\omega \in \mathbb{F}_{2^n}, \widehat{\chi_g}(\omega) \equiv 2^m \pmod{2^m + 1}$.

Semi-bent functions on $\mathbb{F}_{2^n}$ exist for n even or n odd. But we are interested in this paper only in semi-bent functions when n even. Such functions are defined as follows.

DEFINITION 2.4. A Boolean function $f : \mathbb{F}_{2^n} \to \mathbb{F}_2$ (n even) is said to be semi-bent if $\widehat{\chi_f}(\omega) \in \{0, \pm 2^{\frac{n+2}{2}}\}$, for all $\omega \in \mathbb{F}_{2^n}$.

It is well known (see for instance [**4**]) that the algebraic degree of a bent and a semi-bent (with n even) Boolean function defined on $\mathbb{F}_{2^n}$ is at most $\frac{n}{2}$. Consequently, the Hamming weight of all these functions is even. Therefore, the polynomial form of these functions is

$$\forall x \in \mathbb{F}_{2^n}, \quad f(x) = \sum_{j\in\Gamma_n} Tr_1^{o(j)}(a_j x^j) \tag{2.1}$$

where Γ_n, $o(j)$ are defined as above and $a_j \in \mathbb{F}_{2^{o(j)}}$.

From now, $n = 2m$ is an (even) integer.

3. Constructions of semi-bent functions from bent functions

In the sequel, we present several constructions of semi-bent functions involving bent functions.

3.1. Constructions of semi-bent functions on the Galois field $\mathbb{F}_{2^n}$ by considering m-spreads.

First recall that every non-zero element x of $\mathbb{F}_{2^n}$ has a unique decomposition (called the *polar decomposition*) as: $x = yu$ with $y \in \mathbb{F}_{2^m}^\star$ and $u \in U$ where U is the set defined by $\{u \in \mathbb{F}_{2^n} \mid norm(u) = 1\} = \{u \in \mathbb{F}_{2^n} \mid u^{2^m+1} = 1\}$. In the sequel, U will always denote the cyclic group of $(2^m + 1)$-st roots of unity.

An m-spread of $\mathbb{F}_{2^n}$ can be defined as follow.

DEFINITION 3.1. An m-spread of $\mathbb{F}_{2^n}$ is a set of pairwise supplementary m-dimensional subspaces of $\mathbb{F}_{2^n}$ whose union equals $\mathbb{F}_{2^n}$.

In a joint work of the author with Carlet [**6**], semi-bent functions on $\mathbb{F}_{2^n}$ such that their restrictions to the elements of an m-spread have degree at most 1 have been investigated. As far as we know, the only m-spread in the literature is the set $\{u\mathbb{F}_{2^m}, u \in U\}$ where $U := \{u \in \mathbb{F}_{2^n} \mid u^{2^m+1} = 1\}$ (viewed in $\mathbb{F}_{2^n}$) and its images by the linear automorphisms. Note that such an m-spread viewed in bivariate representation (that is, viewed in $\mathbb{F}_{2^n} \approx \mathbb{F}_{2^m} \times \mathbb{F}_{2^m}$) corresponds to the sets $\{E_a, E_\infty\}$ where $E_a := \{(x, ax)\,;\, x \in \mathbb{F}_{2^m}\}$ and $E_\infty := \{(0, y)\,;\, y \in \mathbb{F}_{2^m}\}$.

First let $\mathcal{C}_n$ be the set of Boolean functions $f : \mathbb{F}_{2^n} \to \mathbb{F}_2$ such that the restriction to $u\mathbb{F}_{2^m}^\star$ is constant for every $u \in U$ with $f(0) = 0$. And let $\mathcal{L}_n$ be the set of Boolean functions on $\mathbb{F}_{2^n}$ such that the restriction to $u\mathbb{F}_{2^m}^\star$ is linear for every $u \in U$ with $f(0) = 0$. Note that $f \in \mathcal{C}_n$ means that there exists a Boolean function $b : U \to \mathbb{F}_2$ such that $f(uy) = b(u)$, $\forall u \in U, \forall y \in \mathbb{F}_{2^m}^\star$ with $f(0) = 0$. While $f \in \mathcal{L}_n$ means that there exist a mapping $a : U \to \mathbb{F}_{2^m}$ and a Boolean function $b : U \to \mathbb{F}_2$ such that $f(uy) = Tr_1^m(a(u)y)$, $\forall u \in U, \forall y \in \mathbb{F}_{2^m}^\star$.

In univariate form, functions of $\mathcal{C}_n$ are the so-called Dillon-like functions built with *Dillon-like exponents.* The name of Dillon-like exponent has been introduced in [**20**]. Such exponents are of the form $s(2^m-1)$. In [**22**], the second author has proved the following statement.

PROPOSITION 3.2. *([**22**]) Let $n=2m$. Let f be a Boolean function defined on $\mathbb{F}_{2^n}$ such that $f(0)=0$. The three assertions are equivalent:*

(1) $f(x)=\sum_i Tr_1^{o(d_i)}(a_ix^{d_i})$ *with* $\forall i, d_i \equiv 0 \pmod{2^m-1}$;
(2) $\forall u \in U$, *the restriction of f to $u\mathbb{F}_{2^m}^\star$ is constant (that is, $f(uy)=f(u), \forall y \in \mathbb{F}_{2^m}^\star$);*
(3) $\forall \omega \in \mathbb{F}_{2^n}$ *the restriction of f to $\omega\mathbb{F}_{2^m}^\star$ is constant (that is, $f(\omega y)=f(\omega), \forall y \in \mathbb{F}_{2^m}^\star$).*

The following statement has been proved in a joint work of the second author with Carlet [**5**].

PROPOSITION 3.3. *([**5**]) Let f be a Boolean function over $\mathbb{F}_{2^n}$ and $f(t)=\sum_{d=0}^{2^n-1} a_dt^d$ its univariate representation. Then the restrictions of f to the vectorspaces $\omega\mathbb{F}_{2^m}$, $\omega \in \mathbb{F}_{2^n}^\star$, are all linear if and only if the only exponents d such that $a_d \neq 0$ are congruent to powers of 2 modulo 2^m-1, more precisely, $d \equiv 2^j \pmod{2^m-1}$ for some j, $0 \le j \le m-1$.*

The exponents d in the previous proposition are currently called the *Niho exponents* since they were first studied by Niho in his thesis [**26**]. Moreover, it is well known that a Niho exponent d (always understood modulo 2^n-1) can be written in normalized form as $d=(2^m-1)s+1$ with $0<s<2^m-1$ (note that $d \equiv -2s+1$ modulo 2^m+1). In univariate form, functions of $\mathcal{L}_n$ are called Niho functions since they are constructed via Niho exponents. Now, let us introduce the following notation.

NOTATION 3.4. Denote by $\mathcal{D}_n$ the bent functions in $\mathcal{C}_n$ and by $\mathcal{N}_n$ the bent functions in $\mathcal{L}_n$.

According to the discussion above, $\mathcal{D}_n$ is the set of bent functions f such that $f(x)=\sum_i Tr_1^{o(d_i)}(a_ix^{d_i})$ with $\forall i, d_i \equiv 0 \pmod{2^m-1}$ and $f(0)=0$. While $\mathcal{N}_n$ is the set of bent functions f such that $f(x)=\sum_i Tr_1^{o(d_i)}(a_ix^{d_i})$ with $\forall i, d_i = (2^m-1)s_i+1, 2 \le s_i \le 2^m$ and $f(0)=0$.

A list of the known functions in $\mathcal{D}_n$ can be found in [**25**] with additional functions in [**17**]. A list of the known functions in $\mathcal{N}_n$ can be found for instance in [**20**].

In the following, semi-bent functions on $\mathbb{F}_{2^n}$ such that their restrictions to the elements of the m-spread $u\mathbb{F}_{2^m}$ are affine, are revisited. We introduce the following notation.

NOTATION 3.5.

$\mathcal{A}_n := \{f : \mathbb{F}_{2^n} \to \mathbb{F}_2$ s.t the restriction to $u\mathbb{F}_{2^m}^\star$ is affine for every $u \in U\}$.

Note that $f \in \mathcal{A}_n$ means that there exists a mapping $a : U \to \mathbb{F}_{2^m}$ and a Boolean function $b : U \to \mathbb{F}_2$ such that $f(uy)=Tr_1^m(a(u)y)+b(u)$, $\forall u \in U, \forall y \in \mathbb{F}_{2^m}^\star$.

We denote by $f_{a,b}$ (where $a : U \to \mathbb{F}_{2^m}$ and $b : U \to \mathbb{F}_2$) a function in $\mathcal{A}_n$. Therefore, we have the following natural decomposition:

$$f_{a,b} = f_{a,0} + f_{0,b}$$

where $f_{a,0}$ is a Boolean function defined on $\mathbb{F}_{2^n}$ such that its restrictions to $u\mathbb{F}_{2^m}^\star$ ($u \in U$) are linear and $f_{0,b}$ is a Boolean function on $\mathbb{F}_{2^n}$ such that its restrictions to $u\mathbb{F}_{2^m}^\star$ ($u \in U$) are constant.

REMARK 3.6. $f_{a,b} \in \mathcal{A}_n$ if and only if $1+f_{a,b} \in \mathcal{A}_n$. Indeed, if $f_{a,b} \in \mathcal{A}_n$ then, $\forall u \in U$, $\forall y \in \mathbb{F}_{2^m}^\star$, we have

$$1+f_{a,b}(uy) = Tr_1^m(a(u)y) + b'(u)$$

with $b'(u) := b(u)+1$, which means that $1+f_{a,b} \in \mathcal{A}_n$. The converse is trivial, and we have $1+f_{a,b} = f_{a,b+1}$.

NOTATION 3.7. For $\epsilon \in \{0,1\}$, set

$$\mathcal{A}_n^\epsilon := \{f \in \mathcal{A}_n \mid f_{a,b}(0) = \epsilon\}.$$

We have

$$\mathcal{A}_n^\epsilon = \mathcal{A}_n^0 \cup \mathcal{A}_n^1 = \mathcal{A}_n^0 \cup (1+\mathcal{A}_n^0)$$

where $1+\mathcal{A}_n^0$ is the complement of functions in $\mathcal{A}_n^0$. In the following, we are interested in identifying the functions in $\mathcal{A}_n$ which are semi-bent. Since semi-bentness is affine invariant, it suffices to study the semi-bent functions in $\mathcal{A}_n^0$.

The Walsh transform of a function in $\mathcal{A}_n^0$ can be expressed as follows.

PROPOSITION 3.8. *Let $f_{a,b}$ be a function in $\mathcal{A}_n^0$. Then the Walsh transform of $f_{a,b}$ equals* $\widehat{\chi_{f_{a,b}}}(\omega) = 1 - \sum_{u\in U}(-1)^{b(u)} + 2^m \sum_{u\in U \mid a(u)+Tr_m^n(\omega u)=0}(-1)^{b(u)}$, $\forall \omega \in \mathbb{F}_{2^n}$.

PROOF. Let $f_{a,b} \in \mathcal{A}_n^0$. For all $\omega \in \mathbb{F}_{2^n}$, we have (using the polar decomposition and the properties of trace functions)

$$\begin{aligned}
(3.1)\quad \widehat{\chi_{f_{a,b}}}(\omega) &= 1 + \sum_{x\in\mathbb{F}_{2^n}^\star} (-1)^{f_{a,b}(x)+Tr_1^n(\omega x)} \\
&= 1 + \sum_{u\in U}\sum_{y\in\mathbb{F}_{2^m}^\star} (-1)^{f_{a,b}(uy)+Tr_1^n(\omega uy)} \\
&= 1 + \sum_{u\in U}\sum_{y\in\mathbb{F}_{2^m}^\star} (-1)^{Tr_1^m(a(u)y)+b(u)+Tr_1^m(Tr_m^n(\omega u)y)} \\
&= 1 + \sum_{u\in U}\Big(\sum_{y\in\mathbb{F}_{2^m}} (-1)^{Tr_1^m(a(u)y)+b(u)+Tr_1^m(Tr_m^n(\omega u)y)} - (-1)^{b(u)}\Big) \\
&= 1 - \sum_{u\in U}(-1)^{b(u)} + \sum_{u\in U}\sum_{y\in\mathbb{F}_{2^m}} (-1)^{Tr_1^m(a(u)y)+b(u)+Tr_1^m(Tr_m^n(\omega u)y)} \\
&= 1 - \sum_{u\in U}(-1)^{b(u)} + \sum_{u\in U}(-1)^{b(u)} \sum_{y\in\mathbb{F}_{2^m}} (-1)^{Tr_1^m((a(u)+Tr_m^n(\omega u))y)}.
\end{aligned}$$

But $\sum_{y\in\mathbb{F}_{2^m}} \chi(Tr_1^m\Big((a(u)+Tr_m^n(\omega u))y\Big)$

$$= \begin{cases} 2^m & \text{if } a(u)+Tr_m^n(\omega u) = 0 \\ 0 & \text{otherwise} \end{cases}$$

that is, $\sum_{y\in\mathbb{F}_{2^m}}(-1)^{Tr_1^m\left((a(u)+Tr_m^n(\omega u))y\right)} = 2^m\delta_0(a(u)+Tr_m^n(\omega u))$, where δ_0 is such that $\delta(x)=1$ if $x=0$ and, 0 otherwise.

$$\widehat{\chi_{f_{a,b}}}(\omega) = 1-\sum_{u\in U}(-1)^{b(u)} + \sum_{u\in U}(-1)^{b(u)}2^m\delta_0(a(u)+Tr_m^n(\omega u)) \tag{3.2}$$
$$= 1-\sum_{u\in U}(-1)^{b(u)} + 2^m \sum_{u\in U\mid a(u)+Tr_m^n(\omega u)=0}(-1)^{b(u)}.$$

□

Next we provide an alternative direct proof of the following theorem (Corollary 5, [6]) which identifies in particular all the semi-bent functions in $\mathcal{A}_n$. The reader can notice that the theorem has been obtained in [6] by applying Theorem 1 in [6]. Moreover, the statement concerns only the functions whose restrictions to the m-spreads are affine but not constant and not linear (in fact, it is proved in [5] that there exist no semi-bent functions whose restrictions to the m-spreads are linear).

THEOREM 3.9. *Let $n=2m$ with $m>2$. A semi-bent function in $\mathcal{A}_n$ can be written as the sum of a (bent) function in $\mathcal{D}_n$ and a (bent) function in $\mathcal{N}_n$, where $\mathcal{A}_n$, $\mathcal{D}_n$ and $\mathcal{N}_n$ are defined as in Notation* 3.5 *and Notation* 3.4.

PROOF. According to the discussion above, it suffices to treat the case of semi-bent functions in $\mathcal{A}_n^0$. So, let $f_{a,b}\in\mathcal{A}_n^0$.
According to Proposition 3.8,

$$\widehat{\chi_{f_{a,b}}}(\omega) = 1-\sum_{u\in U}(-1)^{b(u)} + 2^m\sum_{u\in E_\omega}(-1)^{b(u)}, \forall\omega\in\mathbb{F}_{2^n}$$

where $E_\omega := \{u\in U \mid a(u)+Tr_m^n(\omega u)=0\}$.
Now, $f_{a,b}$ is semi-bent if and only if $\widehat{\chi_{f_{a,b}}}(\omega)\in\{0,\pm 2^{m+1}\}, \forall\omega\in\mathbb{F}_{2^n}$, which implies that $\widehat{\chi_{f_{a,b}}}(\omega)\equiv 0 \pmod{2^m}, \forall\omega\in\mathbb{F}_{2^n}$, that is, $\sum_{u\in U}(-1)^{b(u)}\equiv 1 \pmod{2^m}$. Therefore, $\sum_{u\in U}(-1)^{b(u)}\in\{1, 1+2^m, 1-2^m\}$ (since the multiplicative group U is of order 2^m+1). Three cases have to be considered.
Recall the following decomposition: $f_{a,b}=f_{a,0}+f_{0,b}$, where $f_{a,0}$ (resp. $f_{0,b}$) is such that its restrictions to each multiplicative coset $u\mathbb{F}_{2^m}^\star$ $(u\in U)$ is constant (resp. linear).
• Case 1: $\sum_{u\in U}(-1)^{b(u)}=1$.
The function $f_{0,b}$ is such that its restrictions to the multiplicative cosets $u\mathbb{F}_{2^n}^\star$ are constant for every $u\in U$. Hence, for every $\omega\in\mathbb{F}_{2^n}$ the restriction of $f_{0,b}$ to $\omega\mathbb{F}_{2^m}^\star$ is constant (that is, $f_{0,b}(\omega y)=f_{0,b}(\omega), \forall y\in\mathbb{F}_{2^m}^\star$). Indeed, if $\omega\in\mathbb{F}_{2^n}^\star$, then using the polar decomposition, we have: $\omega=uz$ with $u\in U$ and $z\in\mathbb{F}_{2^m}^\star$. Hence, $\forall y\in\mathbb{F}_{2^m}^\star$, $f_{0,b}(\omega y)=f_{0,b}(uzy)=f_{0,b}(u)=f_{0,b}(uz)=f_{0,b}(\omega)$.
Consider the polynomial form of $f_{0,b}$: $f_{0,b}(x)=\sum_{i=1}^{2^n-2}Tr_1^{o(i)}(a_ix^i)+a_{2^n-1}x^{2^n-1}$ (since $f_{0,b}(0)=0$). Then (since $y\in\mathbb{F}_{2^m}^\star\subset\mathbb{F}_{2^n}^\star$ and $y^i=1$ for $i\equiv 0 \pmod{2^m-1}$),

we have: $\forall\omega \in \mathbb{F}_{2^n}$, $\forall y \in \mathbb{F}_{2^m}^\star$,

$$f_{0,b}(\omega y) = \sum_{i=1}^{2^n-2} Tr_1^{o(i)}(a_i\omega^i y^i) + a_{2^n-1}\omega^{2^n-1} \tag{3.3}$$
$$= \sum_{i=1|i\equiv 0 \pmod{2^m-1}}^{2^n-2} Tr_1^{o(i)}(a_i\omega^i) + a_{2^n-1}\omega^{2^n-1}$$
$$+ \sum_{i=1|i\not\equiv 0 \pmod{2^m-1}}^{2^n-2} Tr_1^{o(i)}(a_i\omega^i y^i) + a_{2^n-1}\omega^{2^n-1}.$$

Now, note that for $y = 1$, we have

$$f_{0,b}(\omega) = \sum_{i=1|i\equiv 0 \pmod{2^m-1}}^{2^n-2} Tr_1^{o(i)}(a_i\omega^i + a_{2^n-1}\omega^{2^n-1})$$
$$+ \sum_{i=1|i\not\equiv 0 \pmod{2^m-1}}^{2^n-2} Tr_1^{o(i)}(a_i\omega^i) + a_{2^n-1}\omega^{2^n-1}.$$

But we have $f_{0,b}(\omega y) + f_{0,b}(\omega) = 0$, $\forall\omega \in \mathbb{F}_{2^n}, \forall y \in \mathbb{F}_{2^m}^\star$. Therefore, $\forall\omega \in \mathbb{F}_{2^n}, \forall y \in \mathbb{F}_{2^m}^\star$, $\sum_{i=1|i\not\equiv 0 \pmod{2^m-1}}^{2^n-2} Tr_1^{o(i)}(a_i(y^i+1)\omega^i) = 0$ Now, using the unicity of the polar decomposition, we obtain $\forall i \in [1, 2^n-2], i \not\equiv 0 \pmod{2^m-1}, a_i(y^i+1) = 0$. In particular, if y equals a primitive element β of $\mathbb{F}_{2^m}$ then $y^i = \beta^i \neq 1$. Hence, $\forall i \in [1, 2^n-2], i \not\equiv 0 \pmod{2^m-1}, a_i = 0$, which proves that the polynomial form of $f_{0,b}$ is: $f_{0,b}(x) = \sum_i Tr_1^{o(d_i)}(a_i x^{d_i})$ with $\forall i, d_i \equiv 0 \pmod{2^m-1}$. At this stage, let us prove that the condition $\sum_{u\in U}(-1)^{b(u)} = 1$ is equivalent to the fact that $f_{0,b}$ is bent. To this end, let us compute the Walsh transform of $f_{0,b}$ for $\omega \in \mathbb{F}_{2^n}$.

$$\widehat{\chi_{f_{0,b}}}(\omega) = \sum_{x\in\mathbb{F}_{2^n}} (-1)^{\sum_i Tr_1^{o(d_i)}(a_i x^{d_i}) + Tr_1^n(\omega x)} \tag{3.4}$$
$$= 1 + \sum_{x\in\mathbb{F}_{2^n}^\star} (-1)^{\sum_i Tr_1^{o(d_i)}(a_i x^{d_i}) + Tr_1^n(\omega x)}$$
$$= 1 + \sum_{u\in U}\sum_{y\in\mathbb{F}_{2^m}^\star} (-1)^{\sum_i Tr_1^{o(d_i)}(a_i y^{d_i} u^{d_i}) + Tr_1^n(\omega y u)}$$
$$= 1 + \sum_{u\in U}\sum_{y\in\mathbb{F}_{2^m}^\star} (-1)^{\sum_i Tr_1^{o(d_i)}(a_i u^{d_i}) + Tr_1^n(\omega y u)}$$
$$= 1 + \sum_{u\in U}(-1)^{\sum_i Tr_1^{o(d_i)}(a_i u^{d_i})} \sum_{y\in\mathbb{F}_{2^m}^\star} (-1)^{Tr_1^n(\omega y u)}$$
$$= 1 + \sum_{u\in U}(-1)^{\sum_i Tr_1^{o(d_i)}(a_i u^{d_i})} \sum_{y\in\mathbb{F}_{2^m}} (-1)^{Tr_1^n(\omega y u)}$$
$$- \sum_{u\in U}(-1)^{\sum_i Tr_1^{o(d_i)}(a_i u^{d_i})}.$$

Firstly, if $\omega = 0$ then,

$$\widehat{\chi_f}(0) = 1 + 2^m\sum_{u\in U}(-1)^{\sum_i Tr_1^{o(d_i)}(a_i u^{d_i})} - \sum_{u\in U}(-1)^{\sum_i Tr_1^{o(d_i)}(a_i u^{d_i})}.$$

Hence, the condition $\sum_{u\in U}(-1)^{b(u)}=1$ is equivalent to $\widehat{\chi_f}(0)=2^m$ that is, $f_{0,b}$ bent. Otherwise, if $\omega\neq 0$then, $\sum_{y\in\mathbb{F}_{2^m}}\chi(Tr_1^n(\omega yu))=\sum_{y\in\mathbb{F}_{2^m}}\chi(Tr_1^m(Tr_m^n(\omega u)y))$

$$=\begin{cases} 2^m & \text{if } Tr_m^n(\omega u)=0, \text{ that is, if } u^{2^m-1}=\omega^{1-2^m}\\ 0 & \text{otherwise.}\end{cases}$$

Since $x\mapsto x^{2^m-1}$ is a permutation of U then, $\sum_{y\in\mathbb{F}_{2^m}}\chi(Tr_1^n(\omega yu))$

$$=\begin{cases} 2^m & \text{if } u=\omega^{-1}\\ 0 & \text{otherwise.}\end{cases}$$

Therefore, $\widehat{\chi_f}(\omega)=1-\sum_{u\in U}(-1)^{\sum_i Tr_1^{o(d_i)}(a_iu^{d_i})}+2^m(-1)^{f(\omega^{-1})}$. Hence, $f_{0,b}$ is bent if and only if $\sum_{u\in U}(-1)^{\sum_i Tr_1^{o(d_i)}(a_iu^{d_i})}=1$ that is, $\sum_{u\in U}(-1)^{b(u)}=1$. Therefore, we have proved that the condition on $b(u)$ implies that $f_{0,b}$ is bent.

Moreover, for every $\omega\in\mathbb{F}_{2^n}$, $\widehat{\chi_{f_{a,b}}}(\omega)=2^m\sum_{u\in E_\omega}(-1)^{b(u)}\in\{0,\pm 2^{m+1}\}$. Hence, $\sum_{u\in E_\omega}(-1)^{b(u)}\equiv 0 \pmod 2$. But $\sum_{u\in E_\omega}(-1)^{b(u)}\equiv\sum_{u\in E_\omega}1 \pmod 2$. Thus, $\#E_\omega\equiv 0 \pmod 2$ that is, $\#E_\omega$ is even for every $\omega\in\mathbb{F}_{2^n}$. Set $\#E_\omega=2\kappa(\omega)$ with $\kappa(w)\in\mathbb{Z}$. Now, we have $f_{a,0}(uy)=Tr_1^m(a(u)y)$. Hence, $\widehat{\chi_{f_{a,0}}}(\omega)=1-\sum_{u\in U}(-1)^0+2^m\sum_{u\in E_\omega}(-1)^0$, that is, $\widehat{\chi_{f_{a,0}}}(\omega)=1-\#U+2^m\#E_\omega=2^m(\#E_\omega-1)$. Therefore, $\widehat{\chi_{f_{a,0}}}(\omega)=2^m(2\kappa(\omega)-1)\equiv -2^m \pmod{2^{m+1}}\equiv 2^m \pmod{2^{m+1}}$. Thus, according to Lemma 2.3, $f_{a,0}$ is bent and then, $f_{a,0}$ belongs to $\mathcal{L}_n$.

• Case 2: $\sum_{u\in U}(-1)^{b(u)}=1+2^m$.

Since $\#U=2^m+1$, the mapping $u\mapsto b(u)$ vanishes on U. Thus $f_{a,b}=f_{a,0}$. Now, recall that we have $\widehat{\chi_{f_{a,0}}}=2^m(\#E_\omega-1),\forall\omega\in\mathbb{F}_{2^n}$. Since $f_{a,b}$ is semi-bent then, $2^m(\#E_\omega-1)\in\{0,\pm 2^{m+1}\}$, that is, $\#E_\omega\in\{1,3\}$. Hence, $\widehat{\chi_{f_{a,0}}}\geq 0$ which implies, according to Lemma 2.1, that $f_{a,0}$ is linear and its Walsh spectrum equals $\{0,2^n\}$. This contradicts the semi-bentness of $f_{a,0}=f_{a,b}$. The Case 2 is thus excluded.

• Case 3: $\sum_{u\in U}(-1)^{b(u)}=1-2^m$.

One can prove that there exists a unique $u_o\in U$ such that $b(u_0)=0$ and $b(u)=1$ for every $u\in U\backslash\{u_0\}$. Indeed, for $\epsilon\in\{0,1\}$ denote by B_ϵ the set $\{u\in U\mid b(u)=\epsilon\}$. We have $\sum_{u\in U}(-1)^{b(u)}=\sum_{u\in B_0}(-1)^{b(u)}+\sum_{u\in B_1}(-1)^{b(u)}=1-2^m=\#B_0-\#B_1$. Hence, $\#B_0=1$ and $\#B_1=2^m$ (since $\#B_0+\#B_1=\#U=2^m+1$).

The function $u\in U\mapsto b(u)$ thus equals $\mathbf{1}_U+\mathbf{1}_{u_0\mathbb{F}_{2^m}}$ and $f_{a,b}=f_{a,0}+\mathbf{1}_U+\mathbf{1}_{u_0\mathbb{F}_{2^m}}$. Now, we have $\widehat{\chi_{f_{a,b}}}(\omega)=2^m+2^m\sum_{u\in E_\omega}(-1)^{b(u)},\forall\omega\in\mathbb{F}_{2^n}$ (where E_ω is defined as above). Since $f_{a,b}$ is semi-bent, one has necessarily $2^m(1+\sum_{u\in E_\omega}(-1)^{b(u)})\in\{0,\pm 2^{m+1}\}$, that is $1+\sum_{u\in E_\omega}(-1)^{b(u)}\in\{0,\pm 2\}$. But $(-1)^{b(u)}\equiv 1 \pmod 2$. Hence $\#E_\omega$ is odd (in particular, $\#E_\omega\geq 1$). But $f_{a,0}(uy)=Tr_1^m(a(u)y)$, hence $\widehat{\chi_{f_{a,0}}}(\omega)=1-\sum_{u\in U}(-1)^0+2^m\sum_{u\in E_\omega}(-1)^0=2^m(\#E_\omega-1)\geq 0$. According to Lemma 2.1, we deduce that the function $f_{a,0}$ is linear. Thus, $f_{a,b}=f_{a,0}+\mathbf{1}_U+\mathbf{1}_{u_0\mathbb{F}_{2^m}}$ with $f_{a,b}$ semi-bent and $f_{a,0}+\mathbf{1}_U$ is an affine function. Therefore the function $g:=\mathbf{1}_{u_0\mathbb{F}_{2^m}}$ is semi-bent. Thus its Hamming weight $wt(g)\in\{2^{n-1}-2^m,2^{n-1}+2^m,2^{n-1}\}$ (since $\widehat{\chi_g}(0)=2^n-2wt(g)\in\{0,2^{m+1},-2^{m+1}\}$). But (using the definition of g) we have $wt(g)=2^m$. We conclude that for $m>2$ the function g can not be semi-bent. The proof follows. □

3.2. A construction of semi-bent functions via the indirect sum. In 2004, Carlet [**3**] has introduced a secondary construction of bent functions that he called *indirect sum*. This construction generalizes the well-known direct sum given by Dillon and Rothaus [**9, 27**] and is defined as follows.

DEFINITION 3.10. Let $n = r + s$ where r and s are positive integers. Let f_1, f_2 be Boolean functions defined on $\mathbb{F}_{2^r}$ and g_2, g_2 be two Boolean functions defined on $\mathbb{F}_{2^s}$. Define h as follows (that is, h is the concatenation of the four functions f_1, $f_1 \oplus 1$, f_2 and $f_2 \oplus 1$, in an order controled by $g_1(y)$ and $g_2(y)$):

$$\forall (x, y) \in \mathbb{F}_{2^r} \times \mathbb{F}_{2^s}, \quad h(x, y) = f_1(x) + g_1(y) + (f_1(x) + f_2(x))(g_1(y) + g_2(y)).$$

This construction was used in [**3**] to construct bent functions from bent functions in lower dimension. In the following, we show that the indirect sum could be used to construct semi-bent functions from both bent and semi-bent functions in lower dimension. More precisely, we prove the following result which can be viewed as a secondary-like construction [1].

THEOREM 3.11. *Let $n = r + s$ with r and s two even intergers. Let h be defined as in Definition 3.10. Suppose that f_1 and f_2 are semi-bent on $\mathbb{F}_{2^r}$ and that g_1 and g_2 are bent on $\mathbb{F}_{2^s}$. Then h is semi-bent on $\mathbb{F}_{2^n}$.*

PROOF. Set $r = 2\rho$ and $s = 2\sigma$. Let's compute the Walsh transform of h for every $(a, b) \in \mathbb{F}_{2^r} \times \mathbb{F}_{2^s}$. We have

$$\widehat{\chi_h}(a,b) = \sum_{x \in \mathbb{F}_{2^r}} \sum_{y \in \mathbb{F}_{2^s}} \chi(f_1(x) + g_1(y) + (f_1(x) + f_2(x))(g_1(y) + g_2(y)) + Tr_1^r(ax) + Tr_1^s(by)).$$

Now, one can split the sum depending whether $g_1(y) + g_2(y)$ is equal to 1 or not :

$$\begin{aligned} \widehat{\chi_h}(a,b) &= \sum_{x \in \mathbb{F}_{2^r}} \sum_{y \in \mathbb{F}_{2^s} \mid g_1(y) + g_2(y) = 1} \chi(f_2(x) + g_1(y) + Tr_1^r(ax) + Tr_1^s(by)) \\ &+ \sum_{y \in \mathbb{F}_{2^s} \mid g_1(y) + g_2(y) = 0} \chi(f_1(x) + g_1(y) + Tr_1^r(ax) + Tr_1^s(by)). \end{aligned}$$

Now, note that the indicator of the set $\{y \in \mathbb{F}_{2^s} \mid g_1(y) + g_2(y) = 1\}$ can be written as $\frac{1 - \chi(g_1(y) + g_2(y))}{2}$. Similarly, one can write the indicator of the set $\{y \in \mathbb{F}_{2^s} \mid g_1(y) + g_2(y) = 0\}$ as $\frac{1 + \chi(g_1(y) + g_2(y))}{2}$. Hence,

$$\widehat{\chi_h}(a,b) = \widehat{\chi_{f_1}}(a) \left(\frac{\widehat{\chi_{g_1}}(b) + \widehat{\chi_{g_2}}(b)}{2} \right) + \widehat{\chi_{f_2}}(a) \left(\frac{\widehat{\chi_{g_1}}(b) - \widehat{\chi_{g_2}}(b)}{2} \right).$$

Now, if g_1 and g_2 are bent, then

$$\left(\frac{\widehat{\chi_{g_1}}(b) - \widehat{\chi_{g_2}}(b)}{2} \right) \left(\frac{\widehat{\chi_{g_1}}(b) + \widehat{\chi_{g_2}}(b)}{2} \right) = \frac{1}{4} \left((\widehat{\chi_{g_1}}(b))^2 - (\widehat{\chi_{g_2}}(b))^2 \right) = 0$$

and thus only the two following situations can occur

$$\frac{\widehat{\chi_{g_1}}(b) - \widehat{\chi_{g_2}}(b)}{2} = 0 \text{ and } \frac{\widehat{\chi_{g_1}}(b) + \widehat{\chi_{g_2}}(b)}{2} = \pm 2^\sigma$$

or

$$\frac{\widehat{\chi_{g_1}}(b) - \widehat{\chi_{g_2}}(b)}{2} = \pm 2^\sigma \text{ and } \frac{\widehat{\chi_{g_1}}(b) + \widehat{\chi_{g_2}}(b)}{2} = 0.$$

Now f_1 and f_2 being semi-bent : $\widehat{\chi_{f_1}}(a) \in \{0, \pm 2^{\rho+1}\}$ and $\widehat{\chi_{f_2}}(a) \in \{0, \pm 2^{\rho+1}\}$. Therefore $\widehat{\chi_h}(a, b) \in \{0, \pm 2^{\rho+\sigma+1}\}$ proving that h is semi-bent. □

[1] As opposed to "secondary constructions" which means constructions of new functions from ones having the same properties.

REMARK 3.12. Obviously, the roles of f_1 and f_2 can be exchanged with those of g_1 and g_2. This means that one can exchange the property of bentness and semi-bentness in Theorem 3.11, that is, suppose that f_1 and f_2 are bent and that g_1 and g_2 are semi-bent.

3.3. A construction of semi-bent functions from bent functions by field extension.

Another kind of construction of semi-bent functions from bent functions is given by the simple following statement.

PROPOSITION 3.13. *Let n be an even positive integer. Let f be a Boolean function over $\mathbb{F}_{2^n}$. For $\delta \in \mathbb{F}_4$, we define a Boolean function f_δ over $\mathbb{F}_{2^{n+2}} \simeq \mathbb{F}_{2^n} \times \mathbb{F}_4$ by*

$$f_\delta(y,z) = f(y) + Tr_1^2(\delta z), \forall y \in \mathbb{F}_{2^n}, z \in \mathbb{F}_4.$$

If f is bent over $\mathbb{F}_{2^n}$ then f_δ is semi-bent over $\mathbb{F}_{2^{n+2}}$.

PROOF. Let us compute the Walsh transform at every $\omega := (\omega', \omega_1) \in \mathbb{F}_{2^n} \times \mathbb{F}_4$.

$$\begin{aligned} \widehat{\chi_{f_\delta}}(\omega) &= \sum_{y\in\mathbb{F}_{2^n}} \sum_{z\in\mathbb{F}_{2^2}} (-1)^{f_\delta(y,z)+Tr_1^n(\omega' y)+Tr_1^2(\omega_1 z)} \\ &= \sum_{y\in\mathbb{F}_{2^n}} (-1)^{f(y)+Tr_1^n(\omega' y)} \sum_{z\in\mathbb{F}_{2^2}} (-1)^{Tr_1^2(z(\omega_1+\delta)} \\ &= \widehat{\chi_f}(\omega') \sum_{z\in\mathbb{F}_{2^2}} (-1)^{Tr_1^2(z(\omega_1+\delta)}. \end{aligned} \tag{3.5}$$

Now, since f is bent then $\widehat{\chi_f}(\omega') = \pm 2^{\frac{n}{2}}$. On the other hand,

$$\sum_{z\in\mathbb{F}_{2^2}} (-1)^{Tr_1^2(z(\omega_1+\delta)} = \begin{cases} 4 & \text{if } \omega_1 = \delta \\ 0 & \text{otherwise.} \end{cases}$$

Hence, $\widehat{\chi_{f_\delta}}(\omega) \in \{0, 2^{\frac{n+2}{2}+1}, -2^{\frac{n+2}{2}+1}\}$ proving that f_δ is semi-bent on $\mathbb{F}_{2^n} \times \mathbb{F}_4$. □

4. Conclusion

A lot of research has been devoted to designing constructions of bent functions. This paper investigates constructions of semi-bent functions. To this end, bent functions are exploited to produce new semi-bent functions and thereby extend the list of known primary constructions of semi-bent functions in even dimension.

References

[1] Marat Burnashev and Ilya Dumer, *Error exponents for recursive decoding of Reed-Muller codes on a binary-symmetric channel*, IEEE Trans. Inform. Theory **52** (2006), no. 11, 4880–4891, DOI 10.1109/TIT.2006.883557. MR2300362 (2007m:94234)

[2] Anne Canteaut, Claude Carlet, Pascale Charpin, and Caroline Fontaine, *On cryptographic properties of the cosets of $R(1,m)$*, IEEE Trans. Inform. Theory **47** (2001), no. 4, 1494–1513, DOI 10.1109/18.923730. MR1830095 (2002h:94048)

[3] Claude Carlet, *On the secondary constructions of resilient and bent functions*, Coding, cryptography and combinatorics, Progr. Comput. Sci. Appl. Logic, vol. 23, Birkhäuser, Basel, 2004, pp. 3–28. MR2090638 (2005d:94092)

[4] Claude Carlet, *Boolean Functions for Cryptography and Error Correcting Codes*, Chapter of the monography "Boolean Models and Methods in Mathematics, Computer Science, and Engineering" published by Cambridge University Press, Yves Crama and Peter L. Hammer (eds.), 2010, pp. 257–397.

[5] Claude Carlet and Sihem Mesnager, *On Dillon's class H of bent functions, Niho bent functions and o-polynomials*, J. Combin. Theory Ser. A **118** (2011), no. 8, 2392–2410, DOI 10.1016/j.jcta.2011.06.005. MR2834182 (2012j:06030)

[6] Claude Carlet and Sihem Mesnager, *On semibent Boolean functions*, IEEE Trans. Inform. Theory **58** (2012), no. 5, 3287–3292, DOI 10.1109/TIT.2011.2181330. MR2952547

[7] Pascale Charpin, Enes Pasalic, and Cédric Tavernier, *On bent and semi-bent quadratic Boolean functions*, IEEE Trans. Inform. Theory **51** (2005), no. 12, 4286–4298, DOI 10.1109/TIT.2005.858929. MR2243159 (2007c:94110)

[8] Seongtaek Chee, Sangjin Lee, and Kwangjo Kim, *Semi-bent functions*, Advances in cryptology—ASIACRYPT '94 (Wollongong, 1994), Lecture Notes in Comput. Sci., vol. 917, Springer, Berlin, 1995, pp. 107–118. MR1376378

[9] J. Dillon, *Elementary Hadamard difference sets*, PhD dissertation, University of Maryland.

[10] Ilya Dumer, *Recursive decoding and its performance for low-rate Reed-Muller codes*, IEEE Trans. Inform. Theory **50** (2004), no. 5, 811–823, DOI 10.1109/TIT.2004.826632. MR2090592 (2005e:94283)

[11] R. Gold, *Maximal recursive sequences with 3-valued recursive crosscorrelation functions*, IEEE Trans. Inform. Theory 14 (1), 1968, pp. 154–156.

[12] Tor Helleseth, *Some results about the cross-correlation function between two maximal linear sequences*, Discrete Math. **16** (1976), no. 3, 209–232. MR0429323 (55 #2341)

[13] Tor Helleseth, *Correlation of m-sequences and related topics*, Sequences and their applications (Singapore, 1998), Springer Ser. Discrete Math. Theor. Comput. Sci., Springer, London, 1999, pp. 49–66. MR1843080 (2002f:94030)

[14] Tor Helleseth and P. Vijay Kumar, *Sequences with low correlation*, Handbook of coding theory, Vol. I, II, North-Holland, Amsterdam, 1998, pp. 1765–1853. MR1667957

[15] K. Khoo, G. Gong, and D. R. Stinson, *A new family of Gold-like sequences*, IEEE Trans. Inform. Theory Lausanne, Switzerland, 2002, p. 181.

[16] Khoongming Khoo, Guang Gong, and Douglas R. Stinson, *A new characterization of semi-bent and bent functions on finite fields*, Des. Codes Cryptogr. **38** (2006), no. 2, 279–295, DOI 10.1007/s10623-005-6345-x. MR2197473 (2006i:11146)

[17] Nian Li, Tor Helleseth, Xiaohu Tang, and Alexander Kholosha, *Several new classes of bent functions from Dillon exponents*, IEEE Trans. Inform. Theory **59** (2013), no. 3, 1818–1831, DOI 10.1109/TIT.2012.2229782. MR3030755

[18] M. Matsui, *Linear cryptanalysis method for DES cipher.*, Proceedings of EUROCRYPT'93, Lecture Notes in Computer Science 765, 1994, pp. 386–397.

[19] W. Meier and O. Staffelbach, *Fast correlation attacks on stream ciphers.*, Advances in Cryptology, EUROCRYPT'88, Lecture Notes in Computer Science 330, 1988, pp. 301–314.

[20] Sihem Mesnager, *Bent and hyper-bent functions in polynomial form and their link with some exponential sums and Dickson polynomials*, IEEE Trans. Inform. Theory **57** (2011), no. 9, 5996–6009, DOI 10.1109/TIT.2011.2124439. MR2857949 (2012m:94353)

[21] Sihem Mesnager, *Semibent functions from Dillon and Niho exponents, Kloosterman sums, and Dickson polynomials*, IEEE Trans. Inform. Theory **57** (2011), no. 11, 7443–7458, DOI 10.1109/TIT.2011.2160039. MR2883560 (2012m:94352)

[22] Sihem Mesnager, *Bent functions from Spreads. Preprint.*, 2013.

[23] Sihem Mesnager, *Semi-bent functions from oval polynomials*, Proceedings of Fourteenth International Conference on Cryptography and Coding, Oxford, United Kingdom, IMACC 2013, LNCS 8308, Springer, Heidelberg, 2013, pp. 1–15.

[24] Sihem Mesnager and Gérard Cohen, *On the link of some semi-bent functions with Kloosterman sums*, Coding and cryptology, Lecture Notes in Comput. Sci., vol. 6639, Springer, Heidelberg, 2011, pp. 263–272, DOI 10.1007/978-3-642-20901-7_17. MR2834706

[25] Sihem Mesnager and Jean-Pierre Flori, *Hyperbent functions via Dillon-like exponents*, IEEE Trans. Inform. Theory **59** (2013), no. 5, 3215–3232, DOI 10.1109/TIT.2013.2238580. MR3053406

[26] Y. Niho, *Multi-valued cross-correlation functions between two maximal linear recursive sequences*, Ph.D. dissertation, Univ. Sothern Calif., Los Angeles, 1972.

[27] O. S. Rothaus, *On "bent" functions*, J. Combinatorial Theory Ser. A **20** (1976), no. 3, 300–305. MR0403988 (53 #7797)

[28] G. Sun and C.Wu, *Construction of Semi-Bent Boolean Functions in Even Number of Variables*, Chinese Journal of Electronics, vol 18, No 2, 2009.

[29] Xian-Mo Zhang and Yuliang Zheng, *The nonhomomorphicity of Boolean functions*, Selected areas in cryptography (Kingston, ON, 1998), Lecture Notes in Comput. Sci., vol. 1556, Springer, Berlin, 1999, pp. 280–295, DOI 10.1007/3-540-48892-8_22. MR1715812 (2000f:94038)

[30] Xian-Mo Zhang and Yuliang Zheng, *Relationships between bent functions and complementary plateaued functions*, Lecture Notes in Computer Science, vol 1787, 1999, pp. 60–75.

Ecole Nationale Supérieure des Télécommunications -Telecom-Paristech, UMR 5141, CNRS, France.

E-mail address: cohen@telecom-paristech.fr

Department of Mathematics, University of Paris VIII and University of Paris XIII. CNRS UMR 7539 LAGA (Laboratoire Analyse, Géometrie et Applications), Sorbonne Paris Cité, 2 rue de la liberté, F-93526 Saint-Denis Cedex, France.

E-mail address: smesnager@univ-paris8.fr

Contemporary Mathematics
Volume **625**, 2014
http://dx.doi.org/10.1090/conm/625/12497

Some remarks on multiplicity codes

Swastik Kopparty

To Ilya Dumer, on the occasion of his 60th birthday

ABSTRACT. Multiplicity codes are algebraic error-correcting codes generalizing classical polynomial evaluation codes, and are based on evaluating polynomials *and their derivatives*. This small augmentation confers upon them better *local decoding*, *list-decoding* and *local list-decoding* algorithms than their classical counterparts. We survey what is known about these codes, present some variations and improvements, and finally list some interesting open problems.

1. Introduction

Reed-Solomon codes and Reed-Muller codes are classical families of error-correcting codes which have been widely influential in coding theory, combinatorics and theoretical computer science. These codes are based on evaluations of polynomials: a codeword of one of these codes is obtained by evaluating a polynomial over a finite field $\mathbb{F}_q$ of degree at most d at all points in $\mathbb{F}_q^m$.

Multiplicity codes are a family of recently-introduced algebraic error-correcting codes based on evaluations of polynomials and their derivatives. Specifically, a codeword of a multiplicity code is obtained by evaluating a polynomial of degree at most d, along with all its derivatives of order $< s$, at all points in $\mathbb{F}_q^m$.

The $s = 1$ versions of multiplicity codes are thus the classical Reed-Solomon ($m = 1$) and Reed-Muller ($m \geq 1$) codes. We will see that by allowing s to be larger than 1, in many senses general multiplicity codes go beyond their $s = 1$ counterparts.

Multiplicity codes with $m = 1$ (i.e., based on univariate polynomials) were first considered by Rosenbloom and Tsfasman [**RT97**], who studied them for the purposes of producing optimal codes for the "M metric" (now known as the Rosenbloom-Tsfasman metric). They were also studied by Nielsen [**Nie01**], who showed that they admit list-decoding algorithms upto the Johnson bound, similar to the Reed-Solomon codes.

Multiplicity codes with general m, s were defined by Kopparty, Saraf and Yekhanin [**KSY11**]. The main result of [**KSY11**] was that for every $\epsilon, \alpha > 0$, for all k, there

---

2010 *Mathematics Subject Classification.* Primary 94B40; Secondary 94B35.
Key words and phrases. Multiplicity codes, Reed-Solomon codes, Reed-Muller codes, error-correcting codes, polynomials, derivatives, sublinear-time algorithms.
Research supported in part by a Sloan Fellowship and NSF CCF-1253886.

are multiplicity codes of dimension k, rate $1-\alpha$, and which are locally decodable from a constant fraction of errors with in just $O_{\epsilon,\alpha}(k^{\epsilon})$ time. Prior to [**KSY11**], codes with nontrivial local decoding algorithms were known only at rate $R < 1/2$, and achieving local decoding complexity $O(k^{\epsilon})$ required the code to have rather small rate $R = \epsilon^{(1/\epsilon)}$ (the codes that were known to achieve these parameters were the Reed-Muller codes). It should be noted that more recent results have shown how to construct codes achieving parameters similar to those of multiplicity codes using significantly different ideas: Guo-Kopparty-Sudan [**GKS13**], Guo [**Guo13**] and Hemenway-Ostrovsky-Wooters [**HOW13**].

Subsequently, Guruswami-Wang [**GW11**] and Kopparty [**Kop12**] studied the list-decoding of univariate multiplicity codes, and showed that there are sequences of univariate multiplicity codes of rate R, list-decodable from $1-R-\epsilon$ fraction errors in polynomial time (achieving the so-called list-decoding capacity, thus providing another route to such codes after the breakthrough results of Parvaresh-Vardy [**PV05**] and Guruswami-Rudra [**GR08**]).

Global decoding of multivariate multiplicity codes was also considered in [**Kop12**]. There it was shown that multivariate multiplicity codes can be decoded upto half their minimum distance in polynomial time, and can be list-decoded from the Johnson bound in polynomial time.

The primary purpose of this paper is to survey the state of the art algorithms for dealing with multiplicity codes. Along the way we note some variations and improvements. Specifically:

(1) We give an improved local decoding algorithm for multiplicity codes. The original local decoding algorithm of [**KSY11**] for multiplicity codes worked as follows: in order to recover the correct value of the multiplicity codeword at a point $\mathbf{a} \in \mathbb{F}_q^m$, one would take $s^{O(m)}$ random lines in $\mathbb{F}_q^m$ passing through $\mathbf{a}$, query the codeword on all those lines, and use the answer to decode the correct value at $\mathbf{a}$. Our improved local decoding algorithm is based on queries only $\exp(m)$ random lines through $\mathbf{a}$.

   This new algorithm is based on two new ideas. First, we show that one can extract much more information from each line about the correct value at $\mathbf{a}$ than what the previous algorithm took advantage of. Second, we use a more sophisticated way of combining information from the different lines. For the previous algorithm, the problem of combining information from the various lines through $\mathbf{a}$ to recover the correct value of the codeword at $\mathbf{a}$ amounted to the problem of decoding a Reed-Muller code. In the new algorithm, this problem turns out to be a case of decoding a multiplicity code!

(2) The above framework admits a number of variations that could potentially be interesting for their own sake.

   One variation leads to a "polynomial rate" constant-query error-correction scheme as follows: a message $\sigma \in \Sigma_0^n$, where $|\Sigma_0| = \exp(n)$, gets encoded into a codeword $c \in \Sigma^n$, where $\log|\Sigma| = n^{\epsilon} \cdot \log|\Sigma_0|$, such that even if a constant fraction of the coordinates of c are corrupted, for any given[1] $i \in [n]$ one can recover σ_i with high probability using only

[1]We use $[n]$ to denote the set $\{1, 2, \ldots, n\}$.

$O(1)$ queries into c. Such large alphabet error-correction schemes were considered by Beimel and Ishai [**BI01**].

Another variation allows local correction for some low rate multiplicity codes using only m lines, with a much simpler local correction algorithm.

(3) Using ideas from the above improvements, we give a new algorithm for (global) decoding of multivariate multiplicity codes. The original approach of [**Kop12**] was based on a family of $s^{O(m)}$ space filling curves that passed through all the points of $\mathbb{F}_q^m$. The new algorithm uses only $\exp(m)$ many curves. The property of the $s^{O(m)}$ curves used in [**Kop12**] was "algebraic repulsion": no nonzero polynomial $P(X_1, \ldots, X_m)$ of moderate degree can vanish on all these curves. The family of curves that we use in this paper can be smaller because we require a weaker property: no nonzero polynomial $P(X_1, \ldots, X_m)$ of moderate degree can vanish on all these curves *with high multiplicity.*

(4) We observe that encoding and unique decoding algorithms for multiplicity codes can be implemented in near-linear time (i.e., they run in time $O(n \cdot (\log n)^{O(1)})$) . For $m = 1$, this follows from algorithms nearly identical to the ones from the classical univariate ($s = m = 1$) case, and for general m it follows by refining a reduction to the $m = 1$ case given in [**Kop12**].

(5) We gather a number of open questions and possible future research directions for the study of multiplicity codes.

Organization of this paper: In the next section we formally define multiplicity codes and state their basic properties. In Section 3 we discuss decoding algorithms for univariate multiplicity codes. In Section 4 we discuss decoding algorithms for multivariate multiplicity codes. In Section 5 we discuss encoding algorithms. We conclude with some discussion and open questions.

2. Multiplicity Codes

We begin with some general preliminaries on codes, polynomials and derivatives, and then move on to state the basic definitions and results about multiplicity codes.

2.1. Codes. Let Σ be a finite set and let n be an integer. We will work with Σ^n equipped with the (normalized) Hamming metric Δ, defined by:

$$\Delta(x, y) = \Pr_{i \in [n]}[x_i \neq y_i].$$

A code of length n over the alphabet Σ is a subset $\mathcal{C}$ of Σ^n. The rate of the code is defined to be:

$$R = \frac{\log_{|\Sigma|} |\mathcal{C}|}{n}.$$

The minimum distance of the code $\mathcal{C}$ is defined to be the smallest value δ of $\Delta(c, c')$ for distinct elements c, c' of $\mathcal{C}$.

Encoding. If $\mathcal{C} \subseteq \Sigma^n$ is a code, an encoding map for $\mathcal{C}$ is a bijection $E : \Sigma_0^k \to \mathcal{C}$ for some integer k. Often $\Sigma_0 = \Sigma$, but it need not be. It will be important that this map E is efficiently computable and efficiently invertible.

Unique Decoding. In the problem of **unique decoding** the code $\mathcal{C}$ from η-fraction errors, where $\eta \leq \delta/2$, we are given as input $r \in \Sigma^n$, and we wish to compute the unique $c \in \mathcal{C}$ (if any) such that $\Delta(r,c) < \eta$. The uniqueness follows from our condition relating η and δ.

List-Decoding. In the problem of **list-decoding** the code $\mathcal{C}$ from η-fraction errors, we are given as input $r \in \Sigma^n$, and we wish to compute the set

$$\mathcal{L} = \{c \in \mathcal{C} \mid \Delta(r,c) < \eta\}.$$

The maximum possible value of $|\mathcal{L}|$ as r varies over all elements of Σ^n is called the **list-size** for list-decoding $\mathcal{C}$ from η fraction errors.

Local Correction and Local Decoding. In the problem of **locally correcting** the code $\mathcal{C}$ from η-fraction errors, where $\eta \leq \delta/2$, we are given oracle access to a string $r \in \Sigma^n$, and given as input $i \in [n]$, and we wish to compute c_i for the unique $c \in \mathcal{C}$ (if any) such that $\Delta(r,c) < \eta$. The query complexity of such a local correction algorithm is the number of queries made to r; both the query complexity and time complexity could potentially be sublinear in n (and indeed this is the interesting case).

For **local decoding**, we deal with a code $\mathcal{C}$ along with an encoding map $E : \Sigma_0^k \to \mathcal{C}$. In the problem of local decoding $(\mathcal{C}, E)$ from η-fraction errors, where $\eta \leq \delta/2$, we are given oracle access to $r \in \Sigma^n$, and input $i \in [k]$, and we wish to compute x_i for the unique $x \in \Sigma_0^k$ (if any) such that $\Delta(r, E(x)) < \eta$. The query complexity of such a local decoding algorithm is the number of queries made to r; again, both the query complexity and time complexity could potentially be sublinear in n (and indeed this is the interesting case).

The difference between local decoding and local correction is that in local decoding, we are trying to recover symbols of the original message, while in local correction, we are trying to recover symbols of the codeword.

2.2. Polynomials and Derivatives. For a vector $\mathbf{i} = \langle i_1, \ldots, i_m \rangle$ of non-negative integers, its *weight*, denoted $\mathrm{wt}(\mathbf{i})$, equals $\sum_{j=1}^m i_j$.

For a field $\mathbb{F}$, let $\mathbb{F}[X_1, \ldots, X_m] = \mathbb{F}[\mathbf{X}]$ be the ring of polynomials in the variables $X_1, \ldots, X_m$ with coefficients in $\mathbb{F}$. For a vector of non-negative integers $\mathbf{i} = \langle i_1, \ldots, i_m \rangle$, let $\mathbf{X}^{\mathbf{i}}$ denote the monomial $\prod_{j=1}^m X_j^{i_j} \in \mathbb{F}[\mathbf{X}]$.

We now define derivatives and the multiplicity of vanishing at a point.

Definition 2.1 ((Hasse) Derivative). For $P(\mathbf{X}) \in \mathbb{F}[\mathbf{X}]$ and non-negative vector $\mathbf{i}$, the $\mathbf{i}$th *(Hasse) derivative* of P, denoted $P^{(\mathbf{i})}(\mathbf{X})$, is the coefficient of $\mathbf{Z}^{\mathbf{i}}$ in the polynomial $\tilde{P}(\mathbf{X}, \mathbf{Z}) \stackrel{\text{def}}{=} P(\mathbf{X} + \mathbf{Z}) \in \mathbb{F}[\mathbf{X}, \mathbf{Z}]$.

Thus,

$$P(\mathbf{X} + \mathbf{Z}) = \sum_{\mathbf{i}} P^{(\mathbf{i})}(\mathbf{X})\mathbf{Z}^{\mathbf{i}}. \tag{1}$$

We will need some basic properties of the Hasse derivative (see [**HKT08**]).

Proposition 2.2 (Basic properties of Hasse derivatives). *Let $P(\mathbf{X}), Q(\mathbf{X}) \in \mathbb{F}[\mathbf{X}]^m$ and let $\mathbf{i}$, $\mathbf{j}$ be vectors of nonnegative integers. Then:*

(1) $P^{(\mathbf{i})}(\mathbf{X}) + Q^{(\mathbf{i})}(\mathbf{X}) = (P+Q)^{(\mathbf{i})}(\mathbf{X})$.
(2) $(P \cdot Q)^{(\mathbf{i})}(\mathbf{X}) = \sum_{0 \leq \mathbf{e} \leq \mathbf{i}} P^{(\mathbf{e})}(\mathbf{X}) \cdot Q^{(\mathbf{i}-\mathbf{e})}(\mathbf{X})$.
(3) $\left(P^{(\mathbf{i})}\right)^{(\mathbf{j})}(\mathbf{X}) = \binom{\mathbf{i}+\mathbf{j}}{\mathbf{i}} P^{(\mathbf{i}+\mathbf{j})}(\mathbf{X})$.

DEFINITION 2.3 (Multiplicity). For $P(\mathbf{X}) \in \mathbb{F}[\mathbf{X}]$ and $\mathbf{a} \in \mathbb{F}^m$, the *multiplicity* of P at $\mathbf{a} \in \mathbb{F}^m$, denoted $\mathsf{mult}(P, \mathbf{a})$, is the largest integer M such that for every non-negative vector $\mathbf{i}$ with $\mathrm{wt}(\mathbf{i}) < M$, we have $P^{(\mathbf{i})}(\mathbf{a}) = 0$ (if M may be taken arbitrarily large, we set $\mathsf{mult}(P, \mathbf{a}) = \infty$).

Next, we state a basic bound on the total number of zeroes (counting multiplicity) that a polynomial can have on a product set S^m. An elementary proof of this lemma can be found in [**DKSS09**].

LEMMA 2.4. *Let $P \in \mathbb{F}[\mathbf{X}]$ be a nonzero polynomial of total degree at most d. Then for any finite $S \subseteq \mathbb{F}$,*

$$\sum_{\mathbf{a} \in S^m} \mathsf{mult}(P, \mathbf{a}) \leq d \cdot |S|^{m-1}.$$

In particular, for any integer $s > 0$,

$$\Pr_{\mathbf{a} \in S^m} [\mathsf{mult}(P, \mathbf{a}) \geq s] \leq \frac{d}{s|S|}.$$

2.3. Multiplicity Codes. Finally, we come to the definition of multiplicity codes.

DEFINITION 2.5 (Multiplicity code [**KSY11**]). Let s, d, m be nonnegative integers and let q be a prime power. Let $\Sigma = \mathbb{F}_q^{\binom{m+s-1}{m}} = \mathbb{F}_q^{\{\mathbf{i}:\mathrm{wt}(\mathbf{i})<s\}}$. For $P(X_1, \ldots, X_m) \in \mathbb{F}_q[X_1, \ldots, X_m]$, we define *the order s evaluation of P at* $\mathbf{a}$, denoted $P^{(<s)}(\mathbf{a})$, to be the vector $\langle P^{(\mathbf{i})}(\mathbf{a}) \rangle_{\mathrm{wt}(\mathbf{i})<s} \in \Sigma$.

The multiplicity code of order-s evaluations of degree-d polynomials in m variables over $\mathbb{F}_q$ is defined as follows. The code is over the alphabet Σ, and has length q^m (where the coordinates are indexed by elements of $\mathbb{F}_q^m$). For each polynomial $P(\mathbf{X}) \in \mathbb{F}_q[X_1, \ldots, X_m]$ with $\deg(P) \leq d$, there is a codeword in $\mathcal{C}$ given by:

$$\mathrm{Enc}_{s,d,m,q}(P) = \langle P^{(<s)}(\mathbf{a}) \rangle_{\mathbf{a} \in \mathbb{F}_q^m} \in (\Sigma)^{q^m}.$$

Technically speaking, we have only defined the multiplicity code as a subset of $\Sigma^{\mathbb{F}_q^m}$, without specifying an encoding map. We postpone the choice of a good encoding map to a later section.

LEMMA 2.6 (Rate and distance of multiplicity codes [**KSY11**]). *Let $\mathcal{C}$ be the multiplicity code of order s evaluations of degree d polynomials in m variables over $\mathbb{F}_q$. Then $\mathcal{C}$ has minimum distance at least $\delta = 1 - \frac{d}{sq}$ and rate $\frac{\binom{d+m}{m}}{\binom{s+m-1}{m} q^m}$, which is at least*

$$\left(\frac{s}{m+s}\right)^m \cdot \left(\frac{d}{sq}\right)^m \geq \left(1 - \frac{m^2}{s}\right)(1-\delta)^m.$$

We usually think of m and s as large constants (significantly smaller than q), and in light of the above parameters, having $s \gg m^2$ is particularly interesting. For the rest of this paper, when we speak of near-linear time algorithms, this assumes that m and s are constants, and that q and the blocklength q^m tend to ∞.

One can easily convert such codes into codes over a constant sized (and even binary) alphabet via concatenation, while preserving the local decoding/correction properties. For details, see [**KSY11**].

3. Decoding Univariate Multiplicity Codes

We begin by discussing decoding of univariate multiplicity codes.

3.1. Unique Decoding. The classic Berlekamp-Welch algorithm for decoding Reed-Solomon codes up to half the minimum distance has a simple generalization to the case of univariate multiplicity codes. This generalization was first discovered by Nielsen [**Nie01**][2]. In fact, Nielsen showed how to do list-decoding of univariate multiplicity codes, discussed next.

Let us set the problem up. Recall that the alphabet for this code is $\mathbb{F}_q^s$. Thus the received word is a function $r : \mathbb{F}_q \to \mathbb{F}_q^s$. Abusing notation, we view this as a tuple of s functions $r^{(i)} : \mathbb{F}_q \to \mathbb{F}_q$ for $0 \leq i < s$. We wish to find the unique $P(X)$ such that $\Delta(\mathrm{Enc}_{s,d,1,q}(P), r) < \delta/2$.

The algorithm tries to find an error-locator polynomial $E(X)$ and another polynomial $N(X)$, such that $N(X) = E(X) \cdot P(X)$.

- Search for nonzero polynomials $E(X)$, $N(X)$ of degrees at most $(sq-d)/2$, $(sq+d)/2$ respectively such that for each $x \in \mathbb{F}_q$, we have the following equations:

$$\begin{aligned} N(x) &= E(x) r^{(0)}(x) \\ N^{(1)}(x) &= E(x) r^{(1)}(x) + E^{(1)}(x) r^{(0)}(x) \\ &\cdots \\ N^{(s-1)}(x) &= \sum_{i=0}^{s-1} E^{(i)}(x) r^{(s-1-i)}(x) \end{aligned} \tag{2}$$

  This is a collection of sq homogeneous linear equations in $(sq-d)/2 + 1 + (sq+d)/2 + 1 > sq$ unknowns (the coefficients of E and N). Thus a nonzero solution $E(X), N(X)$ exists. Take any such nonzero solution.
- Given $E(X)$, $N(X)$ as above, output $\frac{N(X)}{E(X)}$.

The analysis proceeds by showing that $N(X) - P(X)E(X)$, which is a degree $(sq+d)/2$ polynomial, has $> (sq+d)/2s$ zeroes of multiplicity $\geq s$, and is thus the zero polynomial. This implies that $P(X) = N(X)/E(X)$, and so $P(X)$ is the output of the algorithm, as desired.

[2]Nielsen's theorem analyzes the decoding radius in terms of the m-metric, and implies the decoding algorithms for the Hamming metric considered here.

3.1.1. *Unique decoding in near-linear time.* In this subsection we describe how to implement the above algorithm in near-linear time. The presentation follows the description of a near-linear time implementation of the Berlekamp-Welch algorithm in Sudan's lecture notes [**Sud01**].

Let $R(X)$ be the unique polynomial of degree at most $sq-1$ such that for each $\alpha \in \mathbb{F}_q$ and $i < s$,

$$R^{(<s)}(\alpha) = r^{(<s)}(\alpha).$$

Such an $R(X)$ can be found in near-linear time by the classical Hermite interpolation algorithm of Chin [**Chi76**]. If $E(X)$ and $N(X)$ satisfy the equations (2), then we have that $N(X) - E(X)R(X)$ vanishes at each $x \in \mathbb{F}_q$ with multiplicity at least s. Thus:

$$N(X) = E(X)R(X) - C(X) \cdot (X^q - X)^s,$$

for some $C(X) \in \mathbb{F}_q[X]$. Equivalently,

$$\frac{N(X)}{E(X)(X^q - X)^s} = \frac{R(X)}{(X^q - X)^s} - \frac{C(X)}{E(X)}.$$

Thus we are looking for $C(X), E(X)$ such that:

(1) $\deg(E(X)) \le (sq - d)/2$,
(2) the rational function $\frac{C(X)}{E(X)}$ approximates the rational function $\frac{R(X)}{(X^q - X)^s}$, in the sense that the numerator of their difference $N(X) = R(X)E(X) - C(X)(X^q - X)^s$ has degree at most $(sq + d)/2$.

This problem can be solved in near-linear time via Strassen's continued fraction algorithm [**Str81**]. In fact, one can minimize the degree of $N(X)$ subject to the constraint that $\deg(E(X)) \le (sq - d)/2$.

Finally, the division step can also be performed in near-linear time. This completes the description of the near-linear time implementation of the unique decoder for univariate multiplicity codes.

3.2. List-Decoding. We now discuss the list-decoding of univariate multiplicity codes. Here we consider the problem of decoding from a fraction of errors which may be larger than half the minimum distance δ.

By the Johnson bound, we know that for list-decoding univariate multiplicity codes from $(1 - \sqrt{1-\delta})$-fraction errors, the list-size is at most $\mathrm{poly}(q)$ (this only uses the fact that the distance of the code is $\ge \delta$). It is thus reasonable to ask whether there is a polynomial time algorithm to list-decode univariate multiplicity codes from $(1 - \sqrt{1-\delta})$-fraction error.

In [**Nie01**], Nielsen gave such an algorithm. His algorithm generalizes the Guruswami-Sudan algorithm for list-decoding Reed-Solomon codes, and is also based on interpolation and root-finding.

Given a received word $r : \mathbb{F}_q \to \mathbb{F}_q^s$, one first interpolates a low-degree bivariate polynomial $Q(X, Y) \in \mathbb{F}_q[X, Y]$ such that for each $\alpha \in \mathbb{F}_q$, the polynomial $Q(X, \sum_{j=0}^{s-1} r^{(j)}(\alpha)(X - \alpha)^j)$ vanishes with high multiplicity at $X = \alpha$. One then shows that every $P(X) \in \mathbb{F}_q[X]$ of degree at most d with $\Delta(\mathrm{Enc}_{s,d,1,q}(P), r) \le 1 - \sqrt{1-\delta}$, we have $Q(X, P(X)) = 0$. Finally, one can find all polynomials $P(X)$ satisfying this latter equation.

Recently Guruswami-Wang [**GW11**] and Kopparty [**Kop12**] independently found improved results for list-decoding univariate multiplicity codes over *prime fields*.

The main result of [**GW11**] is that order s univariate multiplicity codes of distance δ over prime fields can, for every integer $0 \leq t < s$, be list-decoded from η_t fraction errors with list-size at most $q^{O(s)}$, where:

$$\eta_t = \frac{t+1}{t+2}\left(\delta - \frac{t}{s-t}\right).$$

For $t = 0$, the algorithm boils down to Nielsen's version of the Berlekamp-Welch algorithm for unique-decoding multiplicity codes.

The main result of [**Kop12**] is that order s univariate multiplicity codes of distance δ over prime fields can, for every integer $0 \leq t < s$ be list-decoded from η_t' fraction errors with list-size at most $q^{O(ts)}$, where:

$$\eta_t' = 1 - \left(\left(1 - \frac{t}{s-t}\right) \cdot (1-\delta)\right)^{\frac{t+1}{t+2}}.$$

For $t = 0$, the algorithm boils down to Nielsen's version of the Guruswami-Sudan algorithm for list-decoding univariate multiplicity codes.

Both these algorithms are based on deriving an order t differential equation of the form:

$$Q(X, P(X), P^{(1)}(X), \ldots, P^{(t-1)}(X)) = 0$$

from the received word r, such that every P whose encoding is close to r must satisfy this differential equation. In the algorithm of [**GW11**] this differential equation is a linear differential equation, and in the algorithm of [**Kop12**] this equation is a polynomial differential equation. These differential equations are then solved using Hensel-lifting / power series. See [**GW11**] and [**Kop12**] for the details. The decoding radius η_r' is always greater than η_r, but the algorithm and analysis of [**Kop12**] are also more involved than that of [**GW11**].

It is well known that the maximimum fraction of errors η from which a code of rate R and block-length n can be list-decoded from while still having poly(n) list-size is $1 - R - \epsilon$ (for arbitrarily small $\epsilon > 0$). A code which achieves this is said to achieve list-decoding capacity. The first constructions of codes which achieved list-decoding capacity came from the breakthrough results of Parvaresh-Vardy [**PV05**] and Guruswami-Rudra [**GR08**]. The above-mentioned results of [**GW11**] and [**Kop12**] show that univariate multiplicity codes over prime fields achieve list-decoding capacity for every $R \in (0, 1)$. This follows by noting that for univariate multiplicity codes, $R = 1 - \delta$, and that for every δ, if we take r to be a very large constant, and s to be a much larger constant, then the above decoding radii η_r and $\eta_{r'}$ approach $\delta = 1 - R$.

4. Decoding Multivariate Multiplicity Codes

4.1. Local Correction. We begin by discussing local correction algorithms for multiplicity codes. When coupled with a *systematic* encoding map (which we discuss in the next section), this also gives local decoding algorithms for multiplicity codes.

4.1.1. *Preliminaries on Restrictions and derivatives.* We first consider the relationship between the derivatives of a multivariate polynomial P and its restrictions to a line. Fix $\mathbf{a}, \mathbf{b} \in \mathbb{F}_q^m$, and consider the polynomial $Q(T) = P(\mathbf{a} + \mathbf{b}T)$.

- **The relationship of $Q(T)$ with the derivatives of P at a:** By the definition of Hasse derivatives,

$$Q(T) = \sum_{\mathbf{i}} P^{(\mathbf{i})}(\mathbf{a})\mathbf{b}^{\mathbf{i}} T^{\mathrm{wt}(\mathbf{i})}.$$

  Grouping terms, we see that:

$$\sum_{\mathbf{i}|\mathrm{wt}(\mathbf{i})=j} P^{(\mathbf{i})}(\mathbf{a})\mathbf{b}^{\mathbf{i}} = \text{coefficient of } T^j \text{ in } Q(T). \tag{3}$$

- **The relationship of the derivatives of Q at t with the derivatives of P at $\mathbf{a} + t\mathbf{b}$:** Let $t \in \mathbb{F}_q$. By the definition of Hasse derivatives, we get the following two identities:

$$P(\mathbf{a} + \mathbf{b}(t + R)) = Q(t + R) = \sum_j Q^{(j)}(t)R^j.$$

$$P(\mathbf{a} + \mathbf{b}(t + R)) = \sum_{\mathbf{i}} P^{(\mathbf{i})}(\mathbf{a} + \mathbf{b}t)(\mathbf{b}R)^{\mathbf{i}}.$$

  Thus,

$$Q^{(j)}(t) = \sum_{\mathbf{i}|\mathrm{wt}(\mathbf{i})=j} P^{(\mathbf{i})}(\mathbf{a} + \mathbf{b}t)\mathbf{b}^{\mathbf{i}}. \tag{4}$$

  In particular, $Q^{(j)}(t)$ is simply a linear combination of the various $P^{(\mathbf{i})}(\mathbf{a}+\mathbf{b}t)$ (over different $\mathbf{i}$).

We now apply these observations to the derivatives of P. For each nonnegative tuple $\mathbf{e} \in \mathbb{Z}^m$, consider the polynomial $Q_{\mathbf{e}}(T) = P^{(\mathbf{e})}(\mathbf{a} + \mathbf{b}T)$.

- **The relationship of $Q_{\mathbf{e}}(T)$ with the derivatives of P at a:**

(5)

$$\sum_{\mathbf{i}|\mathrm{wt}(\mathbf{i})=j} (P^{(\mathbf{e})})^{(\mathbf{i})}(\mathbf{a})\mathbf{b}^{\mathbf{i}} = \sum_{\mathbf{i}|\mathrm{wt}(\mathbf{i})=j} \binom{\mathbf{e}+\mathbf{i}}{\mathbf{e}} P^{(\mathbf{e}+\mathbf{i})}(\mathbf{a})\mathbf{b}^{\mathbf{i}} = \text{coefficient of } T^j \text{ in } Q_{\mathbf{e}}(T).$$

  In particular, knowing $Q_{\mathbf{e}}(T)$ gives us several linear relations between the evaluations of the derivatives of P at $\mathbf{a}$.

- **The relationship of the derivatives of $Q_{\mathbf{e}}$ at t with the derivatives of P at $\mathbf{a} + t\mathbf{b}$:** Let $t \in \mathbb{F}_q$. We get

$$Q_{\mathbf{e}}^{(j)}(t) = \sum_{\mathbf{i}|\mathrm{wt}(\mathbf{i})=j} (P^{(\mathbf{e})})^{(\mathbf{i})}(\mathbf{a} + \mathbf{b}t)\mathbf{b}^{\mathbf{i}} = \sum_{\mathbf{i}|\mathrm{wt}(\mathbf{i})=j} \binom{\mathbf{e}+\mathbf{i}}{\mathbf{e}} P^{(\mathbf{e}+\mathbf{i})}(\mathbf{a} + \mathbf{b}t)\mathbf{b}^{\mathbf{i}}. \tag{6}$$

  In particular, $Q_{\mathbf{e}}^{(j)}(t)$ is simply a linear combination of evaluations, at $\mathbf{a} + \mathbf{b}t$, of the various derivatives of P.

4.1.2. *The Local Correction Algorithm.* We now give our local correction algorithm which corrects $\delta_0 < \frac{\delta}{8}$ fraction errors. The $\gamma = 0$, $c = 1$ case of this algorithm is the orignal local correction algorithm of [**KSY11**]. Increasing γ reduces the query complexity from $s^{O(m)}$ to $\exp(m)$, while reducing the fraction of correctable errors by a negligible amount.

Main Local Correction Algorithm:
Input: received word $r : \mathbb{F}_q^m \to \Sigma$, point $\mathbf{a} \in \mathbb{F}_q^m$. Abusing notation again, we will write $r^{(\mathbf{i})}(\mathbf{a})$ when we mean the $\mathbf{i}$ coordinate of $r(\mathbf{a})$.

(1) Set $\gamma = 1 - \frac{(1-\delta)}{1-8\delta_0} = \frac{\delta - 8\delta_0}{1-8\delta_0}$. Set $c = \gamma \cdot s + 1$.
(2) **Pick a set B of directions:** Pick $\mathbf{z}, \mathbf{y}_1, \mathbf{y}_2, \ldots \mathbf{y}_m \in \mathbb{F}_q^m$ independently and uniformly at random. Let $S \subset \mathbb{F}_q$ be any set of size $\lceil \frac{5s}{c} \rceil$. Define

$$B = \{\mathbf{z} + \sum_{j=1}^{m} \alpha_j \mathbf{y}_j \mid \alpha_j \in S\}.$$

(3) **Recover $P^{(\mathbf{e})}(\mathbf{a}+\mathbf{b}T)$ for directions $\mathbf{b} \in B$:** For each $\mathbf{e}$ with $\mathrm{wt}(\mathbf{e}) < c$ and each $\mathbf{b} \in B$, consider the function $\ell_{\mathbf{b},\mathbf{e}} : \mathbb{F}_q \to \mathbb{F}_q^{s-\mathrm{wt}(\mathbf{e})}$ given by

$$(\ell_{\mathbf{b},\mathbf{e}}(t))_j = \sum_{\mathbf{i} \mid \mathrm{wt}(\mathbf{i})=j} \binom{\mathbf{e}+\mathbf{i}}{\mathbf{e}} r^{(\mathbf{e}+\mathbf{i})}(\mathbf{a}+\mathbf{b}t)\mathbf{b}^{\mathbf{i}}, \tag{7}$$

for each $0 \le j < s - \mathrm{wt}(\mathbf{e})$. Via a univariate multiplicity code decoding algorithm, find the unique polynomial $Q_{\mathbf{b},\mathbf{e}}(T) \in \mathbb{F}_q[T]$ of degree at most $d - \mathrm{wt}(\mathbf{e})$ (if any), such that

$$\Delta(\mathrm{Enc}_{s-\mathrm{wt}(\mathbf{e}),d-\mathrm{wt}(\mathbf{e}),1,q}(Q_{\mathbf{b},\mathbf{e}}), \ell_{\mathbf{b},\mathbf{e}}) < 2\delta_0.$$

(4) **Decode a constant degree multiplicity code to recover $P^{(<s)}(\mathbf{a})$:** Denote the coefficient of T^j in $Q_{\mathbf{b},\mathbf{e}}(T)$ by $v_{j,\mathbf{b},\mathbf{e}} \in \mathbb{F}_q$. If $j < 0$, we define $v_{j,\mathbf{b},\mathbf{e}} = 0$.
For each j' with $0 \le j' < s$, find the unique homogeneous degree j' polynomial $R_{j'}(\mathbf{X}) \in \mathbb{F}_q[\mathbf{X}]$ such that for at least $1/3$ of the $\mathbf{b} \in B$, for all $\mathbf{e}$ with $\mathrm{wt}(\mathbf{e}) < c$, we have:

$$R_{j'}^{(\mathbf{e})}(\mathbf{b}) = v_{j'-\mathrm{wt}(\mathbf{e}),\mathbf{b},\mathbf{e}}.$$

Note that this is a constant degree multiplicity code decoding problem.
If such an $R_{j'}$ does not exist, or is not unique, the algorithm outputs FAIL.
For each $\mathbf{i}$ with $\mathrm{wt}(\mathbf{i}) < s$, define $u_{\mathbf{i}}$ to equal the coefficient of $\mathbf{X}^{\mathbf{i}}$ in $R_{\mathrm{wt}(\mathbf{i})}(\mathbf{X})$.
(5) Output the vector $\langle u_{\mathbf{i}} \rangle_{\mathrm{wt}(\mathbf{i})<s}$.

We quickly comment on the running time and query complexity. The running time consists of $|S|^m$ instances of decoding univariate multiplicity codes over $\mathbb{F}_q$, as well as on instance of decoding a degree-s m-variate order-c multivariate multiplicity code with evaluation points being S^m. Thus, if m, s are constant, the running time is near-linear in q, which is near-linear in $n^{1/m}$, where n is the block-length of the code. The query complexity is $|S|^m \cdot q$, which equals $(\frac{5}{\gamma})^m \cdot n^{1/m}$. For $\delta = \Omega(1)$ and $\delta_0 < \delta/10$ (say), the query complexity equals $\exp(m) \cdot n^{1/m}$.

4.1.3. *Analysis of the Local Correction Algorithm.* We now analyze the above local correction algorithm.

THEOREM 4.1. *Let* $P(\mathbf{X}) \in \mathbb{F}_q[\mathbf{X}]$ *be such that* $\Delta(\mathrm{Enc}_{s,d,m,q}(P), r) < \delta_0$. *Let* $\mathbf{a} \in \mathbb{F}_q^m$.

With high probability, the local correction algorithm above outputs $P^{(<s)}(\mathbf{a})$.

PROOF. Let $E = \{x \in \mathbb{F}_q^m \mid P^{(<s)}(x) \neq r^{(<s)}(x)\}$ be the error set. We have $|E| < \delta_0 \cdot q^m$.

Let $L_{\mathbf{b}} = \{\mathbf{a} + t\mathbf{b} \mid t \in \mathbb{F}_q\}$ be the line through $\mathbf{a}$ in direction $\mathbf{b}$. We call $\mathbf{b}$ bad if $|L_{\mathbf{b}} \cap E| \geq 4 \cdot \delta_0 \cdot q$.

Note that at most 1/4 of all the lines are bad.

CLAIM 4.2. With high probability, we have:

(1) at most 1/3 of the $\mathbf{b} \in B$ are bad,
(2) $|B| = |S|^m$,

These basic probability/linear-algebra facts are well known, and we omit the proofs.

Henceforth we assume that both these events happen.

CLAIM 4.3. If $\mathbf{b}$ is good, then for every $\mathbf{e}$ with $\mathrm{wt}(\mathbf{e}) < c$, we have:

$$Q_{\mathbf{b},\mathbf{e}}(T) = P^{(\mathbf{e})}(\mathbf{a} + \mathbf{b}T).$$

PROOF. The univariate multiplicity code of order $s - \mathrm{wt}(\mathbf{e})$ evaluations of degree $d - \mathrm{wt}(\mathbf{e})$ polynomials has minimum distance at least $1 - \frac{d}{(s-c+1)q} = 1 - \frac{1-\delta}{1-\gamma}$ which, by choice of γ, is $\geq 8 \cdot \delta_0$.

If $\mathbf{b}$ is good, then we know that $|L_{\mathbf{b}} \cap E| < 4 \cdot \delta_0 \cdot q$. By Equations (7) and (6), we conclude that $P^{(\mathbf{e})}(\mathbf{a} + \mathbf{b}T)$ (which has degree $d - \mathrm{wt}(\mathbf{e})$) satisfies:

$$\Delta(\mathrm{Enc}_{s-\mathrm{wt}(\mathbf{e}),d-\mathrm{wt}(\mathbf{e}),1,q}(P^{(\mathbf{e})}(\mathbf{a} + \mathbf{b}T)), \ell_{\mathbf{b},\mathbf{e}}) \leq \frac{|L_{\mathbf{b}} \cap E|}{q} < 4 \cdot \delta_0,$$

which is less than half the minimum distance of the univariate multiplicity code of order $s - \mathrm{wt}(\mathbf{e})$ evaluations of degree $d - \mathrm{wt}(\mathbf{e})$ polynomials.

Thus $P^{(\mathbf{e})}(\mathbf{a}+\mathbf{b}T)$ is the unique such polynomial found in Step 3, and so $Q_{\mathbf{b},\mathbf{e}}(T) = P^{(\mathbf{e})}(\mathbf{a} + \mathbf{b}T)$. □

For each integer $0 \leq j' < s$, define the polynomial:

$$\tilde{R}_{j'}(\mathbf{X}) = \sum_{\mathbf{i}' \mid \mathrm{wt}(\mathbf{i}')=j'} P^{(\mathbf{i}')}(\mathbf{a})\mathbf{X}^{\mathbf{i}'}.$$

CLAIM 4.4. If $\mathbf{b}$ is good, then for all $\mathbf{e}$ with $\mathrm{wt}(\mathbf{e}) < c$, we have:

$$\tilde{R}_{j'}^{(\mathbf{e})}(\mathbf{b}) = v_{j'-\mathrm{wt}(\mathbf{e}),\mathbf{b},\mathbf{e}},$$

PROOF. We have:

$$\begin{aligned}\tilde{R}^{(\mathbf{e})}_{j'}(\mathbf{X}) &= \sum_{\mathbf{i}' \mid \mathrm{wt}(\mathbf{i}')=j'} \binom{\mathbf{i}'}{\mathbf{e}} P^{(\mathbf{i}')}(\mathbf{a})\mathbf{X}^{\mathbf{i}'-\mathbf{e}} \\ &= \sum_{\mathbf{i}' \mid \mathrm{wt}(\mathbf{i}')=j', \mathbf{i}' \geq \mathbf{e}} \binom{\mathbf{i}'}{\mathbf{e}} P^{(\mathbf{i}')}(\mathbf{a})\mathbf{X}^{\mathbf{i}'-\mathbf{e}} \\ &= \sum_{\mathbf{i} \mid \mathrm{wt}(\mathbf{i})=j} \binom{\mathbf{e}+\mathbf{i}}{\mathbf{e}} P^{(\mathbf{e}+\mathbf{i})}(\mathbf{a})\mathbf{X}^{\mathbf{i}},\end{aligned}$$

where $j = j' - \mathrm{wt}(\mathbf{e})$.

Thus,

$$\begin{aligned}\tilde{R}^{(\mathbf{e})}_{j'}(\mathbf{b}) &= \sum_{\mathbf{i} \mid \mathrm{wt}(\mathbf{i})=j} \binom{\mathbf{e}+\mathbf{i}}{\mathbf{e}} P^{(\mathbf{e}+\mathbf{i})}(\mathbf{a})\mathbf{b}^{\mathbf{i}} \\ &= \text{coeff. of } T^j \text{ in } Q_{\mathbf{b},\mathbf{e}}(T) \quad \text{(by Equation (5) and Claim 4.3, since } \mathbf{b} \text{ is good)} \\ &= v_{j'-\mathrm{wt}(\mathbf{e}),\mathbf{b},\mathbf{e}}.\end{aligned}$$

□

Thus $\tilde{R}_{j'}(\mathbf{X})$ satisfies the conditions required of Step 4 of the algorithm.

Let us now show that no other polynomial can satisfy these conditions. Suppose there was some other solution $\overline{R}_{j'}(\mathbf{X})$. Then the difference $(\tilde{R}_{j'} - \overline{R}_{j'})(\mathbf{X})$ would be a nonzero polynomial of degree $< s$, that vanishes with multiplicity at least c, at $\geq \frac{1}{3}$ of the points of B. But this cannot be, since B is an affine one-to-one image of the set S^m, and the fraction of points of S^m on which a nonzero polynomial of degree $< s$ can vanish with multiplicity $\geq c$ is at most $\frac{s}{c|S|} = \frac{1}{5} < \frac{1}{3}$. Thus $\tilde{R}_{j'}$ is the unique solution found in Step 4.

Finally, we notice that our definition of $R_{j'}$ implies that for every $\mathbf{i}$, we have $u_{\mathbf{i}} = P^{(\mathbf{i})}(\mathbf{a})$, as desired. □

4.1.4. *Variations.* The above algorithm allows a number of variations that may be useful in different contexts.

Let $\mathbf{a} \in \mathbb{F}_q^m$. Suppose $r : \mathbb{F}_q^m \to \Sigma$ is a received word, and suppose $P(\mathbf{X}) \in \mathbb{F}_q[\mathbf{X}]$ is a polynomial of degree at most d such that $\Delta(\mathrm{Enc}_{s,d,m,q}(P), r) < \delta_0$. Let $\gamma = \frac{\delta - 8\delta_0}{1 - 8\delta_0}$, and let $c = \gamma s + 1$.

Let $\mathbf{a} \in \mathbb{F}_q^m$. For each integer $0 \leq j' < s$, define the polynomial:

$$\tilde{R}_{j'}(\mathbf{X}) = \sum_{\mathbf{i}' \mid \mathrm{wt}(\mathbf{i}')=j'} P^{(\mathbf{i}')}(\mathbf{a})\mathbf{X}^{\mathbf{i}'}.$$

Suppose $\mathbf{b} \in \mathbb{F}_q^m$ is good (meaning that the line $L_{\mathbf{a},\mathbf{b}}$ has $< 4\delta_0 q$ errors on it). As we saw in the above analysis, by querying all the points of the line $L_{\mathbf{a},\mathbf{b}}$, we can compute $\tilde{R}^{(\mathbf{e})}_{j'}(\mathbf{b})$, for every $j' < s$ and every $\mathbf{e}$ such that $\mathrm{wt}(\mathbf{e}) < c$.

(1) Suppose we are only interested in recovering $P^{(<c)}(\mathbf{a})$. Then this can be recovered by querying the points of just one line! Indeed, if we pick $\mathbf{b}$ at random, then with high probability $\mathbf{b}$ is good, and then by querying $L_{\mathbf{a},\mathbf{b}}$

we can compute $\tilde{R}^{(\mathbf{e})}_{j'}(\mathbf{b})$ for every $j' < s$ and every $\mathbf{e}$ such that $\mathrm{wt}(\mathbf{e}) < c$. Note that for every $\mathbf{e}$ with $\mathrm{wt}(\mathbf{e}) < c$, we have $\tilde{R}^{(\mathbf{e})}_{\mathrm{wt}(\mathbf{e})}(\mathbf{b}) = P^{(\mathbf{e})}(\mathbf{b})$. Thus we can compute $P^{(<c)}(\mathbf{a})$ with high probability.

We now describe a coding scheme taking advantage of this. Specializing parameters, if we take $\delta = 1/2$, $\delta_0 = 1/100$, $s = \Omega(m^2)$, $d = (1-\delta)\cdot s\cdot q$, $n = q^m$, $\gamma = \frac{1}{5}$, and so $c > \frac{s}{5}$. Let $\Sigma_0 = \mathbb{F}_q^{\{\mathbf{i}|\mathrm{wt}(\mathbf{i})<c\}}$, and let our space of messages be the space of all functions $f : \mathbb{F}_q^m \to \Sigma_0$.

Define the encoding of the message f as follows: find any polynomial $P(\mathbf{X}) \in \mathbb{F}_q[\mathbf{X}]$ of degree at most $(c+m)\cdot q$ such that for each $\mathbf{a} \in \mathbb{F}_q^m$, we have $P^{(<c)}(\mathbf{a}) = f(\mathbf{a})$ (that such a polynomial P exists is an interpolatability statement; it follows from the arguments in Appendix A of [**Kop12**]). By choice of parameters, $(c+m)\cdot q < d$. The encoding of f is then defined to be $\mathrm{Enc}_{s,d,m,q}(P) \in \Sigma^n$. Note that $\log|\Sigma| = \binom{m+s-1}{m}\cdot \log q$, and $\log|\Sigma_0| = \binom{m+c-1}{m}\cdot\log q \approx \gamma^m \cdot \log|\Sigma|$. Thus this encoding blows up the bit-length of an alphabet symbol by a factor of 5^m, which is at most a sublinear polynomial in n for $q > 5$.

By the earlier discussion on local correction, this coding scheme has the following interesting property: given any $\mathbf{a} \in \mathbb{F}_q^m$, given oracle access to some $r \in \Sigma^m$ which is δ_0-close to the encoding of f, one can recover the value of $f(\mathbf{a})$ with high probability using only q queries into r. This scheme can be used even for q as small as $O(1)$!

(2) We now describe another local correction algorithm for multiplicity codes. This algorithm queries only m lines, but it only works for multiplicity codes of low rate.

To locally correct the value of $P^{(<s)}(\mathbf{a})$ for $\mathbf{a} \in \mathbb{F}_q^m$, given oracle access to $r : \mathbb{F}_q^m \to \Sigma$, the algorithm works as follows. First pick a set B of m uniformly random directions from $\mathbb{F}_q^m$. With high probability, we will have that B will be a set of m linearly independent vectors, and if the fraction of errors δ_0 is sufficiently small, then all the $\mathbf{b} \in B$ will be good. This means that we can compute $\tilde{R}^{(\mathbf{e})}_{j'}(\mathbf{b})$ for each $\mathbf{b} \in B$, $j' < s$ and $\mathbf{e}$ with $\mathrm{wt}(\mathbf{e}) < c$.

The following lemma implies that this data uniquely determines the polynomial $\tilde{R}_{j'}(\mathbf{X})$ for each $j' < c' = c \cdot \frac{m}{m-1}$, and this enables us to compute $P^{(<c')}(\mathbf{a})$. If $c' \geq s$, then this is proper local correction algorithm, but even if $c' < s$ this algorithm could be of potential interest.

LEMMA 4.5. *Let $c' = c \cdot \frac{m}{m-1}$. Let $R(X_1, \ldots, X_m)$ be a homogeneous polynomial of degree $j' < c'$. Suppose B is a set of m linearly independent vectors in $\mathbb{F}_q^m$, such that* $\mathsf{mult}(R, \mathbf{b}) \geq c$ *for each $\mathbf{b} \in B$.*

Then $R(X_1, \ldots, X_m) = 0$.

PROOF. Without loss of generality, we may assume $B = \{\mathbf{u}_1, \ldots, \mathbf{u}_m\}$, where $\mathbf{u}_i \in \mathbb{F}_q^m$ is 1 in coordinate i and 0 in every other coordinate. The hypothesis $\mathsf{mult}(R, \mathbf{u}_i) \geq c$ implies that for every $\mathbf{i}$ with $\mathrm{wt}(\mathbf{i}) = j'$ and $\mathbf{i}_i > j' - c$, the coefficient of $\mathbf{X}^{\mathbf{i}}$ in $R(\mathbf{X})$ is 0.

Finally, notice that for every $\mathbf{i}$ with $\mathrm{wt}(\mathbf{i}) = j'$, there always exists some $i \in [m]$ for which $\mathbf{i}_i > \frac{j'}{m} > j' - c$.

Thus $R(\mathbf{X}) = 0$. □

4.2. Global Decoding. We now consider decoding of multivariate multiplicity codes in the global sense. In the case of standard polynomial codes (the $s = 1$ case), the best known algorithm (due to Pellikaan-Wu [**PW04**]) for decoding m-variate codes over $\mathbb{F}_q$ works via a reduction to the decoding of 1-variate codes over the bigger field $\mathbb{F}_{q^m}$.

For multiplicity codes with general s, Kopparty [**Kop12**] gave a reduction from m-variate codes over $\mathbb{F}_q$ to *several instances* of decoding a 1-variate code over the big field $\mathbb{F}_{q^m}$. Using the algorithms for decoding univariate multiplicity codes discussed earlier, this gives polynomial time algorithms for unique decoding multivariate multiplicity codes upto half the minimum distance, and list-decoding multivariate multiplicity codes upto the Johnson bound.

Below we describe a variation of the reduction of [**Kop12**]. The key ingredient of that reduction is the construction of a certain special family of "algebraically-repelling" curves.

Abusing notation, we call $\mathfrak{a} \in \mathbb{F}_{q^m}^m$ a basis if its m coordinates form a basis for $\mathbb{F}_{q^m}$ over $\mathbb{F}_q$. To every basis $\mathfrak{a} = (a_1, \ldots, a_m)$, we associate a curve $\gamma_{\mathfrak{a}}(T) \in \mathbb{F}_{q^m}[T]^m$, given by:

$$\gamma_{\mathfrak{a}}(T) = (\mathrm{Tr}(a_1 T), \mathrm{Tr}(a_2 T), \ldots, \mathrm{Tr}(a_m T)).$$

The most interesting feature of this curve is that $\gamma_{\mathfrak{a}}$ is a bijection between $\mathbb{F}_{q^m}$ and $\mathbb{F}_q^m$. See [**Kop12**] for more properties of these curves $\gamma_{\mathfrak{a}}$.

A collection $\mathfrak{a}_1, \ldots, \mathfrak{a}_M \in \mathbb{F}_{q^m}^m$ of bases is said to be in (s, c)-general position if there does not exist a nonzero polynomial $R(\mathbf{X}) \in \mathbb{F}_{q^m}[\mathbf{X}]$ of degree at most s which vanishes at each $\mathfrak{a}_i$ with multiplicity at least c.

The $c = 1$ case of the following lemma was shown in [**Kop12**].

LEMMA 4.6. *Suppose $c \leq s < q$. Let $\mathfrak{a}_1, \ldots, \mathfrak{a}_M \in \mathbb{F}_{q^m}^m$ be bases in (s, c) general position. Let $Q(\mathbf{X}) \in \mathbb{F}_{q^m}[\mathbf{X}]$ have degree $< s \cdot q$. Suppose that for each $i \in [M]$ and each $\mathbf{e}$ with $\mathrm{wt}(\mathbf{e}) < c$, the univariate polynomial $Q^{(\mathbf{e})} \circ \gamma_{\mathfrak{a}_i}(T) = 0$.*

Then $Q(\mathbf{X}) = 0$.

The proof of this lemma is postponed to Section 4.2.2.

Explicit collections $\mathfrak{a}_1, \ldots, \mathfrak{a}_M \in \mathbb{F}_{q^m}^m$ in (s, c) general position with $M = \left(\frac{s}{c}\right)^m$ can be constructed as follows. Take $\mathfrak{a}$ to be a basis of $\mathbb{F}_{q^m}^m$, and $\mathbf{b}_1, \ldots, \mathbf{b}_M$ to be all the elements of a m-dimensional grid of side $\frac{s}{c}$ in $\mathbb{F}_q^m$, and set $\mathfrak{a}_i = \mathfrak{a} + \mathbf{b}_i$.

Via the lemma, this gives us an explicit collection of $\left(\frac{s}{c}\right)^m$ "algebraically-repelling" curves. We now show how these curves can be used for reducing multivariate decoding to univariate decoding. Again, this generalizes the $\gamma = 0$, $c = 1$ case, which was done in [**Kop12**].

The following lemma, relating the derivatives of a multivariate polynomial to the derivatives of its restriction to the curve $\gamma_{\mathfrak{a}}$, will motivate one of the steps of the algorithm.

LEMMA 4.7. *Let $P(X_1, \ldots, X_m) \in \mathbb{F}_q[X_1, \ldots, X_m]$ and let $Q_{\mathbf{e}}(T) \in \mathbb{F}_{q^m}[T]$ be given by $Q_{\mathbf{e}}(T) = P^{(\mathbf{e})} \circ \gamma_{\mathfrak{a}}(T)$. Then for every $t \in \mathbb{F}_{q^m}$ and every $j < q$:*

$$Q_{\mathbf{e}}^{(j)}(t) = \sum_{\mathbf{i}:\mathrm{wt}(\mathbf{i})=j} \binom{\mathbf{i}+\mathbf{e}}{\mathbf{i}} P^{(\mathbf{i}+\mathbf{e})}(\gamma_{\mathfrak{a}}(t))\mathfrak{a}^{\mathbf{i}}.$$

Algorithm for Reducing Multivariate Decoding to Univariate Decoding

(1) Suppose we have an algorithm $\mathcal{A}$ that list-decodes univariate multiplicity codes of distance δ from $\eta(\delta)$-fraction errors.
(2) Let $\gamma = \frac{\delta - \eta^{-1}(\delta_0)}{1-\eta^{-1}(\delta_0)}$, and let $c = \gamma \cdot s + 1$.
(3) Let $M = (\frac{s}{c})^m$. Pick bases $\mathfrak{a}_1, \mathfrak{a}_2, \ldots, \mathfrak{a}_M \in \mathbb{F}_{q^m}^m$ in (s, c)-general position.
(4) For each $i \in [M]$, for each $\mathbf{e}$ with $\mathrm{wt}(\mathbf{e}) < c$, define $\ell_{i,\mathbf{e}} : \mathbb{F}_{q^m} \to \mathbb{F}_{q^m}^{s-c+1}$ as follows. For each j with $0 \leq j < s - c + 1$, let

$$(\ell_{i,\mathbf{e}}(t))_j = \sum_{\mathbf{i}:\mathrm{wt}(\mathbf{i})=j} \binom{\mathbf{i}+\mathbf{e}}{\mathbf{e}} r^{(\mathbf{i}+\mathbf{e})}(\gamma_{\mathfrak{a}_i}(t)) \cdot \mathfrak{a}_i^{\mathbf{i}}.$$

(5) Using algorithm $\mathcal{A}$, compute the set $\mathcal{L}_{i,\mathbf{e}}$ of all $Q(T) \in \mathbb{F}_{q^m}[T]$ of degree at most dq^{m-1} such that $\Delta(\mathrm{Enc}_{s-c+1,dq^{m-1},1,q^m}(Q), \ell_{i,\mathbf{e}}) < \delta_0$.
(6) For every $(Q_1(T), Q_2(T), \ldots, Q_M(T)) \in \prod_{i=1}^M \mathcal{L}_i$, find all $P(X_1, \ldots, X_m) \in \mathbb{F}_q[X_1, \ldots, X_m]$ with $\deg(P) \leq d$ such that for each $i \in [M]$,

$$P \circ \gamma_{\mathfrak{a}_i}(T) = Q_i(T).$$

(This is a system of linear equations over $\mathbb{F}_{q^m}$ with $\binom{d+m}{m}$ variables and $(d+1) \cdot M$ constraints).
(7) Output the list of all such $P(X_1, \ldots, X_m)$.

Before analyzing correctness, let us comment on the running time of this reduction. For constant m, s, we claim that this reduction can be implemented to run in near-linear time. The running time of this reduction depends the size $\mathcal{L}_i$, and on how quickly we can find $P(X_1, \ldots, X_m)$ given $Q_1(T), \ldots, Q_M(T)$. If we happen to know that the $\mathcal{L}_i$ are all of constant size (as it happens whenever the fraction of errors is bounded below the Johnson radius), then the factor of $\prod |\mathcal{L}_i|$ is at most a constant. Thus we only need to show that the latter step can be implemented in near-linear time. Given $Q_1(T), \ldots, Q_M(T)$, we can: (1) evaluate each $Q_i(T)$ and each of its derivatives of order $< s$ at all points of $\mathbb{F}_q^m$, (2) via Lemma 4.7, we know that this suffices to give us $P^{(<s)}(\mathbf{a})$ for each $\mathbf{a} \in \mathbb{F}_q^m$, (3) to interpolate $P(\mathbf{X})$ given $P^{(<s)}(\mathbf{a})$ for each $\mathbf{a} \in \mathbb{F}_q^m$.

Thus, we get that unique decoding of multivariate multiplicity codes upto half the minimum distance can be done in near-linear time. Furthermore, if Nielsen's algorithm for list-decoding univariate multiplicity codes upto the Johnson bound can be implemented in near-linear time, then one can also list-decode multivariate multiplicity codes upto (almost) the Johnson bound in near-linear time.

4.2.1. *Analysis of the decoding algorithm.* Suppose $P(\mathbf{X}) \in \mathbb{F}_q[\mathbf{X}]$ is such that $\Delta(\mathrm{Enc}_{s,d,m,q}(P), r) < \delta_0$. Let $E \subseteq \mathbb{F}_q^m$ be the set of $\mathbf{a} \in \mathbb{F}_q^m$ where $P^{(<s)}(\mathbf{a})$ differs from $r^{(<s)}(\mathbf{a})$.

We first show that $Q_{i,\mathbf{e}}(T) := P^{(\mathbf{e})} \circ \gamma_{\mathfrak{a}_i}(T) \in \mathcal{L}_{i,\mathbf{e}}$. Indeed, by Lemma 4.7, for every $t \in \mathbb{F}_{q^m}$ such that $\gamma_{\mathfrak{a}}(t) \notin E$, and for every $j < s - c + 1$, we have:

$$\begin{aligned} Q_{i,\mathbf{e}}^{(j)}(t) &= \sum_{\mathbf{i}:\mathrm{wt}(\mathbf{i})=j} \binom{\mathbf{i}+\mathbf{e}}{\mathbf{i}} P^{(\mathbf{i}+\mathbf{e})}(\gamma_{\mathfrak{a}_i}(t))\mathfrak{a}_i^{\mathbf{i}} \\ &= \sum_{\mathbf{i}:\mathrm{wt}(\mathbf{i})=j} \binom{\mathbf{i}+\mathbf{e}}{\mathbf{i}} r^{(\mathbf{i}+\mathbf{e})}(\gamma_{\mathfrak{a}_i}(t))\mathfrak{a}_i^{\mathbf{i}} \\ &= \sum_{\mathbf{i}:\mathrm{wt}(\mathbf{i})=j} \binom{\mathbf{i}+\mathbf{e}}{\mathbf{i}} r^{(\mathbf{i}+\mathbf{e})}(\gamma_{\mathfrak{a}_i}(t))\mathfrak{a}_i^{\mathbf{i}} \\ &= \ell_{i,\mathbf{e}}(t). \end{aligned}$$

Thus $\Delta(\mathrm{Enc}_{s-c+1,d,1,q}(Q_{i,\mathbf{e}}), \ell_{i,\mathbf{e}}) \leq \frac{|E|}{q^m} < \delta_0$, and thus $Q_{i,\mathbf{e}}$ is indeed included in the list $\mathcal{L}_i$. The crucial points here are (a) the relative distance of the univariate multiplicity code of order $s-c+1$ evaluations of degree d polynomials has minimum distance $1 - \frac{1-\delta}{1-\gamma}$, and (b) our choice of γ ensures that $\delta_0 \leq \eta(1 - \frac{1-\delta}{1-\gamma})$. Thus algorithm $\mathcal{A}$ is indeed capable of finding $\mathcal{L}_i$ as required by the reduction.

4.2.2. *Algebraically repelling curves.* In this section we prove Lemma 4.7 and Lemma 4.6.

Proof of Lemma 4.7: By definition of derivatives, we have:

$$Q_{\mathbf{e}}(t+W) = \sum_j Q_{\mathbf{e}}^{(j)}(t)W^j,$$

$$P^{(\mathbf{e})}(\gamma_{\mathfrak{a}}(t) + \mathbf{X}) = \sum_{\mathbf{i}} (P^{(\mathbf{e})})^{(\mathbf{i})}(\gamma_{\mathfrak{a}}(t))\mathbf{X}^{\mathbf{i}} = \sum_{\mathbf{i}} \binom{\mathbf{e}+\mathbf{i}}{\mathbf{i}} P^{(\mathbf{e}+\mathbf{i})}(\gamma_{\mathfrak{a}}(t))\mathbf{X}^{\mathbf{i}}.$$

By linearity, $\gamma_{\mathfrak{a}}(t+W) = \gamma_{\mathfrak{a}}(t) + \gamma_{\mathfrak{a}}(W)$. So

$$\begin{aligned} Q_{\mathbf{e}}(t+W) &= P^{(\mathbf{e})} \circ \gamma_{\mathfrak{a}}(t+W) \\ &= P^{(\mathbf{e})}(\gamma_{\mathfrak{a}}(t) + \gamma_{\mathfrak{a}}(W)) \\ &= \sum_{\mathbf{i}} \binom{\mathbf{e}+\mathbf{i}}{\mathbf{i}} P^{(\mathbf{e}+\mathbf{i})}(\gamma_{\mathfrak{a}}(t))(\gamma_{\mathfrak{a}}(W))^{\mathbf{i}} \end{aligned}$$

Taking this equation mod W^q, we get the following equation:

$$\sum_{j<q} Q_{\mathbf{e}}^{(j)}(t)W^j = \sum_{\mathbf{i}:\mathrm{wt}(\mathbf{i})<q} \binom{\mathbf{e}+\mathbf{i}}{\mathbf{i}} P^{(\mathbf{e}+\mathbf{i})}(\gamma_{\mathfrak{a}}(t))(\mathfrak{a}W)^{\mathbf{i}} \qquad \bmod W^q$$

For $j < q$, note that the coefficient of W^j in the right hand side of this equation equals

$$\sum_{\mathbf{i}:\mathrm{wt}(\mathbf{i})=j} \binom{\mathbf{e}+\mathbf{i}}{\mathbf{i}} P^{(\mathbf{e}+\mathbf{i})}(\gamma_{\mathfrak{a}}(t))\mathfrak{a}^{\mathbf{i}}.$$

On the other hand, the coefficient of W^j in the left hand side of the equation equals $Q_{\mathbf{e}}^{(j)}(t)$. We therefore conclude that for each j with $0 \le j < q$, we have

$$Q_{\mathbf{e}}^{(j)}(t) = \sum_{\mathbf{i}:\mathrm{wt}(\mathbf{i})=j} \binom{\mathbf{e}+\mathbf{i}}{\mathbf{i}} P^{(\mathbf{e}+\mathbf{i})}(\gamma_{\mathfrak{a}}(t))\mathfrak{a}^{\mathbf{i}}.$$

□

Proof of Lemma 4.6: We will show that for each $\mathbf{a} \in \mathbb{F}_q^m$, $\mathsf{mult}(P, \mathbf{a}) \ge s$. Then by Lemma 2.4 (recalling that $\deg(P) < sq$), we can conclude that $P(X_1, \dots, X_m) = 0$.

Fix $i \in [M]$ and $\mathbf{e}$ with $\mathrm{wt}(\mathbf{e}) < c$. We have $P^{(\mathbf{e})} \circ \gamma_{\mathfrak{a}_i}(T) = 0$. By Lemma 4.7, we conclude that for every $t \in \mathbb{F}_{q^m}$ and $j < q$:

$$\sum_{\mathbf{i}:\mathrm{wt}(\mathbf{i})=j} \binom{\mathbf{e}+\mathbf{i}}{\mathbf{i}} P^{(\mathbf{e}+\mathbf{i})}(\gamma_{\mathfrak{a}_i}(t))\mathfrak{a}_i^{\mathbf{i}} = 0.$$

Thus for every $\mathbf{a} \in \mathbb{F}_q^m$, every $i \in [M]$, $0 \le j < s$, and $\mathbf{e}$ with $\mathrm{wt}(\mathbf{e}) < c$:

$$\sum_{\mathbf{i}:\mathrm{wt}(\mathbf{i})=j} \binom{\mathbf{e}+\mathbf{i}}{\mathbf{i}} P^{(\mathbf{e}+\mathbf{i})}(\mathbf{a})\mathfrak{a}_i^{\mathbf{i}} = 0. \tag{8}$$

For $0 \le j' < q$, let $R_{\mathbf{a},j'}(\mathbf{Y}) \in \mathbb{F}_q[\mathbf{Y}]$ be the polynomial

$$R_{\mathbf{a},j'}(\mathbf{Y}) = \sum_{\mathbf{i}':\mathrm{wt}(\mathbf{i}')=j'} P^{(\mathbf{i}')}(\mathbf{a})\mathbf{Y}^{\mathbf{i}'}.$$

Then the derivatives of $R_{\mathbf{a},j'}$ are given by:

$$\begin{aligned} R_{\mathbf{a},j'}^{(\mathbf{e})}(\mathbf{Y}) &= \sum_{\mathbf{i}':\mathrm{wt}(\mathbf{i}')=j'} \binom{\mathbf{i}'}{\mathbf{e}} P^{(\mathbf{i}')}(\mathbf{a})\mathbf{Y}^{\mathbf{i}'-\mathbf{e}} \\ &= \sum_{\mathbf{i}:\mathrm{wt}(\mathbf{i})=j'-\mathrm{wt}(\mathbf{e})} \binom{\mathbf{i}+\mathbf{e}}{\mathbf{e}} P^{(\mathbf{i}+\mathbf{e})}(\mathbf{a})\mathbf{Y}^{\mathbf{i}} \end{aligned}$$

Equation (8) says that for each $\mathbf{a} \in \mathbb{F}_q^m$, $i \in [M]$, $\mathbf{e}$ with $\mathrm{wt}(\mathbf{e}) < c$, and $j' < q + \mathrm{wt}(\mathbf{e})$, we have

$$R_{\mathbf{a},j'}^{(\mathbf{e})}(\mathfrak{a}_i) = 0.$$

Thus for every $j < q$ and $\mathbf{a}, i, \mathbf{e}$ as above:

$$\mathsf{mult}(R_{\mathbf{a},j}, \mathfrak{a}_i) \ge c.$$

By the general position hypothesis on $\mathfrak{a}_i$, this implies that for each $j < s$, the polynomial $R_{\mathbf{a},j}(\mathbf{Y})$ is itself identically 0.

But the coefficients of $R_{\mathbf{a},j}(\mathbf{Y})$ are $P^{(\mathbf{i})}(\mathbf{a})$, for $\mathbf{i}$ satisfying $\mathrm{wt}(\mathbf{i}) = j$. Thus $P^{(\mathbf{i})}(\mathbf{a}) = 0$ for each $\mathbf{a}$ and each $\mathbf{i}$ with $\mathrm{wt}(\mathbf{i}) < s$.

Therefore $\mathsf{mult}(P, \mathbf{a}) \ge s$ for each $\mathbf{a} \in \mathbb{F}_q^m$, which implies that $P = 0$, as desired. □

4.3. Local List-Decoding. The list-decoding algorithm for multivariate multiplicity codes from the Johnson radius can be used to give a *local list-decoding* algorithm for multivariate multiplicity codes upto the Johnson radius. Since the definition of local list-decoding is somewhat involved, we refer the reader to the appendix of [**Kop12**] for the algorithm and its analysis.

5. Encoding

In this section we discuss encoding algorithms.

Since multiplicity codes are $\mathbb{F}_q$-linear subspaces of $\Sigma^{\mathbb{F}_q^m}$, it is natural to choose an encoding map which is $\mathbb{F}_q$-linear. One very natural encoding map to consider is the map:

$$E : \mathbb{F}_q^{\binom{d+m}{m}} \to \mathcal{C},$$

which treats its input as a vector of coefficients of monomials for a polynomial $P \in \mathbb{F}_q[\mathbf{X}]$ of degree at most d, and outputs $\mathrm{Enc}_{s,d,m,q}(P)$. It is well known that the task of computing E, namely evaluating a given polynomial and all its derivatives of order at most s at all points of $\mathbb{F}_q^m$ can be performed in near-linear time $O((d^m + q^m) \cdot \binom{m+s}{m} \cdot \log(d^m + q^m))$.

However, for the purposes of local decoding, it will be important to choose the encoding map $E : \Sigma_0^k \to \mathcal{C} \subseteq \Sigma^n$ a bit more carefully. The goal is to have the encoding be *systematic*; i.e., to have the symbols of the message appear as symbols (or parts of symbols) of its encoding. Once we have such an encoding map, a local correction algorithm immediately gives us a local decoding algorithm.

Such a systematic encoding map can be chosen by giving an *interpolating set*. Concretely, for a given s, d, m, q, we want a set $S \subseteq \mathbb{F}_q^m \times \{\mathbf{i} \mid \mathrm{wt}(\mathbf{i}) < s\}$ such that for every $f : S \to \mathbb{F}_q$, there is **exactly one** $P(X_1, \ldots, X_m) \in \mathbb{F}_q[X_1, \ldots, X_m]$ of degree at most d such that for each $(\mathbf{a}, \mathbf{i}) \in S$, $P^{(\mathbf{i})}(\mathbf{a}) = f(\mathbf{a}, \mathbf{i})$.

It is easy to see that such sets S exist, and any such set S must have $|S| = \binom{d+m}{m}$. In order for the so obtained local decoding algorithm to run in sublinear time (assuming the local correction algorithm runs in sublinear time), it will be important that this interpolating set be *explicit*: given message coordinate $i \in [k]$, we should be able to compute, in time $\mathrm{poly}(\log(k))$, the codeword coordinate $j \in [n]$ which contains the ith symbol of the codeword. In [**Kop12**], an explicit such interpolating set was given.

THEOREM 5.1 ([**Kop12**]). *There exist explicit interpolating sets S as above.*

Thus there exist explicit systematic encoding maps for multiplicity codes.

The interpolating sets for multiplicity codes constructed above are in fact simple combinations of interpolating sets for Reed-Muller codes. Furthermore, it is known that there exist interpolating sets for Reed-Muller codes from which polynomial interpolation can be performed in near-linear time. This implies, by inspecting the proof of Theorem 5.1, that the encoding map described above can be computed in near-linear time.

6. Discussion

(1) The improved local decoding algorithm given in Section 4.1.2 had two main ideas over the original local decoding algorithm of [**KSY11**]: getting more information from each line, and robustly combining this information across different lines by decoding a multiplicity code.

The first idea is naturally motivated by an incongruity between the three papers: [**KSY11**] (on local correction of multiplicity codes),[**GKS13**] (on affine invariant codes, local correction of affine invariant codes, and their relationship to bounds on Nikodym sets), and [**DKSS09**] (giving lower bounds on the size of Nikodym sets via the extended method of multiplicities).

A Nikodym set is a set $N \subseteq \mathbb{F}_q^m$ such that for every $\mathbf{a} \in \mathbb{F}_q^m$, there is some line $L \subseteq \mathbb{F}_q^m$ passing through $\mathbf{a}$, such that $L \setminus \{\mathbf{a}\} \subseteq N$ (i.e., the entire line, except possibly $\mathbf{a}$, is contained in N). In [**GKS13**] it was noted that lower bounds on Nikodym sets follow from the existence of algebraic error-correcting codes that can be locally corrected at $\mathbf{a} \in \mathbb{F}_q^m$ by querying a line through $\mathbf{a}$. In [**DKSS09**], the extended method of multiplicities (which in retrospect can be interpreted in the language of multiplicity codes) was used to show a lower bound of $\left(\frac{q}{2}\right)^m$ on the size of Nikodym sets. On the other hand, if one tried to use the original local correction algorithm of [**KSY11**] for multiplicity codes to prove a lower bound on Nikodym sets (via the [**GKS13**] connection), we would get a significantly weaker bound than the bound of [**DKSS09**]. All this suggests that there should be a better algorithm for local correction of multiplicity codes, and (looking at the details) a way to get more information from each line.

The second idea is motivated by a different incongruity: the local correction algorithm of [**KSY11**] combines information from different lines by decoding a Reed-Muller code, but there seems to be no reason for the $s = 1$ case of multiplicity codes (i.e., Reed-Muller codes) to receive preferential treatment amongst all possibilities for s. The new algorithm resolves this incongruity, and lets us use fewer lines for the local decoding.

(2) The decoding algorithm described in Section 4.1.4 based on m linearly independent lines was motivated by another problem in the combinatorial geometry: the joints problem [**GK10, KSS10**]. Alex Vardy and Abdul Basit independently suggested to me that bounds on the joints problem could potentially be improved using the method of multiplicities. The m-line decoding algorithm for multiplicity codes can be used to give a multiplicity-enhanced variant of the [**KSS10**] bound on the joints problem; unfortunately this variant does not improve on the bounds.

(3) The Reed-Muller codes over small fields and in many variables have been very influential. The analogous multiplicity code would be based on order s evaluations of polynomials over $\mathbb{F}_q$ of total degree at most d and individual degree at most $sq - 1$. It would be very interesting to study properties of these codes, and to see if they have any coding/combinatorial applications.

We note a curious example of how things change when the individual degrees are bounded: a polynomial $P(X_1, \ldots, X_m) \in \mathbb{F}_2[X_1, \ldots, X_m]$ of individual degree at most 1 vanishes at a point $\mathbf{a} \in \mathbb{F}_2^m$ with multiplicity at least s if and only if P vanishes at all points of $\mathbb{F}_2^n$ which are at a Hamming distance at most $s-1$ from $\mathbf{a}$.

(4) There are various interesting variations/cousins of Reed-Solomon andReed-Muller codes which are interesting from the coding perspective. These include algebraic-geometric codes, BCH codes, projective Reed-Muller codes, Grassman codes, etc. It would be interesting to investigate multiplicity-based generalizations of these codes. For algebraic-geometric codes, some such investigations were made (in the 1-dimensional case, analogous to Reed-Solomon codes) by Rosenbloom-Tsfasman [**RT97**], Xing [**Xin03**] and Nielsen [**Nie01**].

7. Open Questions

We conclude with a list of some interesting open questions. Some of these questions are open even in the classical $s = 1$ case (Reed-Solomon codes and Reed-Muller codes).

(1) What is the list-decoding radius for univariate multiplicity codes? In other words, what is the largest fraction η of errors from which univariate multiplicity codes of distance δ and block-length n can be list-decoded with poly(n) list-size? For general univariate multiplicity codes, this is only known to be true for $\eta \leq 1 - \sqrt{1-\delta}$, while for univariate multiplicity codes over prime fields with s sufficiently large, it is known for every $\eta < \delta$.

This question is even open for Reed-Solomon codes.

Here are some related questions. Does the answer depend on the field? Does the answer depend on the set of evaluation points? What happens for multivariate multiplicity codes (again, this is open for multivariate Reed-Muller codes too)?

(2) It is an extremely interesting question whether the list-size for list-decoding univariate multiplicity codes over prime fields (as in [**GW11, Kop12**]) needs to be poly(q), or if it can be reduced to a constant independent of q.

It is also extremely interesting to know whether the primality of q is essential for the improved list-decoding of multiplicity codes in [**GW11, Kop12**].

(3) Can Nielsen's algorithm for list-decoding univariate multiplicity codes up to the Johnson bound be implemented to run in near-linear time? This would also immediately imply a near-linear time global algorithm and a faster local algorithm for list-decoding multivariate multiplicity codes up to the Johnson radius. This seems to require some nontrivial adaptation of the ideas of Alekhnovich [**Ale05**], who showed how to list-decode Reed-Solomon codes in near-linear time.

(4) Are multiplicity codes locally testable?

(5) Are there any applications of multiplicity codes to computational complexity theory? Reed-Solomon codes and Reed-Muller codes have found many celebrated applications, and it would be interesting to see if multiplicity codes can improve on any of them.

(6) Are there any practical applications of multiplicity codes? By combining high rate with sublinear-time decoding, and also supporting various other efficent operations, multiplicity codes seem to be theoretically practical. Perhaps they really are practical?

Acknowledgements

Thanks to Alexander Barg and Oleg Musin for organizing an excellent workshop to celebrate Ilya Dumer's birthday, and for encouraging me to write this article. Many, many thanks to Ilya Dumer for being a wonderful friend and mentor for many years, and for getting me hooked on coding theory in the first place.

References

[Ale05] Michael Alekhnovich, *Linear Diophantine equations over polynomials and soft decoding of Reed-Solomon codes*, IEEE Trans. Inform. Theory **51** (2005), no. 7, 2257–2265, DOI 10.1109/TIT.2005.850097. MR2246361 (2007e:94126)

[BI01] Amos Beimel and Yuval Ishai, *Information-theoretic private information retrieval: a unified construction (extended abstract)*, Automata, languages and programming, Lecture Notes in Comput. Sci., vol. 2076, Springer, Berlin, 2001, pp. 912–926, DOI 10.1007/3-540-48224-5_74. MR2066561

[Chi76] Francis Y. Chin, *A generalized asymptotic upper bound on fast polynomial evaluation and interpolation*, SIAM J. Comput. **5** (1976), no. 4, 682–690. MR0495185 (58 #13917)

[DKSS09] Zeev Dvir, Swastik Kopparty, Shubhangi Saraf, and Madhu Sudan, *Extensions to the method of multiplicities, with applications to Kakeya sets and mergers*, 2009 50th Annual IEEE Symposium on Foundations of Computer Science (FOCS 2009), IEEE Computer Soc., Los Alamitos, CA, 2009, pp. 181–190, DOI 10.1109/FOCS.2009.40. MR2648400 (2011d:68128)

[GK10] Larry Guth and Nets Hawk Katz, *Algebraic methods in discrete analogs of the Kakeya problem*, Adv. Math. **225** (2010), no. 5, 2828–2839, DOI 10.1016/j.aim.2010.05.015. MR2680185 (2011h:52019)

[GKS13] A. Guo, S. Kopparty, and M. Sudan, *New affine-invariant codes from lifting*, ITCS, 2013, pp. 529–540.

[GR08] Venkatesan Guruswami and Atri Rudra, *Explicit codes achieving list decoding capacity: error-correction with optimal redundancy*, IEEE Trans. Inform. Theory **54** (2008), no. 1, 135–150, DOI 10.1109/TIT.2007.911222. MR2446745 (2010b:94096)

[Guo13] A. Guo, *High rate locally correctable codes via lifting*, Electronic Colloquium on Computational Complexity (ECCC) **20** (2013), 53.

[GW11] Venkatesan Guruswami and Carol Wang, *Optimal rate list decoding via derivative codes*, Approximation, randomization, and combinatorial optimization, Lecture Notes in Comput. Sci., vol. 6845, Springer, Heidelberg, 2011, pp. 593–604, DOI 10.1007/978-3-642-22935-0_50. MR2863293

[HKT08] J. W. P. Hirschfeld, G. Korchmáros, and F. Torres, *Algebraic curves over a finite field*, Princeton Series in Applied Mathematics, Princeton University Press, Princeton, NJ, 2008. MR2386879 (2008m:14040)

[HOW13] B. Hemenway, R. Ostrovsky, and M. Wootters, *Local correctability of expander codes*, ICALP (1), 2013, pp. 540–551.

[Kop12] S. Kopparty, *List-decoding multiplicity codes*, Electronic Colloquium on Computational Complexity (ECCC), TR12-044, 2012.

[KSS10] Haim Kaplan, Micha Sharir, and Eugenii Shustin, *On lines and joints*, Discrete Comput. Geom. **44** (2010), no. 4, 838–843, DOI 10.1007/s00454-010-9246-3. MR2728035 (2012g:52043)

[KSY11] Swastik Kopparty, Shubhangi Saraf, and Sergey Yekhanin, *High-rate codes with sublinear-time decoding*, STOC'11—Proceedings of the 43rd ACM Symposium on Theory of Computing, ACM, New York, 2011, pp. 167–176, DOI 10.1145/1993636.1993660. MR2931966

[Nie01] R. R. Nielsen, *List decoding of linear block codes*, Ph.D. thesis, Technical University of Denmark, 2001.

[PV05] F. Parvaresh and A. Vardy, *Correcting errors beyond the Guruswami-Sudan radius in polynomial time*, 46th IEEE Symposium on Foundations of Computer Science (FOCS), 2005, pp. 285–294.

[PW04] Ruud Pellikaan and Xin-Wen Wu, *List decoding of q-ary Reed-Muller codes*, IEEE Trans. Inform. Theory **50** (2004), no. 4, 679–682, DOI 10.1109/TIT.2004.825043. MR2065650 (2005g:94127)

[RT97] M. Yu. Rozenblyum and M. A. Tsfasman, *Codes for the m-metric* (Russian, with Russian summary), Problemy Peredachi Informatsii **33** (1997), no. 1, 55–63; English transl., Problems Inform. Transmission **33** (1997), no. 1, 45–52. MR1441528 (98c:94023)

[Str81] V. Strassen, *The computational complexity of continued fractions*, Proceedings of the fourth ACM symposium on Symbolic and algebraic computation, ACM, 1981, pp. 51–67.

[Sud01] M. Sudan, *Notes on rational function interpolation*, Lecture notes for *Algorithmic Introduction to Coding Theory*, 2001.

[Xin03] Chaoping Xing, *Nonlinear codes from algebraic curves improving the Tsfasman-Vlăduţ-Zink bound*, IEEE Trans. Inform. Theory **49** (2003), no. 7, 1653–1657, DOI 10.1109/TIT.2003.813559. MR1985569 (2004f:94116)

Department of Mathematics & Department of Computer Science, Rutgers University.

E-mail address: swastik.kopparty@gmail.com

Contemporary Mathematics
Volume **625**, 2014
http://dx.doi.org/10.1090/conm/625/12498

Multivariate positive definite functions on spheres

Oleg R. Musin

To Ilya Dumer, on the occasion of his 60th birthday

ABSTRACT. In 1942, I. J. Schoenberg proved that a function is positive definite in the unit sphere if and only if this function is a positive linear combination of the Gegenbauer polynomials. In this paper we extend Schoenberg's theorem for multivariate positive definite functions. This extension derives new positive semidefinite constraints for the distance distribution which can be applied for spherical codes.

1. Introduction

Let M be a metric space with a distance function τ. A real continuous function $f(t)$ is said to be positive definite (p.d.) in M if for arbitrary points $p_1, \dots, p_r$ in M, real variables $x_1, \dots, x_r$, and arbitrary r we have

$$\sum_{i=1}^{r} \sum_{j=1}^{r} f(t_{ij})\, x_i x_j \geq 0, \quad t_{ij} = \tau(p_i, p_j),$$

or equivalently, the matrix $\big(f(t_{ij})\big) \succeq 0$, where the sign $\succeq 0$ stands for: "is positive semidefinite".

Let $\mathbb{S}^{n-1}$ denote the unit sphere in $\mathbb{R}^n$, and let φ_{ij} denote the angular distance between points p_i, p_j. Schoenberg [**21**] proved that:

$f(\cos\varphi)$ is p.d. in $\mathbb{S}^{n-1}$ if and only if $f(t) = \sum_{k=0}^{\infty} f_k G_k^{(n)}(t)$ with all $f_k \geq 0$.

Here $G_k^{(n)}(t)$ are the Gegenbauer polynomials.

Schoenberg's theorem has been generalized by Bochner [**6**] to more general spaces. Namely, the following fact holds: *f is p.d. in a 2-point-homogenous space M if and only if $f(t)$ is a nonnegative linear combination of the zonal spherical functions $\Phi_k(t)$* (see details in [**16**, Th. 2], [**10**, Chapter 9]).

Note that the Bochner–Schoenberg theorem is widely used in coding theory and discrete geometry for finding bounds for error-correcting codes, constant weight codes, spherical codes, sphere packings and other packing problems in 2-point-homogeneous spaces (see [**10, 16**–**20**] and many others).

2010 *Mathematics Subject Classification.* Primary 52C35; Secondary 94B75.

Key words and phrases. Positive Define Functions, Schoenberg theorem, spherical codes, Semidefinite Programming.

This work was partially supported by NSF grant DMS - 1101688 and by RFBR grant 13-01-12458.

Consider the following problem: *Let* $Q = \{q_1, \ldots, q_m\}$ *be a set of points in* M. *To describe the class of continuous functions* $F(t, u, v)$ *in* $2m + 1$ *variables with* $t \in \mathbb{R}$, $u, v \in \mathbb{R}^m$, $F(t, u, v) = F(t, v, u)$ *such that for arbitrary points* $p_1, \ldots, p_r$ *in* M *the matrix* $\big(F(t_{ij}, u_i, u_j)\big) \succeq 0$, *where* $t_{ij} = \tau(p_i, p_j)$, $u_i = (\tau(p_i, q_1), \ldots, \tau(p_i, q_m))$.

Denote this class by $\mathrm{PD}(M, Q)$. If $Q = \emptyset$, then $\mathrm{PD}(M, Q)$ is the class of p.d. functions in M.

Schrijver [**22**] improved some upper bounds on binary codes using semidefinite programming. Schrijver's method has been adapted for some non-binary codes (Gijswijt, Schrijver, and Tanaka [**13**]), and for spherical codes (Bachoc and Vallentin [**2–4**]). (See also an interesting review written by Bachoc [**1**] and the recent paper by Cohn and Woo [**9**].)

In fact, by using the stabilizer subgroup of the isometry group this method derives new positive semidefinite constraints which are stronger than constraints given by the Bochner–Schoenberg theorem (see [**2**]). In other words, it shows that constraints for $\{t_{ij}\}$ and $\{u_i\}$ given by $\mathrm{PD}(M, \{q_1\})$ are stronger than constraints in $\mathrm{PD}(M, \emptyset)$ for spherical, Hamming and some other spaces.

Clearly, if $F_1, F_2 \in \mathrm{PD}(M, Q)$, then for any functions α_1, α_2 in m variables we have

$$F(t, u, v) = \alpha_1(u)\alpha_1(v)F_1(t, u, v) + \alpha_2(u)\alpha_2(v)F_2(t, u, v) \in \mathrm{PD}(M, Q). \tag{1.1}$$

In this paper we consider the class $\mathrm{PD}(\mathbb{S}^{n-1}, Q)$. Namely, we show that this class can be generated by (1.1) using as a basis the polynomials $G_k^{(n-m)}$, where $m = \mathrm{card}(Q) \le n - 2$. Actually, the class $\mathrm{PD}(\mathbb{S}^{n-1}, Q)$ becomes "more rich" whenever m is increasing.

The paper is organized as follows: Section 2 introduces multivariate Gegenbauer polynomials and its basic properties. Section 3 extends the Schoenberg theorem. Section 4 extends the Schoenberg theorem to Euclidean spaces. Section 5 discusses applications of the extended Bochner– Schoenberg theorem to spherical codes.

2. Gegenbauer polynomials and multivariate positive definite functions

In this section we introduce polynomials $G_k^{(n,m)}(t, \mathbf{u}, \mathbf{v})$ in $2m+1$ variables with $0 \le m \le n-2$, $t \in \mathbb{R}$, $\mathbf{u} = (u_1, \ldots, u_m)$, $\mathbf{v} = (v_1, \ldots, v_m) \in \mathbb{R}^m$.

2.1. Gegenbauer polynomials. There are many ways to define Gegenbauer (or ultraspherical) polynomials $G_k^{(n)}(t)$ (see [**10, 12, 20, 21**]). $G_k^{(n)}$ are a special case of Jacobi polynomials $P_k^{(\alpha,\beta)}$ with $\alpha = \beta = (n-3)/2$ and the normalization by $G_k^{(n)}(1) = 1$. Also $G_k^{(n)}$ can be defined by the recurrence formula:

$$G_0^{(n)} = 1,\ G_1^{(n)} = t,\ \ldots,\ G_k^{(n)} = \frac{(2k+n-4)\,t\,G_{k-1}^{(n)} - (k-1)\,G_{k-2}^{(n)}}{k+n-3}.$$

Note that for any even $k \ge 0$ (resp. odd) $G_k^{(n)}(t)$ is even (resp. odd). Therefore, G_{2k} and $G_{2\ell+1}$ are orthogonal on $[-1, 1]$. Moreover, all polynomials $G_k^{(n)}$ are orthogonal on $[-1, 1]$ with respect to the weight function $(1-t^2)^{(n-3)/2}$:

$$\int_{-1}^{1} G_k^{(n)}(t)\, G_\ell^{(n)}(t)\, (1-t^2)^{(n-3)/2}\, dt = 0, \quad k \ne \ell. \tag{2.1}$$

Recall the addition theorem for Gegenbauer polynomials:

$$G_k^{(n)}(\cos\theta_1\cos\theta_2+\sin\theta_1\sin\theta_2\cos\varphi)$$

$$=\sum_{s=0}^{k} c_{nks}\, G_{k-s}^{(n+2s)}(\cos\theta_1)\, G_{k-s}^{(n+2s)}(\cos\theta_2)\,(\sin\theta_1)^s\,(\sin\theta_2)^s\, G_s^{(n-1)}(\cos\varphi),$$

where c_{nks} are *positive* coefficients whose values of no concern here (see [**7**,**12**]).

2.2. Basic definitions. Let us denote by $\langle \mathbf{u},\mathbf{v}\rangle$ the inner product of vectors $\mathbf{u},\mathbf{v}\in\mathbb{R}^m$, and by $|\mathbf{v}|$ we denote the norm of $\mathbf{v}$ (i.e. $|\mathbf{v}|^2=\langle\mathbf{v},\mathbf{v}\rangle$).

DEFINITION 2.1. *Let $0\le m\le n-2$, $t\in\mathbb{R}$, $\mathbf{u},\mathbf{v}\in\mathbb{R}^m$ for $m>0$, and $\mathbf{u}=\mathbf{v}=0$ for $m=0$. Then the following polynomial in $2m+1$ variables of degree k in t is well defined:*

$$G_k^{(n,m)}(t,\mathbf{u},\mathbf{v}):=(1-|\mathbf{u}|^2)^{k/2}\,(1-|\mathbf{v}|^2)^{k/2}\,G_k^{(n-m)}\left(\frac{t-\langle\mathbf{u},\mathbf{v}\rangle}{\sqrt{(1-|\mathbf{u}|^2)(1-|\mathbf{v}|^2)}}\right).$$

Note that by definition we have: $G_k^{(n,0)}(t)=G_k^{(n,0)}(t,0,0)=G_k^{(n)}(t)$.

Let $\mathbf{u}=(u_1,\dots,u_n)\in\mathbb{R}^n$. For $0<m\le n$ we denote by $\mathbf{u}^{(m)}$ the vector $(u_1,\dots,u_m)$ in $\mathbb{R}^m$. Let $\mathbf{u}^{(0)}=0$. For $0\le m\le n-2$, $\mathbf{u},\mathbf{v}\in\mathbb{R}^n$ put

$$G_k^{(n,m)}(t,\mathbf{u},\mathbf{v}):=G_k^{(n,m)}(t,\mathbf{u}^{(m)},\mathbf{v}^{(m)}).$$

Remark. Note that these multivariate polynomials are different from the multivariate Jacobi polynomials (see [**11**,**15**]).

DEFINITION 2.2. *We define a matrix $Z_d^m(\mathbf{u},\mathbf{v})$ of size $\binom{m+d}{m}\times\binom{m+d}{m}$ for $\mathbf{u}\in\mathbb{R}^m$, $\mathbf{v}\in\mathbb{R}^m$ by $Z_d^m(\mathbf{u},\mathbf{v}):=(z_d^m(\mathbf{u}))^T\,z_d^m(\mathbf{v})$, where*

$$z_d^m(\mathbf{x})=z_d^m(x_1,\dots,x_m):=(1,x_1,\dots,x_m,x_1^2,x_1x_2,\dots,x_{m-1}x_m,x_m^2,\dots,x_m^d)$$

is the vector of monomials.

Denote the inner product of matrices $A=(a_{ij})$, $B=(b_{ij})$ both of size $d\times d$ by $\langle A,B\rangle$, i.e. $\langle A,B\rangle=\mathrm{Tr}(AB)=\sum_{i,j=1}^{d}a_{ij}b_{ij}$.

DEFINITION 2.3. *Let $f(\mathbf{u},\mathbf{v})$ be a symmetric polynomial in $\mathbf{u},\mathbf{v}\in\mathbb{R}^m$, i.e. $f(\mathbf{u},\mathbf{v})=f(\mathbf{v},\mathbf{u})$. We say that $f(\mathbf{u},\mathbf{v})$ is positive semidefinite and write $f\succeq 0$ if there are a symmetric matrix $H\succeq 0$ and $d\ge 0$ so that $f(\mathbf{u},\mathbf{v})=\langle H,Z_d^m(\mathbf{u},\mathbf{v})\rangle$.*

Let polynomials $f_r\succeq 0$ for all r, $f_r\to f$ as $r\to\infty$, where a function $f(\mathbf{u},\mathbf{v})$ is symmetric in $\mathbf{u},\mathbf{v}$ and continuous on $\{|\mathbf{u}|\le 1,|\mathbf{v}|\le 1\}$. Then we say that $f(\mathbf{u},\mathbf{v})$ is positive semidefinite and write $f\succeq 0$.

2.3. Orthogonality. Let $t\in\mathbb{R}$, $\mathbf{u}=(u_1,\dots,u_m)$, $\mathbf{v}=(v_1,\dots,v_m)\in\mathbb{R}^m$. Denote

$$Q_m(t,\mathbf{u},\mathbf{v}):=\begin{pmatrix}1&\dots&0&u_1&v_1\\ \vdots&\vdots&\ddots&\vdots&\\ 0&\dots&1&u_m&v_m\\ u_1&\dots&u_m&1&t\\ v_1&\dots&v_m&t&1\end{pmatrix}.$$

Let

$$D_m := \{(t, \mathbf{u}, \mathbf{v}) \in \mathbb{R}^{2m+1} : Q_m(t, \mathbf{u}, \mathbf{v}) \succeq 0\}.$$

Since

$$\det(Q_m(t, \mathbf{u}, \mathbf{v})) = (1 - |\mathbf{u}|^2)(1 - |\mathbf{v}|^2) - (t - \langle \mathbf{u}, \mathbf{v}\rangle)^2,$$

we have $(t, \mathbf{u}, \mathbf{v}) \in D_m$ if and only if

$$(t - \langle \mathbf{u}, \mathbf{v}\rangle)^2 \le (1 - |\mathbf{u}|^2)(1 - |\mathbf{v}|^2), \quad |\mathbf{u}| \le 1.$$

Denote $(\det(Q_m(t, \mathbf{u}, \mathbf{v})))^{(n-m-3)/2}$ by $\rho_{n,m}(t, \mathbf{u}, \mathbf{v})$, i.e.

$$\rho_{n,m}(t, \mathbf{u}, \mathbf{v}) := \left((1 - |\mathbf{u}|^2)(1 - |\mathbf{v}|^2) - (t - \langle \mathbf{u}, \mathbf{v}\rangle)^2\right)^{(n-m-3)/2}.$$

THEOREM 2.4. *Let $0 \le m \le n-2$. Let $q(\mathbf{u}, \mathbf{v})$ be any continuous function on $\{(\mathbf{u}, \mathbf{v}) \in \mathbb{R}^{2m} : |\mathbf{u}| \le 1, |\mathbf{v}| \le 1\}$. If $k \ne \ell$, then*

$$\int_{D_m} G_k^{(n,m)}(t, \mathbf{u}, \mathbf{v})\, G_\ell^{(n,m)}(t, \mathbf{u}, \mathbf{v})\, q(\mathbf{u}, \mathbf{v})\, \rho_{n,m}(t, \mathbf{u}, \mathbf{v})\, dt\, d\mathbf{u}\, d\mathbf{v} = 0.$$

PROOF. Let

$$t = s\,((1 - |\mathbf{u}|^2)(1 - |\mathbf{v}|^2))^{1/2} + \langle \mathbf{u}, \mathbf{v}\rangle.$$

In the variables $s, \mathbf{u}, \mathbf{v}$ we have

$$D_m = \{(s, \mathbf{u}, \mathbf{v}) : -1 \le s \le 1,\ |\mathbf{u}| \le 1,\ |\mathbf{v}| \le 1\}.$$

The Jacobian of this change of variables is $((1 - |\mathbf{u}|^2)(1 - |\mathbf{v}|^2))^{1/2}$. Then

$$I = \int_{D_m} G_k^{(n,m)}(t, \mathbf{u}, \mathbf{v})\, G_\ell^{(n,m)}(t, \mathbf{u}, \mathbf{v})\, q(\mathbf{u}, \mathbf{v})\, \rho_{n,m}(t, \mathbf{u}, \mathbf{v})\, dt\, d\mathbf{u}\, d\mathbf{v}$$

$$= I_0 \int_{-1}^{1} G_k^{(n-m)}(s)\, G_\ell^{(n-m)}(s)\, (1 - s^2)^{(n-m-3)/2}\, ds,$$

$$I_0 = \int_{|\mathbf{u}| \le 1} \int_{|\mathbf{v}| \le 1} \left((1 - |\mathbf{u}|^2)(1 - |\mathbf{v}|^2)\right)^{(n-m+k-2)/2} q(\mathbf{u}, \mathbf{v})\, d\mathbf{u}\, d\mathbf{v}.$$

Thus, (2.1) yields $I = 0$. □

Let $e_1, \dots, e_n$ be an orthonormal basis of $\mathbb{R}^n$, and let $(x_1, \dots, x_n)$ be the coordinates expression of a point $x \in \mathbb{R}^n$ in this basis. Then $x^{(m)}$ as well as $G_k^{(n,m)}(\langle x, y\rangle, x, y) = G_k^{(n,m)}(\langle x, y\rangle, x^{(m)}, y^{(m)})$ for $x, y \in \mathbb{R}^n$ are well defined.

LEMMA 2.1. *For any continuous F on D_m and $0 \le m \le n-2$ we have*

$$\int_{(\mathbb{S}^{n-1})^2} F\big(\langle x, y\rangle, x^{(m)}, y^{(m)}\big)\, d\omega_n(x)\, d\omega_n(y)$$

$$= \omega_{n-m}\, \omega_{n-m-1} \int_{D_m} F(t, \mathbf{u}, \mathbf{v})\, r_m(\mathbf{u})\, r_m(\mathbf{v})\, \rho_{n,m}(t, \mathbf{u}, \mathbf{v})\, dt\, d\mathbf{u}\, d\mathbf{v},$$

where $r_m(\mathbf{u}) = \prod_{i=1}^{m-1}(1 - |\mathbf{u}^{(i)}|^2)^{1/2}$ for $m > 1$, $r_0 = r_1 = 1$, and ω_n is the surface area of $\mathbb{S}^{n-1}$ for the standard measure $d\omega_n$.

PROOF. Let $x, y \in \mathbb{S}^{n-1}$. Let $t = \langle x, y\rangle$, $\mathbf{u} = x^{(m)}$, $\mathbf{v} = y^{(m)}$. Let

$$a = \frac{x - \mathbf{u}}{h(\mathbf{u})}, \quad b = \frac{y - \mathbf{v}}{h(\mathbf{v})}, \quad s = \langle a, b\rangle, \quad \text{where } h(\mathbf{u}) = \sqrt{1 - |\mathbf{u}|^2}.$$

Then

$$F\big(\langle x, y\rangle, x^{(m)}, y^{(m)}\big) = F(h(\mathbf{u})\, h(\mathbf{v})\, \langle a, b\rangle + \langle \mathbf{u}, \mathbf{v}\rangle, \mathbf{u}, \mathbf{v}) = F(t, \mathbf{u}, \mathbf{v}).$$

Therefore, we can change the variables $(x, y) \to (a, b, \mathbf{u}, \mathbf{v}) \to (s, \mathbf{u}, \mathbf{v}) \to (t, \mathbf{u}, \mathbf{v})$.

We obviously have $|\mathbf{u}| \le 1$, $|\mathbf{v}| \le 1$, $-1 \le s \le 1$. Clearly, $|a| = |b| = 1$. Then $(a, b) \in (\mathbb{S}^{n-m-1})^2$. Since $|t - \langle \mathbf{u}, \mathbf{v}\rangle| \le h(\mathbf{u})\, h(\mathbf{v})$ we have $(t, \mathbf{u}, \mathbf{v}) \in D_m$.

Consider $(x, y) \to (a, b, \mathbf{u}, \mathbf{v})$. The Jacobian is defined by

$$d\omega_n(x) = k_{n,m}(\mathbf{u})\, d\mathbf{u}\, d\omega_{n-m}(a),$$

$$k_{n,m}(\mathbf{u}) = (1 - |\mathbf{u}|^2)^{(n-m-2)/2} \prod_{i=0}^{m-1} (1 - |\mathbf{u}^{(i)}|^2)^{1/2}.$$

Combining the volume of $\{(a, b) \in (\mathbb{S}^{n-m-1})^2 : \langle a, b\rangle = s\}$ and the Jacobian of $(a, b, \mathbf{u}, \mathbf{v}) \to (s, \mathbf{u}, \mathbf{v})$ we get: $\omega_{n-m}\, \omega_{n-m-1}\, \rho_{n,m}(s, 0, 0)$. The Jacobian of the changing $(s, \mathbf{u}, \mathbf{v}) \to (t, \mathbf{u}, \mathbf{v})$ is $(h(\mathbf{u})\, h(\mathbf{v}))^{-1}$. That completes the proof. $\square$

Combining Lemma 2.1 and Theorem 2.1 we get:

COROLLARY 2.1. *Let* $0 \le m \le n-2$. *Let* $f(\mathbf{u}, \mathbf{v})$ *be any continuous function on* $\{(\mathbf{u}, \mathbf{v}) \in \mathbb{R}^{2m} : |\mathbf{u}| \le 1, |\mathbf{v}| \le 1\}$. *If* $k \ne \ell$, *then*

$$\int_{(\mathbb{S}^{n-1})^2} G_k^{(n,m)}(\langle x, y\rangle, x, y)\, G_\ell^{(n,m)}(\langle x, y\rangle, x, y)\, f\big(x^{(m)}, y^{(m)}\big)\, d\omega_n(x)\, d\omega_n(y) = 0.$$

2.4. The addition theorem.

THEOREM 2.5 (The addition theorem). *Let* $1 \le m \le n-2$. *Then*

$$G_k^{(n,m-1)}(t, \mathbf{u}, \mathbf{v}) = \sum_{s=0}^{k} C_{k-s}^{n,m}(\mathbf{u})\, C_{k-s}^{n,m}(\mathbf{v})\, G_s^{(n,m)}(t, \mathbf{u}, \mathbf{v}),$$

where $C_d^{n,m}(\mathbf{u}) = C_d^{n,m}(u_1, \ldots, u_m)$ *is a polynomial in* $u_1, \ldots, u_m$ *of degree* d.

PROOF. Suppose t, $\mathbf{u} = (u_1, \ldots, u_m)$, $\mathbf{v} = (v_1, \ldots, v_m)$ are such that

$$|\mathbf{u}| < 1, \quad |\mathbf{v}| < 1, \quad (t - \langle \mathbf{u}, \mathbf{v}\rangle)^2 < (1 - |\mathbf{u}|^2)(1 - |\mathbf{v}|^2).$$

Then $\theta_1, \theta_2, \varphi \in (0, \pi)$ are uniquely defined by the following equations:

$$\cos\theta_1 = \frac{u_m}{\sqrt{1 - u_1^2 - \ldots - u_{m-1}^2}}, \qquad \cos\theta_2 = \frac{v_m}{\sqrt{1 - v_1^2 - \ldots - v_{m-1}^2}},$$

$$\cos\varphi = \frac{t - \langle \mathbf{u}, \mathbf{v}\rangle}{\sqrt{(1 - |\mathbf{u}|^2)(1 - |\mathbf{v}|^2)}}.$$

The addition theorem for Gegenbauer polynomials yields:

$$G_k^{(n,m-1)}(t, \mathbf{u}, \mathbf{v}) = \sum_{s=0}^{k} C_{k-s}^{n,m}(\mathbf{u})\, C_{k-s}^{n,m}(\mathbf{v})\, G_s^{(n,m)}(t, \mathbf{u}, \mathbf{v}), \tag{2.2}$$

$$C_{k-s}^{n,m}(\mathbf{u}) := \sqrt{c_{nks}}\, w^{k-s}\, G_{k-s}^{(n+2s)}\left(\frac{u_m}{w}\right), \quad w = \sqrt{1 - u_1^2 - \ldots - u_{m-1}^2}.$$

It's easy to see that $C_{k-s}^{n,m}(u_1, \ldots, u_m)$ is a polynomial of degree $k - s$. Thus, (2.2) holds for all $t, \mathbf{u}, \mathbf{v}$. $\square$

Let $b_s(\mathbf{u}, \mathbf{v}) = C_{k-s}^{n,m}(\mathbf{u})\, C_{k-s}^{n,m}(\mathbf{v})$. Clearly, $b_s \succeq 0$. Then Theorem 2.2 yields

COROLLARY 2.2. *Let* $0 \le m \le \ell \le n-2$. *Let* $\mathbf{u}, \mathbf{v} \in \mathbb{R}^n$. *Then*

$$G_k^{(n,m)}(t,\mathbf{u},\mathbf{v}) = \sum_{s=0}^{k} f_s\big(\mathbf{u}^{(\ell)}, \mathbf{v}^{(\ell)}\big)\, G_s^{(n,\ell)}(t,\mathbf{u},\mathbf{v})$$

with $f_s \succeq 0$ *for all* $0 \le s \le k$.

3. An extension of the Schoenberg theorem

3.1. Schoenberg's theorem. Let $p_1, \dots, p_r$ be points in $\mathbb{S}^{n-1}$, and let $a_1, \dots, a_r$ be any real numbers. Then

$$0 \le |a_1p_1 + \ldots + a_rp_r|^2 = \sum_{i,j} \langle p_i, p_j\rangle a_i a_j,$$

or equivalently the Gram matrix $\big(\langle p_i, p_j\rangle\big)$ is positive semidefinite.

Schoenberg [**21**] extended this property to Gegenbauer polynomials $G_k^{(n)}$. He proved that *for any finite* $X = \{p_1, \dots, p_r\} \subset \mathbb{S}^{n-1}$ *the matrix* $\Big(G_k^{(n)}(\langle p_i, p_j\rangle)\Big)$ *is positive semidefinite.*

Schoenberg proved also that the converse holds: *if* $f(t)$ *is a real polynomial and for any finite* $X \subset \mathbb{S}^{n-1}$ *the matrix* $\big(f(\langle p_i, p_j\rangle)\big) \succeq 0$, *then* $f(t)$ *is a linear combination of* $G_k^{(n)}(t)$ *with nonnegative coefficients.*

3.2. An extension of the direct Schoenberg theorem.

THEOREM 3.1. *Let* $e_1, \dots, e_n$ *be an orthonormal basis of* $\mathbb{R}^n$, *and let* $p_1, \dots, p_r$ *be points in* $\mathbb{S}^{n-1}$. *Then for any* $k \ge 0$ *and* $0 \le m \le n-2$ *the matrix* $\Big(G_k^{(n,m)}(\langle p_i, p_j\rangle, p_i, p_j)\Big)$ *is positive semidefinite.*

PROOF. Actually, this theorem is a simple consequence of the Schoenberg theorem. Indeed, let $\mathbf{v}_i = \langle p_i, e_1\rangle\, e_1 + \ldots + \langle p_i, e_m\rangle\, e_m$, and let $\mathbf{x}_i = p_i - \mathbf{v}_i$. Then $\mathbf{x}_i$ is a vector in $\mathbb{R}^{n-m}$ with the basis $e_{m+1}, \dots, e_n$. Note that $|\mathbf{v}_i|^2 + |\mathbf{x}_i|^2 = 1$.

Let $\mathbf{y}_i = \mathbf{x}_i/|\mathbf{x}_i|$ for $|\mathbf{x}_i| > 0$. In the case $|\mathbf{x}_i| = 0$ put $\mathbf{y}_i = e_n$. Then $\mathbf{y}_i \in \mathbb{S}^{n-m-1}$.

Recall the Schur theorem: *If* $A = \big(a_{ij}\big) \succeq 0$, $B = \big(b_{ij}\big) \succeq 0$, *then* $C = \big(a_{ij}b_{ij}\big) \succeq 0$. Let $h_i = |\mathbf{x}_i|^k = (1 - |\mathbf{v}_i|^2)^{k/2}$, $\mathbf{h} = (h_1, \dots, h_r)$, $A = \mathbf{h}^T\mathbf{h}$. Clearly, $A \succeq 0$. The Schoenberg theorem yields: $B = \Big(G_k^{(n-m)}(\langle \mathbf{y}_i, \mathbf{y}_j\rangle)\Big) \succeq 0$. It is easy to see that $G_k^{(n,m)}(t_{ij}, \mathbf{v}_i, \mathbf{v}_j) = a_{ij}b_{ij}$, where $t_{ij} := \langle p_i, p_j\rangle$. Thus, $C = \Big(G_k^{(n,m)}(t_{ij}, \mathbf{v}_i, \mathbf{v}_j)\Big) \succeq 0$.

Using the addition theorem it is not hard to give a direct proof. Note that $G_k^{(2)}$ is the Chebyshev polynomial of the first kind, i.e. $G_k^{(2)}(\cos\theta) = \cos(k\theta)$. It is easy to see that for any $\varphi_1, \dots, \varphi_r$ the matrix $\Big(\cos(\varphi_i - \varphi_j)\Big) \succeq 0$. From this follows that Schoenberg's theorem holds for $n = 2$. Therefore, we have proved the theorem for $m = n-2$. Put $\ell = n-2$ in Corollary 2.2. That yields the theorem for all $0 \le m \le n-2$. □

Remark 3.1. We see that the second proof (as well as Schoenberg's original proof in [**21**]) is based on the addition theorem for Gegenbauer polynomials. There exists another proof of Schoenberg's theorem which is using the *addition theorem* *for spherical harmonics* (see, for instance, [**20**]). It is possible (see [**2**] for the case $m = 1$) to derive Theorem 3.1 from this theorem. However, this proof looks more complicated than our proof.

Remark 3.2. Bachoc and Vallentin [**2**] derived new upper bounds for spherical codes based on positive semidefinite constraints that are given in [**2**, Corollary 3.4]. In fact, this corollary easily follows from Theorem 3.1 with $m = 1$.

Indeed, let us fix some ℓ, $1 \le \ell \le r$, and take an orthonormal basis $e_1, \dots, e_n$ of $\mathbb{R}^n$ with $e_1 = p_\ell$. Then for $m = 1$ we have $\mathbf{v}_i = t_{i\ell}$. Let

$$(A_\ell)_{ij} = G_k^{(n,1)}(t_{ij}, t_{i\ell}, t_{j\ell}), \quad 0 \le i \le r,\ 1 \le j \le r,$$

$$Y_\ell = W A_\ell W^T, \quad (W)_{ij} = \lambda_i G_i^{(n+2k)}(t_{j\ell}),\ 0 \le i \le d,\ 1 \le j \le r.$$

Theorem 3.1 yields: $A_\ell \succeq 0$. Then $Y_\ell \succeq 0$ and $Y_1 + \dots + Y_r \succeq 0$. These constraints are equivalent to the constraints in [**2**, Corollary 3.4].

COROLLARY 3.1. *Let $0 \le m \le n-2$, $d \ge 0$. Let $e_1, \dots, e_n$ be an orthonormal basis of $\mathbb{R}^n$. Let a polynomial $F(t, \mathbf{u}, \mathbf{v})$ can be represented in the form*

$$F(t, \mathbf{u}, \mathbf{v}) = \sum_{k=0}^{d} f_k(\mathbf{u}, \mathbf{v})\, G_k^{(n,m)}(t, \mathbf{u}, \mathbf{v}),$$

where $f_k \succeq 0$ for all $0 \le k \le d$. Then for any points $p_1, \dots, p_r$ in $\mathbb{S}^{n-1}$ the matrix $\left(F(\langle p_i, p_j\rangle, p_i^{(m)}, p_j^{(m)})\right)$ is positive semidefinite. In other words, $F \in \mathrm{PD}(\mathbb{S}^{n-1}, \{e_1, \dots, e_m\})$.

PROOF. Let $f(\mathbf{u}, \mathbf{v}) \succeq 0$. Then $f(\mathbf{u}, \mathbf{v}) = \langle H, Z_d^m(\mathbf{u}, \mathbf{v})\rangle$ with $H \succeq 0$. From an eigenvalue factorization of H it follows that there exist polynomials $h_i(\mathbf{u})$ such that $f(\mathbf{u}, \mathbf{v}) = \sum_i h_i(\mathbf{u})\, h_i(\mathbf{v})$. Therefore, $\left(f(p_i^{(m)}, p_j^{(m)})\right) \succeq 0$.

We have $A_k = \left(f_k(p_i^{(m)}, p_j^{(m)})\right) \succeq 0$. Theorem 3.1 yields

$$B_k = \left(G_k^{(n,m)}(\langle p_i, p_j\rangle, p_i^{(m)}, p_j^{(m)})\right) \succeq 0.$$

Let $(C_k)_{ij} = (A_k)_{ij}\,(B_k)_{ij}$. The Schur theorem implies: $C_k \succeq 0$. Thus, $\left(F(\langle p_i, p_j\rangle, p_i^{(m)}, p_j^{(m)})\right) = \sum_{k=0}^{d} C_k \succeq 0$. □

3.3. An extension of the converse Schoenberg theorem.

THEOREM 3.2. *Let $0 \le m \le n-2$. Let $e_1, \dots, e_n \in \mathbb{R}^n$ be an orthonormal basis. Let $F(t, \mathbf{u}, \mathbf{v})$ be a polynomial in t and a symmetric polynomial in $\mathbf{u}, \mathbf{v} \in \mathbb{R}^m$. Suppose $F \in \mathrm{PD}(\mathbb{S}^{n-1}, \{e_1, \dots, e_m\})$. Then*

$$F(t, \mathbf{u}, \mathbf{v}) = \sum_{k=0}^{d} f_k(\mathbf{u}, \mathbf{v})\, G_k^{(n,m)}(t, \mathbf{u}, \mathbf{v}) \tag{3.1}$$

with $f_k \succeq 0$ for all $0 \le k \le d = \deg_t(F)$.

PROOF. The polynomial $G_k^{(n,m)}(t, \mathbf{u}, \mathbf{v})$ has degree k in the variable t, so that F has a unique expression in the form (3.1), where $f_k(\mathbf{u}, \mathbf{v})$ is a symmetric polynomial in $\mathbf{u}, \mathbf{v}$ of degree d_k in $\mathbf{u}$. Then $f_k(\mathbf{u}, \mathbf{v}) = \left\langle F_k, Z_{d_k}^m(\mathbf{u}, \mathbf{v})\right\rangle$, where F_k is a symmetric matrix. Now we prove that $F_k \succeq 0$.

Let $f(t,\mathbf{u},\mathbf{v})$, $g(t,\mathbf{u},\mathbf{v})$ be continuous functions on D_m. Denote

$$\{f,g\} := \int_{(\mathbb{S}^{n-1})^2} f(\langle x,y\rangle, x^{(m)}, y^{(m)})\, g(\langle x,y\rangle, x^{(m)}, y^{(m)})\, d\omega_n(x)\, d\omega_n(y).$$

For any continuous function $h(\mathbf{u},\mathbf{v})$ Corollary 2.1 yields:

$$\{F, h\,G_k^{(n,m)}\} = \{f_k\, G_k^{(n,m)}, h\,G_k^{(n,m)}\} = \{f_k\, h, (G_k^{(n,m)})^2\}. \tag{3.2}$$

Let $f(\mathbf{u},\mathbf{v})$, $g(\mathbf{u},\mathbf{v})$ be continuous functions on $\{|\mathbf{u}|\le 1,\ |\mathbf{v}|\le 1\}$. Denote

$$[f,g] := \{fg, (G_k^{(n,m)})^2\}.$$

Clearly, $[\cdot,\cdot]$ is an inner product. Let $\alpha_1, \alpha_2, \ldots$ be an orthonormal basis for $[\cdot,\cdot]$ in the space of real polynomials $\mathbb{R}[\mathbf{u},\mathbf{v}]$. We observe that F_k defines a quadratic form on $\mathbb{R}[\mathbf{u},\mathbf{v}]$. Let us denote by $\tilde{F}_k$ the matrix expression of this quadratic form in the basis $\{\alpha_i\}$. Then there exists a matrix A such that $F_k = A^T\tilde{F}_kA$. Therefore, $\tilde{F}_k \succeq 0$ yields $F_k \succeq 0$.

Let $h(\mathbf{u},\mathbf{v}) = \langle H, Z_{d_k}^m(\mathbf{u},\mathbf{v})\rangle$. It's easy to see that $[f_k, h] = \langle\tilde{F}_k, \tilde{H}\rangle$.

Note that the sum of all entries of a positive semidefinite matrix is nonnegative. This implies: if $f \in \mathrm{PD}(\mathbb{S}^{n-1}, E_m)$, $E_m := \{e_1,\ldots,e_m\}$, then $\{f,1\} \ge 0$. Moreover, the Schur theorem yields: if f and $g \in \mathrm{PD}(\mathbb{S}^{n-1}, E_m)$, then $fg \in \mathrm{PD}(\mathbb{S}^{n-1}, E_m)$. Therefore, $\{f,g\} \ge 0$. Thus, for any $f,g \in \mathrm{PD}(\mathbb{S}^{n-1}, E_m)$ we have $[f,g] \ge 0$.

Let $H \succeq 0$. Then $h(\mathbf{u},\mathbf{v}) \in \mathrm{PD}(\mathbb{S}^{n-1}, E_m)$. We have $F \in \mathrm{PD}(\mathbb{S}^{n-1}, E_m)$. Then (3.2) yields $[f_k, h] = \langle\tilde{F}_k, \tilde{H}\rangle \ge 0$ for any $\tilde{H} \succeq 0$. Thus, $\tilde{F}_k \succeq 0$ and $F_k \succeq 0$. □

Remark 3.3. In [**3**, Proposition 4.12] a statement that is equivalent to Theorem 3.2 was proven with $m = 1$. However, in [**3**, Proposition 4.12] the formula (3.1) is written in terms of an orthogonal basis for $\{\cdot,\cdot\}$.

3.4. The class $\mathrm{PD}(\mathbb{S}^{n-1}, Q)$**.** If $f_r \in \mathrm{PD}(M,Q)$, $f_r \to f$ as $r \to \infty$, and $f(t,\mathbf{u},\mathbf{v})$ is continuous, then also $f \in \mathrm{PD}(M,Q)$. Therefore, Corollary 3.1 and Theorem 3.2 imply

THEOREM 3.3. *Let $0 \le m \le n-2$. Let $e_1,\ldots,e_n$ be an orthonormal basis of $\mathbb{R}^n$. Then $F \in \mathrm{PD}(\mathbb{S}^{n-1}, \{e_1,\ldots,e_m\})$ if and only if*

$$F(t,\mathbf{u},\mathbf{v}) = \sum_{k=0}^{\infty} f_k(\mathbf{u},\mathbf{v})\, G_k^{(n,m)}(t,\mathbf{u},\mathbf{v}),$$

where for any $k \ge 0$ a function $f_k(\mathbf{u},\mathbf{v})$ is positive semidefinite.

Now we consider $\mathrm{PD}(\mathbb{S}^{n-1}, Q)$ for any $Q = \{q_1,\ldots,q_m\} \subset \mathbb{S}^{n-1}$ with $0 \le m \le n-2$. Theorem 3.3 yields

COROLLARY 3.2. *Let $0 \le m \le n-2$. Let $Q = \{q_1,\ldots,q_m\} \subset \mathbb{S}^{n-1}$ with $\mathrm{rank}(Q) = m$. Let $e_1,\ldots,e_m$ be an orthonormal basis of the linear space with the basis $q_1,\ldots,q_m$, and let L_Q denotes the linear transformation of coordinates. Then $F \in \mathrm{PD}(\mathbb{S}^{n-1}, Q)$ if and only if*

$$F(t,\mathbf{u},\mathbf{v}) = \sum_{k=0}^{\infty} f_k(\mathbf{u},\mathbf{v})\, G_k^{(n,m)}(t, L_Q(\mathbf{u}), L_Q(\mathbf{v})),$$

where $f_k(\mathbf{u},\mathbf{v}) \succeq 0$ for all $k \ge 0$.

Remark 3.4. It is not hard to describe $\mathrm{PD}(\mathbb{S}^{n-1}, Q)$ with $\mathrm{rank}(Q) \geq n-1$.

First, consider the case $Q = \{e_1, \ldots, e_{n-1}\}$, where $\{e_i\}$ is an orthonormal basis. Let $p_1, \ldots, p_r \in \mathbb{S}^{n-1}$, $\mathbf{u}_i = \langle p_i, e_1\rangle e_1 + \ldots + \langle p_i, e_{n-1}\rangle e_{n-1}$, and $t_{ij} = \langle p_i, p_j\rangle$. Obviously, $t_{ij} - \langle \mathbf{u}_i, \mathbf{u}_j\rangle = w_i w_j$, $w_i = \langle p_i, e_n\rangle$. Then

$$\Big(t_{ij} - \langle \mathbf{u}_i, \mathbf{u}_j\rangle\Big) = \mathbf{w}^T \mathbf{w} \succeq 0, \quad \mathbf{w} = (w_1, \ldots, w_r). \tag{3.3}$$

Since $w_i^2 = 1 - |\mathbf{u}_i|^2$, we have for all $i = 1, \ldots, r$, $j = 1, \ldots, r$:

$$(t_{ij} - \langle \mathbf{u}_i, \mathbf{u}_j\rangle)^2 = (1 - |\mathbf{u}_i|^2)(1 - |\mathbf{u}_j|^2). \tag{3.4}$$

Let $H_2(t, \mathbf{u}, \mathbf{v}) := (t - \langle \mathbf{u}, \mathbf{v}\rangle)^2 - (1 - |\mathbf{u}|^2)(1 - |\mathbf{v}|^2)$. Since $H_2(t_{ij}, \mathbf{u}_i, \mathbf{u}_j) = 0$, we have $F \in \mathrm{PD}(\mathbb{S}^{n-1}, \{e_1, \ldots, e_{n-1}\})$ if and only if

$$F(t, \mathbf{u}, \mathbf{v}) = f_0(\mathbf{u}, \mathbf{v}) + f_1(\mathbf{u}, \mathbf{v})\,(t - \langle \mathbf{u}, \mathbf{v}\rangle) + R(t, \mathbf{u}, \mathbf{v})\, H_2(t, \mathbf{u}, \mathbf{v}),$$

where $f_0 \succeq 0$, $f_1 \succeq 0$, and $R(t, \mathbf{u}, \mathbf{v})$ is any continuous function on D_m.

Let $Q = \{q_1, \ldots, q_{n-1}\} \subset \mathbb{S}^{n-1}$ with $\mathrm{rank}(Q) = n-1$. Let L be a linear transformation of the basis $q_1, \ldots, q_{n-1}$ to an orthonormal basis $e_1, \ldots, e_{n-1}$ of $\mathbb{R}^{n-1}$. Then $F \in \mathrm{PD}(\mathbb{S}^{n-1}, Q)$ if and only if

$$F(t, \mathbf{u}, \mathbf{v}) = f_0(\mathbf{u}, \mathbf{v}) + f_1(\mathbf{u}, \mathbf{v})\,(t - \langle L(\mathbf{u}), L(\mathbf{v})\rangle) + R(t, \mathbf{u}, \mathbf{v})\, H_2(t, L(\mathbf{u}), L(\mathbf{v})),$$

with $f_0 \succeq 0$, $f_1 \succeq 0$, and any $R(t, L(\mathbf{u}), L(\mathbf{v})) \in C(D_m)$.

In the case $\mathrm{rank}(Q = \{q_1, \ldots, q_m\}) \geq n-1$ consider all $Q_i \subset Q$ with $\mathrm{rank}(Q_i) = n-1$. Let a linear transformation $L_i : \mathbb{R}^m \to \mathbb{R}^{n-1}$ is defined by $L_i(q) = q$ if $q \in Q_i$, $L_i(q) = 0$ if $q \in Q$ and $\mathrm{rank}(Q_i \cup \{q\}) = n$. Denote by N the number of distinct Q_i. It is not hard to prove that $F \in \mathrm{PD}(\mathbb{S}^{n-1}, Q)$ iff

$$F(t, \mathbf{u}, \mathbf{v}) = \sum_{i=1}^{N} f_i(\mathbf{u}, \mathbf{v})\, F_i(t, L_i(\mathbf{u}), L_i(\mathbf{v})),$$

where for all i: $f_i(\mathbf{u}, \mathbf{v}) \succeq 0$, $F_i(t, L_i(\mathbf{u}), L_i(\mathbf{v})) \in \mathrm{PD}(\mathbb{S}^{n-1}, Q_i)$.

4. Positive definite functions in $\mathbb{R}^n$

Direct extensions of the Bochner–Schoenberg theorem and finding bounds on sphere packings in $\mathbb{R}^n$ are not so straightforward because this space is not compact. Different indirect ways of deriving bounds on sphere packings in $\mathbb{R}^n$ were suggested in the literature [**8, 14, 16**].

Here we note that the multivariate Gegenbauer polynomials defined above enable one to define p.d. functions in $\mathbb{R}^n$ as follows:

$$H_k^{(n,m)}(t, x, y, \mathbf{u}, \mathbf{v}) := (xy)^{k/2}\, G_k^{(n,m)}(t', \mathbf{u}', \mathbf{v}'),$$

where $0 \leq m \leq n-2$, $t, x, y \in \mathbb{R}$, $\mathbf{u}, \mathbf{v} \in \mathbb{R}^m$ for $m > 0$ and $\mathbf{u} = \mathbf{v} = 0$ for $m = 0$, and

$$t' = t/\sqrt{xy}, \quad \mathbf{u}' = \mathbf{u}/\sqrt{x}, \quad \mathbf{v}' = \mathbf{v}/\sqrt{y}.$$

The positive semidefiniteness of the polynomials $G_k^{(n,m)}$ (Theorem 3.1) implies the following result.

THEOREM 4.1. *Let $e_1, \ldots, e_n \in \mathbb{R}^n$ be an orthonormal basis, and let $p_1, \ldots, p_N$ be points in $\mathbb{R}^n$. Then for any $k \geq 0$ and $0 \leq m \leq n-2$ the matrix (g_{ij}), where*

$$g_{ij} = H_k^{(n,m)}(\langle p_i, p_j\rangle, |p_i|^2, |p_j|^2, p_i, p_j),$$

is positive semidefinite.

REMARK 4.1. *It is not hard to find Euclidean analogs of Theorems 3.2, 3.3 and Corollary 3.2*

Theorem 4.1 gives a family of positive-semidefinite constraints for distance distributions of points in Euclidean spaces. For instance, consider the simple case of $m = 0$. Now for any matrix $A = \left(a_{ij}\right)$ of size $N \times N$ we have a matrix $H_k^{(n)}(A)$ which is defined by

$$\left(H_k^{(n)}(A)\right)_{ij} = (a_{ii}a_{jj})^{k/2}\, G_k^{(n)}\left(a_{ij}/\sqrt{a_{ii}a_{jj}}\right).$$

COROLLARY 4.1. *If A is a symmetric positive semidefinite matrix with* rank$(A) \leq n$*, then for any positive integer k we have*

$$H_k^{(n)}(A) \succeq 0.$$

PROOF. It is a well-known fact: *A symmetric matrix A of size $N \times N$ is the Gram matrix of N vectors in $\mathbb{R}^n$ if and only if $A \succeq 0$ and* rank$(A) \leq n$. Therefore, there are vectors $p_1, \ldots, p_N$ in $\mathbb{R}^n$ such that $A = \left(\langle p_i, p_j\rangle\right)$. Then

$$\left(H_k^{(n)}(A)\right)_{ij} = H_k^{(n,0)}(\langle p_i, p_j\rangle, |p_i|^2, |p_j|^2, 0, 0).$$

Thus Theorem 4.1 yields $H_k^{(n)}(A) \succeq 0$. □

For small k it is not hard to give explicit expressions for $H_k^{(n)}(A)$. Clearly, $H_1^{(n)}(A) = A$. Since $G_2^{(n)}(t) = (nt^2 - 1)/(n-1)$, we have

$$\left(H_2^{(n)}(A)\right)_{ij} = \frac{na_{ij}^2 - a_{ii}a_{jj}}{n-1}.$$

Then

$$H_2^{(n)}(A) = \frac{nA_2 - \mathbf{a}^T\mathbf{a}}{n-1},$$

where

$$(A_2)_{ij} := a_{ij}^2, \quad \mathbf{a} := (a_{11}, a_{22}, \ldots, a_{NN}).$$

So if $A \succeq 0$ and rank$(A) \leq n$, then

$$nA_2 \succeq \mathbf{a}^T\mathbf{a}.$$

5. Upper bounds for spherical codes

In this section we set up upper bounds for spherical codes which are based on multivariate p.d. functions. These bounds extend the famous Delsarte's bound. Note that for the case $m = 1$ this bound is the Bachoc - Vallentin bound [**4**].

DEFINITION 5.1. *Consider a vector $J = (j_1, \ldots, j_d)$. Split the set of numbers $\{j_1, \ldots, j_d\}$ into maximal subsets $I_1, \ldots, I_k$ with equal elements. That means, if $I_r = \{j_{r_1}, \ldots, j_{r_s}\}$, then $j_{r_1} = \ldots = j_{r_s} = a_r$ and all other $j_\ell \neq a_r$. Without loss of generality it can be assumed that $i_1 = |I_1| \geq \ldots \geq i_k = |I_k| > 0$. (Note that we have $i_1 + \ldots + i_k = d$.) Denote by $\psi(J)$ the vector $\omega = (i_1, \ldots, i_k)$.*

Let

$$W_d := \{\omega = (i_1, \ldots, i_k) : i_1 + \ldots + i_k = d,\ i_1 \geq \ldots \geq i_k > 0,\ i_1, \ldots, i_k \in \mathbb{Z}\}.$$

Let $\omega \in W_d$. Denote

$$\tilde{q}_\omega(N) := \#\{J = (j_1, \ldots, j_d) \in \{1, \ldots, N\}^d : \psi(J) = \omega\},$$

$$q_\omega(N) := \frac{\tilde{q}_\omega(N)}{N}.$$

It is not hard to see that $q_\omega(N)$ is a polynomial of degree $d-1$ for $\omega \in W_d$ and

$$\sum_{\omega \in W_d} q_\omega(N) = N^{d-1}.$$

DEFINITION 5.2. *For any vector* $\mathbf{x} = \{x_{ij}\}$ *with* $1 \le i < j \le d$ *denote by* $A(\mathbf{x})$ *a symmetric* $d \times d$ *matrix* (a_{ij}) *with all* $a_{ii} = 1$ *and* $a_{ji} = a_{ij} = x_{ij}$, $i < j$.

Let $0 < \theta < \pi$ *and*

$$X(\theta) := \{\mathbf{x} = \{x_{ij}\} : x_{ij} \in [-1, \cos\theta] \text{ or } x_{ij} = 1,\ 1 \le i < j \le d\}.$$

Now for any $\mathbf{x} = \{x_{ij}\} \in X(\theta)$ *we define a vector* $J(\mathbf{x}) = (j_1, \ldots, j_d)$ *such that* $j_k = k$ *if there are no* $i < k$ *with* $x_{ik} = 1$, *otherwise* $j_k = i$, *where* i *is the minimum index with* $x_{ik} = 1$.

Let $\omega \in W_d$. *Denote*

$$D_\omega(\theta) := \{\mathbf{x} \in X(\theta) : \psi(J(\mathbf{x})) = \omega \text{ and } A(\mathbf{x}) \succeq 0\}.$$

Let $f(\mathbf{x})$ *be a real function in* $\mathbf{x}$, *and let*

$$B_\omega(\theta, f) := \sup_{\mathbf{x} \in D_\omega(\theta)} f(\mathbf{x}).$$

Note that the assumption $A(\mathbf{x}) \succeq 0$ implies existence of unit vectors $p_1, \ldots, p_d$ such that $A(\mathbf{x})$ is the Gram matrix of these vectors, i.e. $x_{ij} = \langle p_i, p_j \rangle$. Moreover, if $x_{ij} = 1$, then $p_i = p_j$. In particular, $D_{(d)}(\theta) = \{(1, \ldots, 1)\}$ and therefore $B_{(d)}(\theta, f) = f(1, \ldots, 1)$.

DEFINITION 5.3. *Let* $\mathbf{x} = \{x_{ij}\}$, *where* $1 \le i < j \le m+2 \le n$, *and let* $A(\mathbf{x}) \succeq 0$. *Then there exist* $P = \{p_1, \ldots, p_{m+2}\} \subset \mathbb{S}^{n-1}$ *such that* $x_{ij} = \langle p_i, p_j \rangle$. *Let* $F(\mathbf{x})$ *be a continuous function in* $\mathbf{x}$ *with* $F(\tilde{\mathbf{x}}_{k\ell}) = F(\mathbf{x})$ *for all* $\tilde{\mathbf{x}}_{k\ell}$ *that can be obtained by interchanging two points* p_k *and* p_ℓ *in* P. *We say that* $F(\mathbf{x}) \in \mathrm{PD}_m^n$ *if for all* $\mathbf{x}$ *with* $A(\mathbf{x}) \succeq 0$ *we have* $\tilde{F}(x_{12}, \mathbf{u_1}, \mathbf{u_2}) \in \mathrm{PD}(\mathbb{S}^{n-1}, Q(\mathbf{x}))$, *where* $\mathbf{u_i} = (x_{i3}, \ldots, x_{i,m+2})$, $Q(\mathbf{x}) = \{p_3, \ldots, p_{m+2}\}$, *and* $\tilde{F}(x_{12}, \mathbf{u_1}, \mathbf{u_2}) = F(\mathbf{x})$.

For the classical case $m = 0$ Schoenberg's theorem says that $f \in \mathrm{PD}_0^n$ if and only if $f(t) = \sum_k f_k G_k^{(n)}(t)$ with all $f_k \ge 0$. Theorem 3.3 (see also [**2**–**4**]) yields $F(\mathbf{x}) \in \mathrm{PD}_1^n$ if and only if

$$F(x_{12}, x_{13}, x_{23}) = \sum_k f_k(x_{12}, x_{13}, x_{23})\, G_k^{(n,1)}(x_{12}, x_{13}, x_{23}),$$

where $f_k \succeq 0$ for all k, and

$$F(x_{12}, x_{13}, x_{23}) = F(x_{13}, x_{12}, x_{23}) = F(x_{23}, x_{13}, x_{12}).$$

Using Corollary 3.2 it is possible to describe the class of functions PD_m^n, $m \ge 2$.

Let C be an N-element subset of the unit sphere $\mathbb{S}^{n-1} \subset \mathbb{R}^n$. It is called an (n, N, θ) *spherical code* if every pair of distinct points (c, c') of C have inner product $\langle c, c' \rangle$ at most $\cos\theta$.

THEOREM 5.4. *Let* $f_0 > 0$, $0 \le m \le n-2$, *and* $F(\mathbf{x}) = f(\mathbf{x}) - f_0 \in \mathrm{PD}_m^n$. *Then an* (n, N, θ) *spherical code satisfies*

$$f_0 N^{m+1} \le \sum_{\omega \in W_{m+2}} B_\omega(\theta, f)\, q_\omega(N).$$

PROOF. Let C be an (n, N, θ) spherical code. Define

$$S = \sum_{\mathbf{c}=(c_1,\ldots,c_d)\in C^d} f(\{\langle c_i, c_j\rangle\}), \quad d = m+2.$$

Then

$$S = \sum_{\omega\in W_d} \sum_{\mathbf{c}:\psi(\mathbf{c})=\omega} f(\{\langle c_i, c_j\rangle\}) \leq \sum_{\omega\in W_d} B_\omega(\theta, f)\, \tilde{q}_\omega(N).$$

On the other hand, since $F \in \mathrm{PD}_m^n$ we have

$$\sum_{\mathbf{c}\in C^d} F(\{\langle c_i, c_j\rangle\}) \geq 0.$$

Thus

$$S = \sum_{\mathbf{c}\in C^d} (f_0 + F(\{\langle c_i, c_j\rangle\})) \geq f_0\, N^d.$$

□

It is easy to see for $m = 0$ that $q_{(2)}(N) = 1$, $q_{(1,1)}(N) = N-1$, and $B_{(2)}(\theta, f) = f(1)$. Therefore, from Theorem 5.1 we have

$$f_0 N \leq f(1) + B_{(1,1)}(\theta, f)(N-1).$$

Suppose $B_{(1,1)}(\theta, f) \leq 0$, i.e. $f(t) \leq 0$ for all $t \in [-1, \cos\theta]$. Thus for (n, N, θ) spherical code we obtain

$$N \leq \frac{f(1)}{f_0}.$$

This upper bound is called Delsarte's bound.

The Bachoc-Vallentin bound [**4**, Theorem 4.1] is the bound in Theorem 5.1 for $m = 1$ and $B_{(1,1,1)}(\theta, f) \leq 0$. Indeed, let $B_{(2,1)}(\theta, f) \leq B$. Since $q_{(3)}(N) = 1$, $q_{(2,1)}(N) = 3(N-1)$, and $B_{(3)}(\theta, f) = f(1,1,1)$, we have

$$f_0 N^2 \leq f(1,1,1) + 3(N-1)B.$$

Let us consider Theorem 5.1 also for the case $m = 2$ with $B_{(1,1,1,1)}(\theta, f) \leq 0$. Let $B_{(3,1)}(\theta, f) \leq B_1$, $B_{(2,2)}(\theta, f) \leq B_2$, and $B_{(2,1,1)}(\theta, f) \leq B_3$. Then

$$f_0 N^3 \leq f(1,1,1,1,1,1) + 4(N-1)B_1 + 3(N-1)B_2 + 6(N-1)(N-2)B_3.$$

Let $f(\mathbf{x})$ be a polynomial of degree d. Then the assumptions in Theorem 5.1 can be written as positive semidefinite constraints for the coefficients of F (see for details [**2**–**4**, **13**, **22**]). Actually, the bound given by Theorem 5.1 can be obtained as a solution of an SDP (semidefinite programming) optimization problem. In [**2**, **3**] using numerical solutions of the SDP problem for the case $m = 1$ has obtained new upper bounds for the kissing numbers and for the one-sided kissing numbers in several dimensions $n \leq 10$.

However, the dimension of the corresponding SDP problem is growth so fast whenever d and m are increasing that this problem can be treated numerically only for relatively small d and small m. It is an interesting problem to find (explicitly) suitable polynomials F for Theorem 5.1 and using it to obtain new bounds for spherical codes.

References

[1] C. Bachoc, Semidefinite programming, harmonic analysis and coding theory, preprint, arXiv:0909.4767, 2009.

[2] Christine Bachoc and Frank Vallentin, *New upper bounds for kissing numbers from semidefinite programming*, J. Amer. Math. Soc. **21** (2008), no. 3, 909–924, DOI 10.1090/S0894-0347-07-00589-9. MR2393433 (2009c:52029)

[3] Christine Bachoc and Frank Vallentin, *Semidefinite programming, multivariate orthogonal polynomials, and codes in spherical caps*, European J. Combin. **30** (2009), no. 3, 625–637, DOI 10.1016/j.ejc.2008.07.017. MR2494437 (2010d:90065)

[4] Christine Bachoc and Frank Vallentin, *Optimality and uniqueness of the* $(4, 10, 1/6)$ *spherical code*, J. Combin. Theory Ser. A **116** (2009), no. 1, 195–204, DOI 10.1016/j.jcta.2008.05.001. MR2469257 (2010f:94337)

[5] Alexander Barg and Oleg R. Musin, *Codes in spherical caps*, Adv. Math. Commun. **1** (2007), no. 1, 131–149, DOI 10.3934/amc.2007.1.131. MR2262773 (2007m:94257)

[6] S. Bochner, *Hilbert distances and positive definite functions*, Ann. of Math. (2) **42** (1941), 647–656. MR0005782 (3,206d)

[7] Billie Chandler Carlson, *Special functions of applied mathematics*, Academic Press [Harcourt Brace Jovanovich, Publishers], New York-London, 1977. MR0590943 (58 #28707)

[8] Henry Cohn and Noam Elkies, *New upper bounds on sphere packings. I*, Ann. of Math. (2) **157** (2003), no. 2, 689–714, DOI 10.4007/annals.2003.157.689. MR1973059 (2004b:11096)

[9] Henry Cohn and Jeechul Woo, *Three-point bounds for energy minimization*, J. Amer. Math. Soc. **25** (2012), no. 4, 929–958, DOI 10.1090/S0894-0347-2012-00737-1. MR2947943

[10] J. H. Conway and N. J. A. Sloane, *Sphere packings, lattices and groups*, 3rd ed., Grundlehren der Mathematischen Wissenschaften [Fundamental Principles of Mathematical Sciences], vol. 290, Springer-Verlag, New York, 1999. With additional contributions by E. Bannai, R. E. Borcherds, J. Leech, S. P. Norton, A. M. Odlyzko, R. A. Parker, L. Queen and B. B. Venkov. MR1662447 (2000b:11077)

[11] A. W. Davis, *Spherical functions on the Grassmann manifold and generalized Jacobi polynomials. I*, Linear Algebra Appl. **289** (1999), no. 1-3, 75–94, DOI 10.1016/S0024-3795(98)10178-7. Linear algebra and statistics (Istanbul, 1997). MR1670974 (2000i:33015)

[12] A. Erdélyi, editor, Higher Transcendental Function, McGraw-Hill, NY, 3 vols, 1953, Vol. II, Chap. XI.

[13] Dion Gijswijt, Alexander Schrijver, and Hajime Tanaka, *New upper bounds for nonbinary codes based on the Terwilliger algebra and semidefinite programming*, J. Combin. Theory Ser. A **113** (2006), no. 8, 1719–1731, DOI 10.1016/j.jcta.2006.03.010. MR2269550 (2007g:94105)

[14] D. V. Gorbachev, *An extremal problem for entire functions of exponential spherical type, which is connected with the Levenshteĭn bound for the density of a packing of* $\mathbb{R}^n$ *by balls* (Russian, with English and Russian summaries), Izv. Tul. Gos. Univ. Ser. Mat. Mekh. Inform. **6** (2000), no. 1, Matematika, 71–78. MR2018751 (2004h:41009)

[15] Alan T. James and A. G. Constantine, *Generalized Jacobi polynomials as spherical functions of the Grassmann manifold*, Proc. London Math. Soc. (3) **29** (1974), 174–192. MR0374523 (51 #10723)

[16] G. A. Kabatjanskiĭ and V. I. Levenšteĭn, *Bounds for packings on the sphere and in space* (Russian), Problemy Peredači Informacii **14** (1978), no. 1, 3–25. MR0514023 (58 #24018)

[17] Oleg R. Musin, *The kissing number in four dimensions*, Ann. of Math. (2) **168** (2008), no. 1, 1–32, DOI 10.4007/annals.2008.168.1. MR2415397 (2009e:52036)

[18] Oleg R. Musin, *The one-sided kissing number in four dimensions*, Period. Math. Hungar. **53** (2006), no. 1-2, 209–225, DOI 10.1007/s10998-006-0033-0. MR2286472 (2007j:52019)

[19] O. R. Musin, *Bounds for codes by semidefinite programming*, Tr. Mat. Inst. Steklova **263** (2008), no. Geometriya, Topologiya i Matematicheskaya Fizika. I, 143–158, DOI 10.1134/S0081543808040111; English transl., Proc. Steklov Inst. Math. **263** (2008), no. 1, 134–149. MR2599377 (2011c:94085)

[20] Florian Pfender and Günter M. Ziegler, *Kissing numbers, sphere packings, and some unexpected proofs*, Notices Amer. Math. Soc. **51** (2004), no. 8, 873–883. MR2145821 (2006a:52015)

[21] I. J. Schoenberg, *Positive definite functions on spheres*, Duke Math. J. **9** (1942), 96–108. MR0005922 (3,232c)

[22] Alexander Schrijver, *New code upper bounds from the Terwilliger algebra and semidefinite programming*, IEEE Trans. Inform. Theory **51** (2005), no. 8, 2859–2866, DOI 10.1109/TIT.2005.851748. MR2236252 (2007a:94147)

Department of Mathematics, University of Texas at Brownsville, Brownsville, Texas 78520 — and — Inst. for Problems of Inform. Trans., RAS, Moscow, Russia

E-mail address: oleg.musin@utb.edu

Selected Published Titles in This Series

625 **Alexander Barg and Oleg R. Musin, Editors,** Discrete Geometry and Algebraic Combinatorics, 2014

616 **G. L. Litvinov and S. N. Sergeev, Editors,** Tropical and Idempotent Mathematics and Applications, 2014

615 **Plamen Stefanov, András Vasy, and Maciej Zworski, Editors,** Inverse Problems and Applications, 2014

614 **James W. Cogdell, Freydoon Shahidi, and David Soudry, Editors,** Automorphic Forms and Related Geometry, 2014

613 **Stephan Stolz, Editor,** Topology and Field Theories, 2014

612 **Patricio Cifuentes, José García-Cuerva, Gustavo Garrigós, Eugenio Hernández, José María Martell, Javier Parcet, Keith M. Rogers, Alberto Ruiz, Fernando Soria, and Ana Vargas, Editors,** Harmonic Analysis and Partial Differential Equations, 2014

611 **Robert Fitzgerald Morse, Daniela Nikolova-Popova, and Sarah Witherspoon, Editors,** Group Theory, Combinatorics, and Computing, 2014

610 **Pavel Etingof, Mikhail Khovanov, and Alistair Savage, Editors,** Perspectives in Representation Theory, 2014

609 **Dinh Van Huynh, S. K. Jain, Sergio R. López-Permouth, S. Tariq Rizvi, and Cosmin S. Roman, Editors,** Ring Theory and Its Applications, 2014

608 **Robert S. Doran, Greg Friedman, and Scott Nollet, Editors,** Hodge Theory, Complex Geometry, and Representation Theory, 2014

607 **Kiyoshi Igusa, Alex Martsinkovsky, and Gordana Todorov, Editors,** Expository Lectures on Representation Theory, 2014

606 **Chantal David, Matilde Lalín, and Michelle Manes, Editors,** Women in Numbers 2, 2013

605 **Omid Amini, Matthew Baker, and Xander Faber, Editors,** Tropical and Non-Archimedean Geometry, 2013

604 **José Luis Montaña and Luis M. Pardo, Editors,** Recent Advances in Real Complexity and Computation, 2013

603 **Azita Mayeli, Alex Iosevich, Palle E. T. Jorgensen, and Gestur Ólafsson, Editors,** Commutative and Noncommutative Harmonic Analysis and Applications, 2013

602 **Vyjayanthi Chari, Jacob Greenstein, Kailash C. Misra, K. N. Raghavan, and Sankaran Viswanath, Editors,** Recent Developments in Algebraic and Combinatorial Aspects of Representation Theory, 2013

601 **David Carfi, Michel L. Lapidus, Erin P. J. Pearse, and Machiel van Frankenhuijsen, Editors,** Fractal Geometry and Dynamical Systems in Pure and Applied Mathematics II, 2013

600 **David Carfi, Michel L. Lapidus, Erin P. J. Pearse, and Machiel van Frankenhuijsen, Editors,** Fractal Geometry and Dynamical Systems in Pure and Applied Mathematics I, 2013

599 **Mohammad Ghomi, Junfang Li, John McCuan, Vladimir Oliker, Fernando Schwartz, and Gilbert Weinstein, Editors,** Geometric Analysis, Mathematical Relativity, and Nonlinear Partial Differential Equations, 2013

598 **Eric Todd Quinto, Fulton Gonzalez, and Jens Gerlach Christensen, Editors,** Geometric Analysis and Integral Geometry, 2013

597 **Craig D. Hodgson, William H. Jaco, Martin G. Scharlemann, and Stephan Tillmann, Editors,** Geometry and Topology Down Under, 2013

596 **Khodr Shamseddine, Editor,** Advances in Ultrametric Analysis, 2013

For a complete list of titles in this series, visit the AMS Bookstore at **www.ams.org/bookstore/conmseries/**.